T0290997

FINANCING INVESTMENT IN WATER SECURITY

FINANCING INVESTMENT IN WATER SECURITY

Recent Developments and Perspectives

Edited by

XAVIER LEFLAIVE
OECD Environment Directorate

KATHLEEN DOMINIQUE
OECD Environment Directorate

GUY J. ALAERTS
IHE-Delft International Institute for Water Education

ELSEVIER

Elsevier
Radarweg 29, PO Box 211, 1000 AE Amsterdam, Netherlands
The Boulevard, Langford Lane, Kidlington, Oxford OX5 1GB, United Kingdom
50 Hampshire Street, 5th Floor, Cambridge, MA 02139, United States

ISBN: 978-0-12-822847-0

For Information on all Elsevier publications visit our website at
https://www.elsevier.com/books-and-journals

Publisher: Candice Janco
Acquisitions Editor: Louisa Munro
Editorial Project Manager: Chris Hockaday
Production Project Manager: Sruthi Satheesh
Cover Designer: Victoria Pearson
Cover Photograph: Pascal Kobeh, http://www.pascalkobeh.com, Instagram: @pascal_kobeh

Typeset by Aptara, New Delhi, India

Contents

Contributors

Guy J. Alaerts
IHE-Delft International Institute for Water Education

Richard Ashley
Emeritus Professor University of Sheffield, Department of Civil and Structural Engineering, United Kingdom

Martin S. Baker
Counsel, Dentons

Bert De Bièvre
Fondo para la Protección del Agua (FONAG), Quito, Ecuador

Maarten Blokland
Water Finance Facility, Kenya

Silvia Cardascia
Asian Development Bank

Lorena Coronel
AquaNature, Quito, Ecuador

Adam Davis
Ecosystem Investment Partners, Baltimore, Maryland

Kathleen Dominique
OECD Environment Directorate

Coral Fernandez-Illescas
Asian Development Bank

James Gebhardt
Office of Water, US Environmental Protection Agency, Washington, D.C.

Hein Gietema
CSC Strategy & Finance, Rotterdam, The Netherlands

Dick van Ginhoven
Water Finance Facility, Kenya

Bruce Horton
Environmental Policy Consulting and Stantec, United Kingdom

Guy Hutton
Senior Adviser for Water, Sanitation and Hygiene, UNICEF New York, NY, United States

Sara L. Johnson
Ecological Restoration Business Association, McLean, Virginia

Xavier Leflaive
OECD Environment Directorate

Mireille J. Martini
OECD Environment Directorate

Ng'ang'a Mbage
Kenya Pooled Water Fund

Alex Money
Smith School for Enterprise and the Environment, University of Oxford

Alex Mourant
Office of Water, US Environmental Protection Agency, Washington, D.C.

Dean Muruven Nalandhren
Global Policy Lead Freshwater, WWF, Switzerland

Eddy Njoroge
Kenya Pooled Water Fund

Joris van Oppenraaij
Water Finance Facility, Kenya

Nele-Frederike Rosenstock
European Commission DG Environment

Alex Simalabwi
Global Water Partnership

Mark Smith
International Water Management Institute, Colombo, Sri Lanka

Jean Pierre Sweerts
Water Finance Facility, Kenya

Roy Torkelson
Water Finance Facility, Kenya

Elaine Wu
J.P. Morgan

Rick C. Ziegler
Office of International and tribal Affairs (OITA), US Environmental Protection Agency, Washington, D.C.

General introduction

Xavier Leflaive[a], Guy J. Alaerts[b] and Kathleen Dominique[a]
[a]OECD Environment Directorate
[b]IHE-Delft International Institute for Water Education

A new conversation on financing water

Financing investment in water security is a major global challenge and is increasingly recognized as an issue at the heart of sustainable development and financial sustainability, as well as peace and international security. The global recognition of the role of water security in broad development, economic, environmental, and financial agendas emerged about 20 years ago. Since then, this conversation has gained dramatically in urgency and in depth driven by the global policy discourses on climate change and the UN Sustainable Development Goals. Still, water is hardly on the radar for most financiers and financial policymakers. Whether or not water risks are financially material today (or have the potential to be in the future) is the subject of debate. This volume clearly argues in favor of a further deepening of this debate, toward the growing attention to the opportunities and challenges related to water in the sphere of finance.

The recognition of financing water as a major global challenge first gained traction at the international level within the water community itself. Michel Camdessus, former Managing Director of the International Monetary Fund (IMF), was seminal in putting finance on the agenda of the water community at the 2003 World Water Forum in Kyoto. Angel Gurría, then Secretary-General of the OECD, reinforced this message at the subsequent Forum in Mexico City in 2006. Since then, finance has become a recurrent theme of the World Water Forums as the critical enabling condition, together with governance, to achieve progress toward increased water security. This was encapsulated in the Sustainable Development Goal 6 (SDG6) on water and sanitation. Traditionally, public finance had always been considered the default option but constrained fiscal resources in many countries strengthened the case to attract more private risk-bearing capital.

While the debate around these issues remained initially within the water community, the global discourse also started emphasizing that water is not an insular sector but is strongly affected by policies and investments in the

Financing Investment in Water Security: Recent Developments and Perspectives.
DOI: https://doi.org/10.1016/B978-0-12-822847-0.00021-1

an insular sector but is strongly affected by policies and investments in the economy, land use, and ecosystems. Hence, sustainable use and management of water resources require policy alignment and collaborative action across several sectors and ministries, raising the expectations regarding financing and adding complexity around the questions of who should and how to finance needed investments. The political visibility of water was further elevated in 2020 when the first G20 Water Dialogue took place under the Presidency of Saudi Arabia. The G20 Leaders' Declaration dedicated, for the first time, a full paragraph to water[1], and the G20 committed to an annual dialogue on water. The background note (authored by OECD, FAO, and IIASA) that informed the G20 Water Dialogue covered notably policy, finance, and innovation.

Importantly, the tenor of the conversation around water financing has shifted. In 2003, the message focused on the financing gap—financing needs far exceed available finance for water-related investments—and reiterated the call for more finance to close this gap: "There is widespread agreement that the flow of funds for water infrastructure has to roughly double, with the increase to come from all sources [public and private]" (Winpenny, 2003, p.11). In many developing countries only a quarter of the calculated financial "need"—to connect people to safe water and improved sanitation—is actually covered, often even less. Subsequent studies and high-level fora, however, have kept highlighting the persistent shortage of "bankable" projects and the need to improve the quality of investment proposals for water-related investments. While this line of thought remains prominent and certainly relevant, more recent analyses also explore whether financial markets and the finance industry are equipped to direct a significant share of financial flows to projects that contribute to water security and other global water-related goals such as those under SDG 6. These goals have social and economic value and may have a solid business case. Yet the nature of these investments and their political economy appear to discourage transactions. The situation is arguably compounded by ill-adapted financial instruments and misaligned incentive systems.

For the financial actors, most water investments are still less attractive and perceived as riskier compared with many alternative investments. Concern started growing that financial assets themselves are increasingly at material

[1] "We acknowledge that affordable, reliable, and safe water, sanitation, and hygiene services are essential for human life and that access to clean water is critical to overcome the pandemic. We welcome the G20 Dialogue on Water as a basis to share best practices and promote innovation, and new technologies, on a voluntary basis, that will foster sustainable, resilient, and integrated water management". (http://www.g20.utoronto.ca/2020/G20_Riyadh_Summit_Leaders_Declaration_EN.pdf).

and systemic risk from water insecurity across the world: under the pressure of demographic, economic development, and climate change. Sustained adequate provision of water for productive purposes such as for agriculture, industrial and energy production, and protection from river or sea flooding, could no longer be taken for granted for the duration of an asset's life.

With real estate and industrial assets threatened, water security and hence its financing, are now emerging as a condition for a healthy financial industry, and for the stability of the financial sector itself. Central banks and regulators are taking a rapidly growing interest in the implications of climate and environmental risks, including water, as exemplified by the 2017 and 2021 Recommendations by the Task Force on the Climate-related Financial Disclosures of the Financial Stability Board, and emerging discussions in the Network for Greening the Financial System. Such issues were topical on the agendas of the 2020 and 2021 meetings of the Roundtable on Financing Water[2]. In a sense, "water" is now increasingly perceived as having a *financial* value too.

A new context for the discourse

The context for the discourse on financing water is changing as well. Three developments deserve particular attention, as they point to valuable directions while identifying further challenges to financing water.

First, we better understand how water management drives growth and sustainable development. The economic case for water management and more investment in water security has been firmly established. In 2015, *Investing in Water, Sustaining Growth* (Sadoff et al., 2015) firmly positioned water security as a condition for sustainable development, at basin and global levels. It documented the cost of inaction and outlined novel perspectives for water-related investments, in particular investment pathways supporting successful development strategies and effective investments in different parts of the world. That same year, the IMF published a landmark staff discussion note (Kochhar et al., 2015) documenting the impact of water challenges on growth and macro-stability; the report also recommended macro-economic policies, such as replacing subsidies for water security by targeted social support.

[2] The Roundtable on Financing Water is a global public-private platform established by the OECD, the Netherlands, the World Water Council and the World Bank, to facilitate increased financing of investments that contribute to water security and sustainable growth. For more information, visit https://www.oecd.org/water/roundtable-on-financing-water.htm.

More recently, the Dutch initiative Water as Leverage[3] explores how investing in water is effective to build resilience to climate change. It documents, and endeavors to address, the challenge of investing the catalytic first millions of USD that can leverage investment at scale. Its *Reflect* report (Ovink et al., 2021) notes that most stakeholders—including development finance institutions—welcome this notion, yet struggle to apply it in practice. In particular, it highlights the impediments in prevailing financing modalities to finance urban resilience and water-related projects.

Second, the global discourses, policies, and initiatives related to climate action and sustainable finance are growing more urgent and intensive and the linkages to water increasingly recognized. While climate mitigation concerns predominantly energy and transport, climate adaptation is primarily about the water cycle, agriculture, and habitats. Water security and resilience, thus, are gaining additional traction in these narratives. Financing climate action for both mitigation and adaptation is a rapidly expanding field of work, on the global agenda and in practice. These recent developments in climate and sustainable finance are opening new opportunities for financing water. However, as IPCC reports observe, of all climate-related finance currently only about 5%–10% is geared to adaptation measures (Gupta et al., 2014). New dedicated funds (such as the Green Climate Fund), financing instruments (e.g., green bonds and Special Purpose Vehicles) and engagement by a widening range of financiers (including, but not limited to institutional and impact investors) can accelerate investments, which will significantly contribute to climate resilience and water security. They are benefiting from institutional and regulatory developments intended to guide conducive investment decisions and avoid investments with high carbon and/or water impacts. Taxonomies and guidelines are being formulated which set criteria that define green, low-carbon, climate-resilient investments, and for financial disclosures. While the European Commission is pioneering this approach, other countries and institutions are proceeding along similar lines.

While promising, such developments also face limitations. Uncertainties and ambiguities may arise in definitions and there is a distinct risk of "green-washing." The size of sustainable finance is still modest compared to main-stream financial flows. While supporting the steady expansion of sustainable finance matters, a major challenge lies in displacing environmentally-harmful investments and re-aligning financing flows toward investments that support environmental objectives in time to achieve a net-zero, resilient world by

[3] For more information and inspiring stories, visit https://www.worldwateratlas.org/curated/water-as-leverage/.

mid-century. And finally, some ambivalence remains because what qualifies as climate-resilient or sustainable finance may not necessarily support other key tenets of the water agenda such as for instance provision of safe water to communities.

Third, at global scale the debate regarding the public versus private roles in water services and regarding the water resource has matured and taken a more pragmatic turn, focusing on their respective comparative advantage rather than as competing solutions to water financing. Deeper analysis and, indeed, experience, with a much wider set of field-tested arrangements suggest that nations and communities can retain critical responsibilities and functions—such as ownership of the resource and of critical infrastructure, and regulatory functions—in the public domain while productively partnering with private parties and initiatives for other functions. In the past, this debate had tended to be narrowed down to two arenas. On one hand, water supply and sewerage (and, rarely, irrigation and river infrastructure) services could be delegated to private operators in concessional or lease constructions. The assumption in the late 1990s that such operators would be distinctly more effective in raising capital for investments did not materialize. After the 2008 financial crisis, the appetite and capability of these operators to scale up also proved constrained. On the other hand, the right of access to water is an intrinsically cultural and political issue. Private ownership systems have remained exceedingly rare (south-western US, Chile) but several nations have initiated market-based or economic-value-based arrangements to allocate entitlements to use water under scarcity conditions to their optimal use (e.g., Australia and some EU countries). Such arrangements are increasingly also applied to, e.g., wetland conservation and management (Davis and Johnson; De Biévre and Coronel; and Muruven, this volume). Still, the secure access to tap or drinking water is often perceived as of "national strategic import" as described by Wu (this volume) for China, where most wastewater treatment and industrial water provision is delegated to private operators in contrast to the public water supply which remains in the hands of local governments. Nonetheless, public utilities in developed and developing economies alike that are technically competent, sizeable and well-governed find it easier to attract finance from capital markets. They do so in the form of (municipal) bonds, loans or other funds, sometimes through a (semi-)public financial intermediary such as for instance the US Environmental Protection Agency (Gebhardt, this volume), the Netherlands Water Bank, and an array of new dedicated funds in Asia (Fernandez-Illescas et al., this volume) and elsewhere. In this volume, discussions on private finance in effect explore the opportunities and conditions to access commercial sources of finance, irrespective of the status of operators of water services.

Ambition and scope of the book

This volume aims to frame water financing as central to the debate and action on sustainable development. It raises the sense of urgency to resolve outstanding challenges in order to scale up investments, faster and more effectively. It does so while reflecting on the new context outlined above. It intends to push the boundaries by taking stock of the diversity of new developments and documenting new opportunities while attempting to better understand and address some of the pervasive bottlenecks.

To do so, the scope of the book is distinctively broad:

- The volume is global in scope and ambition, yet it stresses the critical importance of regional and local specific circumstances and action.
- It endeavors to cover the full spectrum of water-related investments; while water supply and sanitation infrastructure and services feature prominently, other dimensions feature as well, notably irrigation, ecosystem management, and water security at large.
- While it focuses on water, it argues that financing water extends beyond the traditional scope of the water sector, as water is deeply impacted by decisions in different sectors such as energy, land use and agriculture; it is, thus, a matter of interest—and, often, concern—for a wide range of policy communities, financiers, and institutions in charge of regulating financial markets, consumer affairs and the environment.

To deliver on its ambition, this volume builds on three sets of expertise:

- Scholars and other experts who argue convincingly about why financing water matters, and what conditions need to be satisfied to align financing flows with needs and capacities at local, national, and global levels.
- Analysts who assess the financing needs and capacities and describe the policies required to achieve them.
- Practitioners who have put in place financing policies, institutions, financial instruments, and contractual arrangements that lead to transactions and actually have contributed to financing water.

The book is intended for a wide audience from the water and finance communities, as well as from other sectors and policy communities which either heavily depend on water and/or whose activities strongly impact the resource, such as agriculture, industry, and ecosystems and biodiversity management. Within these sectors, the book should be able to provide insight and guidance to those involved in the industries, such as water service providers, agro-businesses, and financiers, but also insurers, credit rating agencies, financial and environmental accountants, and consultancies specializing in climate and sustainable finance taxonomies. This audience would also extend explicitly to regulating institutions on both the water and environmental side (e.g., on tariffs, performance and conservation) and the financing side (capital market regulators; bank, pension funds, and insurance oversight bodies; central banks and treasuries).

The book's structure reflects this ambition. In Part 1, five chapters lay the foundations for a thorough discussion on financing water. They set the scene by describing and analyzing the broader context that affects financing water as well as basic operational aspects. Mark Smith highlights the "systemic mission" of water as regards the SDGs. He emphasizes that the assumption of water abundance is no longer valid and that the political economy determines how water is valued and taken into account by different authorities and diverse stakeholders. Guy Alaerts outlines the mismatch between the systemic and integrated nature of water as resource, and the fragmented institutional landscape. He analyses the diverse water-related asset classes and their distinctive features as relevant for financing. At the same time, he clarifies how financial assets are increasingly exposed and vulnerable to water-related risks. This message provides a good segue into Mireille Martini's chapter, which explains why, under prevailing post-financial-crisis conditions, the financial sector is ill-equipped to consider water and water-related risks. She calls for a revision of financial and economic regulation to help redirect financial flows toward investments that contribute to water security and sustainable growth. Thereafter, Hein Gietema focuses on a key element in transactions; he illustrates how financial structuring can help enhance the risk-return profile of water-related investments over the project lifecycle. Finally, Martin Baker provides a practitioner's reflection on the complexity of drafting contractual arrangements designed to reconcile the different perspectives and priorities of the parties to a transaction.

Part 2 presents evidence of financing needs and capacities, globally and in several of the world's regions. The first two chapters characterize financing needs and capacities globally. Bruce Horton and Richard Ashley look into the drivers of the financing needs and capacity for investments in water supply and sanitation; they make a heuristic distinction between geographical areas with extended coverage, and those in need of further coverage extension. They introduce innovation as an option to mitigate the rising costs of service provision. Guy Hutton compares the most recent projections of water-related financing needs and capacities, taking SDG6 as the overall ambition. The chapter provides some level of disaggregation by world region. It introduces the discussion on affordability of water services, a concept that is often oversimplified and not adequately reflected in policy-making. Then, four chapters characterize financing needs and capacity at regional level, with focus on Africa, Asia, China, and Europe. Projections and analyses, however, suffer from the paucity of comparable and robust data, in particular beyond water supply and sanitation. Still, some distinctive features emerge, such as the benefits of having a harmonized policy framework and data collection effort in Europe, and the magnitude and diversity of financing needs and opportunities across Asia. Elaine Wu documents the latest developments of financing wastewater management in China, which is marked by the introduction of more sophisticated financial instruments such as real estate investment trusts.

Part 3 collates real-life practical examples of financing mechanisms and enabling conditions that facilitate transactions and financing flows to projects that contribute to water security and sustainable growth. Lessons are learned on the prerequisites to make these mechanisms and arrangements deliver and to adapt and scale them up, if and when appropriate. To provide an introduction to the case studies, Kathleen Dominique and Alex Money provide an overview of different types of water-related investments, financing models, and the extent to which they could be adapted to address a diverse set of financing challenges and contexts. Several case studies follow, detailing distinctive water financing approaches.

Jim Gebhardt et al. discuss the history of the public funding provided to water and sanitation infrastructure through the lens of US government programs established to address national water management goals. The chapter documents in detail the two dominant US federal loan programs—notably the Clean Water and Drinking Water State Revolving Funds managed by the states and the Water Infrastructure Finance Innovation Act and reflect on the necessary enabling conditions to deploy these approaches in other contexts. Also drawing on experience in the United States, Adam Davis and Sara L. Johnson focus on mobilizing private capital for large-scale ecological restoration and conservation. They detail the key elements for the market and regulatory requirements that underpin this financing model and how it applies in specific cases.

Exploring financing models in developing countries, the following two chapters detail an approach to financing water supply and sanitation as exemplified in Kenya and an approach to catchment protection deployed in numerous countries around the world. Joris van Oppenraaij and co-authors depict the establishment and operations of the Kenya Pooled Water Fund, which aims to access local capital markets to mobilize water and sanitation infrastructure investments. They recount the key steps and challenges encountered in setting up such a fund, as well as the design features that underpin the approach. Turning to catchment protection, Bert De Bièvre and Lorena Coronel document the experience of Water Funds, most notably via the first of such funds establishment over 20 years ago in Quito Ecuador. The lessons learned provide insights on how such funds can be established and adapted to new contexts. Finally, Dean Muruven explores opportunities to develop bankable and beneficial projects that contribute to sustainable and resilient freshwater ecosystems within the context of broader landscape financing plans.

Cross-cutting messages and ways forward

A strong economic case has failed to translate into financing flows at scale

The work collated in the volume concurs in the observation that the economic case for investing in water generally is compelling, but struggles to be translated into financing flows at the scale commensurate with the challenges (OECD, 2022). This observation has been made before. Since the endorsement of the Sustainable Development Goals by the global community in 2015, it is well acknowledged and documented that water—in addition to being the subject of a distinctive SDG—is a condition to achieve multiple other goals. From that perspective, water is endowed with a "systemic mission" as regards the SDGs (Smith, this volume).

The economic case for financing water, however, still suffers from several caveats. Distinct caveats relate to the difficulties with valuing water (Alaerts, this volume, documents the common misalignment between water value, cost, and price); and the split incentives between the ones who benefit from these investments and the ones who should bear the costs. Additional complexity to assessing the case for investment and designing investment under uncertainty derives from the uncertainties about future water demand and availability and the pronounced vulnerability to risks under a changing climate (Smith, this volume).

All projections of financing needs and capacities concur to stress that an enduring financing gap exists between current levels of finance and needs to meet national, regional, or global goals related to water (SDG6, or regional objectives where they exist) and to sustainable development at large (see most chapters in Part 2 of this volume). They also concur in claiming that this gap, mostly, does not reflect any shortage of money, as financial resources are globally abundant, including in a number of developing countries where local capital markets have been emerging and maturing over the past two decades.

In their own ways, each chapter of the book stresses some of the reasons why finance struggles to reach some of the most needed and valuable water-related investments. Well-recognized causes are poor project definition, weak credit-worthiness of water agencies and municipal utilities, an unattractive risk-return profile due to weak enabling conditions, deficient institutional capacity, and lack of financial structuring (Baker, and Gietema, this volume). However, new analyses also emphasize the issues arising from the prevailing conditions under which the financial sector operates (e.g., prudential regulations), which impede financing flows to activities that are valuable from an economic, social, or environmental perspective and may also undermine investments contributing to water security and sustainable growth. This perspective further explains the well-established insight—that commercial financiers and capital markets currently play a minimal role in financing water in most cases.

As a consequence, there is a role for policies, regulations, and institutions to set enabling conditions that direct financial flows toward investments that contribute to water security and sustainable growth. This volume argues that such enabling conditions include water and environmental policy, as well as a broader range of domains as well, most prominently financial regulation and prudential rules. Aligning needs and capacities will require action from a range of stakeholders, including public budgets, accounting standards and water tariffs, as well as central banks, in their capacity of regulators of the finance industry.

Available data and analytical tools are evolving but still fall short of being fit for purpose

Data are pivotal in making the economic case for water-related investments, in particular, to monitor exposure and vulnerability to water-related risks and the benefits of investing in water security and sustainable growth. The fact that a large share of relevant data on water risks is in private hands (e.g., the insurance industry) can create asymmetries of information that hinder policy development and affect a fair allocation of risk and benefits from investing in water. A combined effort from public authorities at local, national, and international levels is required to produce, standardize, share and update public data on related issues.

Additional sources of information should be designed to inform analytical tools that can document exposure and vulnerability to water risks and the benefits of investing in productive and secure water. Four sets of tools are discussed in this volume, which are taken for granted and seldom questioned in the literature and policy guidance on financing water.

First, cost-benefit analyses fall typically short when assessing the "true" value of water and of water-related investments, in particular in an uncertain future. Investment decisions would benefit from new analytical tools designed to assess the robustness of investment and policy decisions across a range of possible future water regimes (Smith, this volume) and to value flexibility and the capacity to adapt to shifting conditions and unexpected events.

Second, environmental impact assessments (EIA) need to be deepened or redesigned to capture the potentially diverse consequences of (series of) investments in water at different geographical scales. Alaerts (this volume) argues in favor of sectoral EIA. Others recommend to supplement project level analysis by the multicriteria analysis of investment pathways, informed by the values of stakeholders to assess a portfolio of projects in a particular landscape or basin and how these may evolve over time under different scenarios (Brown and Boltz, 2022).

Third, it is increasingly essential that financiers conduct due diligence on the water impacts of their investments (Alaerts, and Gietema, this volume). This matters as each new water appropriation now impinges on already existing ones, triggering cascading, or spill-over effects and affecting the risks and return profiles of other investments. This is in line with the requirements of emerging taxonomies that help define green, climate-resilient, or water-wise sustainable investments. For reasons noted above, reliance on due diligence is also plagued with conceptual, methodological, and institutional challenges.

- How are water impacts defined? How are potential inconsistencies across geographical scales and time horizons considered and addressed?
- How to deal with uncertainties, be they derived from the paucity of data, or from deep uncertainties triggered by climate change and hard-to-predict changing water regimes?
- In what ways are different categories of water users, stakeholders and communities potentially exposed to water-related risks, and how are they involved in the assessment and in the investment decision?

Finally, expanded and more reliable data on exposure and vulnerability to water risks and on the value of water can inform new modeling tools, which can then support decisions across a range of policy domains and investment opportunities (Martini, this volume). One area that is likely to receive more attention in the future is the use of geospatial data, and in particular its potential for improving the allocative efficiency of capital investments into the water sector (Dominique & Money, this volume).

The way the financial markets and industry operate today hampers water finance

There are two broad ways in which financial markets and the prevailing modalities of the finance industry's operation today fail to properly value water-related investments and incorporate water-related risks.

First, while financial assets, such as productive assets and real estate, are increasingly exposed and vulnerable to water risks (Alaerts, this volume), the finance industry seems to only slowly come to terms with these risks (with the notable exception of the insurance industry). It helps to make a distinction between long-tenor investors (institutional, impact investors, etc., but also, increasingly, fund, and asset managers) and short-term financiers. The former ones are ramping up their efforts to come to grips with new sustainability risk categories in the face of environmental "tipping points" that are projected for the next 2–3 decades. Investors with only short-term positions are starting to realize that, even though they are not directly affected by what will be happening after the next decade, a systemic risk is arising with asset value getting destroyed - at times on very short notice. Such a systemic risk may result from assets being located in value chains or locations that unexpectedly may prove particularly vulnerable to water risks; or from all investors getting access to the same information at the same time, triggering a sell-off; or from regulators or governments taking regulatory action on short notice, nationalizing assets, or deciding to intervene in markets to push green strategies or protect national interests.

From that perspective, the voluntary (and in the future possibly mandatory) disclosure of firm-level data on exposure and vulnerability to, and mitigation strategy of water-related risks is a significant development (see the recommendations of the Task Force for on climate-related Financial Disclosure of the Financial Stability Board). However, disclosure still cannot fully ensure alignment of the economic and financial implications of water-related risks or the benefits of investing in water. This is particularly a concern in the case of banks and financial institutions. While regulatory developments under several jurisdictions mandate banks to disclose information about their portfolio's exposure to water-related risks, such a request partly misses the point because, as analyzed by Martini (2022), the transmission of risks caused by floods, droughts, or water pollution into material financial impacts on the financial sector remains minimal. Under prevailing prudential regulations and accounting standards, disclosure remains voluntary. More importantly, banks and financial institutions may arrange to ignore, hedge or outsource these risks. Ground-breaking work on the materiality of water risks for financial institutions points to a new role of financial regulators (including central banks) in setting the incentives correctly (Martini, 2022) and reconciling the economic and financial perception of water-related risks and their sectoral and potentially systemic implications. Until further regulatory initiatives are taken, it is unlikely that the vast majority of financial flows will start to consider and reflect the issues related to water security and sustainable growth.

Second, Martini (this volume) explains how—in several world regions—the combination of prudential regulations and credit deregulation forces banks to "originate and distribute," making financing through markets (bonds) much more common than financing through debt (loans). This shift in financing mechanisms is consequential for a wide range of water-related projects, which tend to be less adapted to meet the expectations of financial markets, because they do often lack a strong revenue-generating capacity. In such a context, opportunities to scale up financing for water—in particular through debt finance—will remain limited, despite repetitive calls for action from the global development community; this situation is compounded by the sometimes poor creditworthiness of the borrowing entity for water-related investments (sovereign, city, utility, etc.). Such opportunities will need to focus on the level of transactions, as a transformation of the financial system is out of reach. Still, the global new drive, accelerating rapidly, for climate and green finance may lead to more conducive environments and create new suitable instruments and arrangements.

Opportunities exist to scale up transactions that contribute to water-related investments

Water projects are diverse and distinct asset classes (Alaerts, this volume) will have different capacities to access finance. The distinct risk-return profile and project attributes of each investment should inform the appropriate financing strategy (Dominique & Money, this volume). Some of these asset classes have fairly straightforward financing cases, in particular when regular revenues can be secured and ring-fenced, and operational risks are well understood (e.g., reservoirs for hydropower generation, or desalination plants for sea or brackish water). Others are less straightforward (e.g., nature-based solutions for flood prevention, or distributed water distribution systems), but financing options are still available.

The understanding that water investments are typically made as part of a broader investment or development plan and ensuring proper project definition can go a long way in making projects more bankable (Bakker, this volume). For instance, the definition of urban water supply, which can encompass surrounding peri-urban areas, partly determines bankability of access to water in rural areas. Similarly, aggregation of fragmented water supply or sanitation can make investing in service provision more attractive for financiers. At the same time, appropriately designed financial structuring can make different steps in the life cycle of a project bankable (Gietema, this volume). Financial structuring allocates risks and revenues in ways that meet the distinctive expectations of different financiers and can lead to the selection of appropriate financing instruments and contractual incentives.

A recurring theme is that operational performance is a precondition to bankability (Horton and Ashley, Rosentock and Leflaive, Fernandez-Illescas et al., this volume). In the case of water supply and sanitation, structural cooperation among utilities leads to better pooling of resources, lower unit costs and higher technical competence, all contributing to operational performance and better planning.

Bankability needs to be reconciled with other policy objectives, including equity

Discussions on financing water have traditionally focused on, on one hand, the bankability of investment proposals and the creditworthiness of the borrower institutions, and, on the other hand, on expanding the pipeline of bankable projects. As one partial solution, a variety of project preparation facilities were established by financial institutions and development agencies to support the borrowers with the development of bankable projects. While this effort is contributing to accelerating transactions, such facilities come with their own disadvantages such as moral hazard, conflicts of interest, and unclear accountability.

Although bankability is central to the financial assessment of a proposal, the concept may need to be augmented or enhanced by factoring in other considerations. In particular, financial structuring would benefit from guidance or a frame regarding how much risk can be transferred to the public sector while revenues accrue to private stakeholders. This raises important questions of equity and fairness, including across time when risks are transferred to subsequent generations.

A particular angle to this issue relates to the cost of water security (OECD, 2013), namely, how that cost is shared across communities, now and in the future, and the resulting affordability of water security for these communities. Affordability is a common, and challenging topic in discussions around water, notably on financing water supply and sanitation services. The notion is equally applicable to other water services, such as access of poor or smallholder farmers to agricultural water to grow their crops and ensure the livelihood of rural communities, and the fair allocation of risks of water excess (flooding, drainage) or scarcity across communities and users. Hutton (this volume) claims that more than 50% of the projected financing needs to achieve SDG6 ought to be spent on the population with the bottom 40% of income. However, this raises severe challenges of affordability and calls to allocate public and development (concessional) finance wisely. Hutton (this volume) convincingly argues that prevailing definitions of affordability are not fit for purpose, based on simplistic understanding of water uses and unrealistic assumptions about households' income. Taking a step back, addressing affordability issues first requires ensuring a fair allocation of risks and revenues, now and in the future, so that the most vulnerable are not disproportionally affected (OECD, 2019). Blended finance can be part of the answer, with public and development finance used to crowd in other sources of finance, while ensuring that risks and revenues are fairly and equitably allocated. Also, social payments targeted at poorer households, and fine-tuned micro-credit arrangements (preferably complemented with technical assistance) can help manage this challenge.

Political economy is pivotal to make financing water happen where it is most valuable

Financing water is not merely about analytics, economics, and finance. The political economy—at local, basin, or larger scales—is fundamental in determining how water is valued (Smith, this volume) and how, in turn, the value of water informs decisions in policy domains that affect water availability and demand, and exposure and vulnerability to water risks. Furthermore, the political economy shapes what are considered appropriate levels of water security, i.e. the level of water-related risks communities are willing and able to assume. It is also foundational to the definition of the other side of the equation, namely, how much a community is willing and able to pay for water security. In practice, the current level of investment in water security reflects power and domination in relation to these issues, who is (and will be) exposed and vulnerable to water risks and who bears the costs, now and in the future. From that perspective, calls for stakeholder consultation and inclusive decision-making are both relevant and naïve. They are relevant as, indeed, inclusiveness contributes to better informed decisions that eventually receive stronger support; naïve, because no institutional arrangement or process will of itself be able to overcome such cases of power and domination. We need new narratives to tell the value of water for different communities, including the ones whose voice is silent or not listened to in financing decisions today.

As mentioned above, these narratives need to go beyond the water community and resonate with institutions and communities that affect water availability and demand, and exposure and vulnerability to water-related risks. For instance, this volume argues that we start to understand better the critical role played by finance regulation, financial institutions, and financial instruments, in directing the systemic transitions required to make water financing available where it creates the most value for communities.

New developments in water economics and beyond need to reframe and advance the on-going conversation with institutions and stakeholders active in agriculture, land use and urban development, energy supply, and climate action. Such developments should also have the ambition to accelerate the process of putting water on the agenda of central banks, which are tasked primarily with long-term financial stability. This is a fundamental and strategic recalibration of the way in which the value of water affects national development. More work is required to ensure that agenda will further shift from voluntary disclosure and monitoring to regulation. For the International Monetary Fund, for example, it would make sense to urge central banks to put water high on their agenda, acknowledging that a growing range of economic objectives and financial assets will be increasingly vulnerable to water-related risks, and that financial markets and institutions cannot remain artificially insulated from such systemic risks.

In addition, multilateral development banks (MDBs) have an opportunity to be transformational, while delivering on their development mandate.

MDBs enjoy the distinctive advantages that they can access global currencies, lend to sovereigns, and have direct and privileged access to Ministries of Finance and Treasuries. Thus, they would be well positioned to mitigate the bias of many Treasuries of developing countries for export-oriented investments that insufficiently reflect the value of their water resources. They can promote an enabling environment that is conducive to direct finance to investments which positively contribute to water security and sustainable growth notably by simultaneously crowding in domestic commercial finance.

Contributions in this volume contribute to a re-evaluation of changes required to align finance with the ambition of water security and sustainable growth. Some required changes are technical and would benefit from innovation in financing techniques and transaction arrangements. Others relate to rules and regulations—the enabling environment—and we better understand that such enabling conditions reach beyond water policies and institutions and encompass, in particular financial regulation. Yet others relate to how communities value water and water security, the social norms that eventually translate such values into economic and financing opportunities. The global community would greatly benefit from paying attention to the multiple options that coexist across societies and cultures to make the best use of available water resources, to avoid building future liabilities, and to secure investments in line with social expectations. While this is probably true for other issues, water—thanks to its ubiquitous economic, social, and cultural importance—provides an opportunity to inform such a dialogue that is inherently global and that would resonate across multiple agendas.

The editors of this volume

Xavier Leflaive leads the Team that works on Resilience, Adaptation to climate change, and Water in the OECD Environment Directorate. The team's work covers water quality (with a focus on contaminants of emerging concern) and financing water – more specifically enabling conditions that facilitate financing investments that contribute on water security and sustainable development. Xavier Leflaive coordinated the Recommendation of the OECD Council on Water, which captures OECD policy guidance on the management water quantity and quality, water-related risks, governance, and finance. He spearheaded regional analyses of water-related financing needs and capacities (in Europe, and Asia and the Pacific). Over the last decade, he has facilitated water policy reforms in Brazil, the Caucasus and Central

Asia, several European countries, Korea and Thailand. Xavier Leflaive studied business administration and social theory in France, Canada, and the UK. He holds a Ph.D. in Social and Political Sciences from the University of Cambridge, UK. He lectures on the intergovernmental organizations and on Financing for Water Security and Sustainable Growth at the Paris School of International Affairs at Sciences Po.

Guy Alaerts has a long track record in water engineering, policy and finance, covering water services, irrigation, river basin and flood management, environmental management, and climate adaptation. After a 3-year development assignment in Indonesia, he was appointed Professor in Public Health Engineering in 1986 at the IHE International Institute for Hydraulic and Environmental Engineering, Delft, The Netherlands, of which he also became Vice Rector. From 1996 through 2015 he was Principal Water Resources Specialist and Programme Manager at the World Bank, Washington, DC. In that capacity, he prepared and implemented numerous operations in Asia, Africa, Latin America and Europe, conducting sectoral and economic analytical work, advising on national policy, and designing and supervising investment programs in water and natural resources. He is currently a Professor of Capacity and Knowledge Development at IHE. He earned his doctorate in engineering and studied business administration at the Catholic University of Leuven, Belgium.

Kathleen Dominique is a Senior Policy Analyst at the OECD and leads the OECD's program of work on financing water, including the Roundtable on Financing Water—a dedicated platform to engage policy makers and the finance and investment community. Since joining the OECD in 2010, Kathleen has led work on water resources allocation, groundwater, water management in the context of climate change adaptation, biodiversity policy reform and country-level reviews. She also coordinates the OECD's Working Party on Biodiversity, Water & Ecosystems. She engages in numerous advisory roles, including advising the Saudi G20 Presidency to support their efforts to launch the G20 Dialogue on Water and contributing to numerous technical expert and advisory groups on finance and water issues. Prior to joining OECD, Kathleen worked in the United States and Nigeria as well as the Paris-based think-tank IDDRI. She lectures on Financing for Water Security and Sustainable Growth at the Paris School of International Affairs at Sciences Po. She has a Master's in Public Policy from Science Po (France) and a Bachelor's in Business Administration from the University of Notre Dame (USA).

Disclaimer

Xavier Leflaive and Kathleen Dominique acknowledge the OECD Environment Directorate for the support provided throughout the process. Anthony Cox, formerly Deputy Director, provided insightful comments on the overall introduction.

Xavier Leflaive, Kathleen Dominique and Mireille Martini contributed to this volume in their own personal capacity. The opinions in this volume do not necessarily reflect those of the OECD, or its Member countries.

References

Brown, C., Boltz, F., 2022. Strategic investment pathways for resilient water systems. In: OECD Environment Working Paper. Organisation for Economic Co-operation and Development, Paris.

Gupta, S., Harnisch, J., Barua, D.C., Chingambo, L., Frankel, P., Garrido Vázquez, R.J., Gómez-Echeverri, L., Haites, E., Huang, Y., Kopp, R., Lefèvre, B., Machado-Filho, H., Massetti, E., 2014. Cross-cutting Investment and Finance Issues. In: Climate Change 2014: Mitigation of Climate Change. Contribution of Working Group III to the Fifth Assessment Report of the Intergovernmental Panel on Climate Change.

Kochhar, K., Pattillo, C., Sun, Y., Suphaphiphat, N., Swiston, A., Tchaidze, R., Clements, B., Fabrizio, S., Flamini, V., Redifer, L., Finger, H. (2015), Is the glass half empty or half full? Issues in managing water challenges and policy instruments. International Monetary Fund, SDN/15/11, Washington, DC.

Martini, M., 2022. Watered down? Investigating the financial materiality of water-related risks in the financial system. In: OECD Environment Working Paper. Organisation for Economic Co-operation and Development, Paris.

OECD, 2013. Water security for better lives. OECD Studies on Water. https://doi.org/10.1787/9789264202405-en.

OECD, 2019. Making blended finance work for water and sanitation: unlocking commercial finance for SDG 6. OECD Studies on Water. https://doi.org/10.1787/5efc8950-en.

OECD, 2022. Financing a water secure future. OECD Studies on Water.

Ovink, H., Schous, J., Declerck, J., van Godtsenhoven, S., Vandenmoortel, B., Naudts, N., Leflaive, X., Lyons, S., Dominique, K., Kempenaar, A., Laeni, N., Brink van den, M., Busscher, T., Arts, J., 2021. Water as Leverage – Reflect. Government of the Netherlands, The Hague. https://drive.google.com/file/d/1s9tNlk9VCLwtXakQ6-POKPlY4W6P112v/preview

Sadoff, C.W., Hall, J.W., Grey, D., Aerts, J.C.J.H., Ait-Kadi, M., Brown, C., Cox, A., Dadson, S., Garrick, D., Kelman, J., McCornick, P., Ringler, C., Rosegrant, M., Whittington, D., Wiberg, D., 2015. Securing water, sustaining growth. Report of the GWP/OECD Task Force on Water Security and Sustainable Growth, University of Oxford.

Winpenny, J., 2003. Financing water for all. Report of the World Panel on Financing Water Infrastructure chaired by Michel Camdessus, World Water Council, Global Water Partnership, Marseilles. ISBN 92-95017-01-3.

PART I

Investing in water and growth: A global perspective

CHAPTER ONE

If not now, when? Converging needs for water security, systemic change, and finance and investment

Mark Smith
International Water Management Institute, Colombo, Sri Lanka

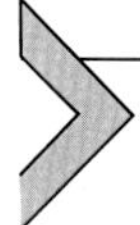

1.1 Water in the economy: multiple objectives and competing needs

"Water is life" is a common refrain, because all life needs water. Water, as a consequence, shapes the characteristics and myriad forms of ecosystems, as water flows and cycles through them and as scarcity and pollution disrupt the intricate webs of life within them. An economy is no different. Demand for water is distributed across economic sectors, and throughout human societies. Just as it shapes ecosystems, water flows and cycles through economies, and whether there is too little water, too much or it is too polluted influences success or failure, profit or loss, and how welfare is distributed among groups of people, across sectors and even between countries. "Water is life" matters not just to biology, but also in the economy, and therefore—in principle, at least (see also Alaerts, This volume)—to the case for investment in water.

Water needs have to be met in every household on Earth, every day. Water and sanitation are a human right that remains unmet for many, with 29% of people worldwide lacking safely managed supply of drinking water in 2017, 40% lacking basic handwashing facilities, and 55% or 4.2 billion people without safe sanitation services (UNICEF and WHO, 2019). Yet, water withdrawals for domestic use are a relatively small proportion of the total, at 10% on average worldwide, while water use in agriculture accounts for about 70% (and higher in agrarian economies and dry regions), and energy generation and industry approximately 20% (FAO, 2020). One set of uses can impact other uses, at the scale of the household or across the whole economy, through changes in the availability of water, its timing or its quality. How

Financing Investment in Water Security: Recent Developments and Perspectives.
DOI: https://doi.org/10.1016/B978-0-12-822847-0.00009-0

water is used and managed therefore creates risks that propagate through the economy, exacerbating or (if managed well) ameliorating risks from drought, flood, and extreme weather. As a consequence of water's potential to cause damage and disruption, "water crisis" was ranked among the top 5 global risks each year for 2015–2020 in the World Economic Forum's annual survey of perceptions of risk by economic and opinion leaders (World Economic Forum, 2020). In tandem with human demand for water and its attendant risks, nature's water needs have to be met, or biodiversity is lost (Tickner et al., 2020; WWF, 2020) and the ecosystem services that people rely on—including for water storage, supply, filtration, and recycling—are degraded, with consequences not just for conservation, but also risks, welfare and jobs (WWAP/UN-Water, 2018).

The intricate connections among water uses and users are, however, changing. Demographic change, economic growth, and shifts in diets each affect patterns of water resource exploitation and water availability for different uses (FAO, 2020). Alongside these shifts, climate change has profound implications for the influence of water in the economy and considerations for investment in water, and for water-related risks to financing and investment in other sectors (Alaerts, 2019). The water cycle is embedded in the global climate and, therefore, as the climate changes, the availability of water, seasonality in water flows and the risks of drought, floods, and extreme events, as well as salinization and pollution, are changing too (Smith et al., 2019). Uncertainty related to water is expanding, which is compromising conventional and long-accepted methods for assessing water risks (Ray and Brown, 2015). Climate change is, as a consequence, changing demand for water and capacity to meet demand, as well as capabilities for managing water risks across society's goals and across sectors (Smith et al., 2019).

The many imperatives for water are woven into the term "water security," which has gained increasing acceptance over the last decade, and which Grey and Sadoff (2007) defined as achieving and sustaining "the availability of an acceptable quantity and quality of water for health, livelihoods, ecosystems, and production, coupled with an acceptable level of water-related risks to people, environments and economies." From the perspective of water security, the core challenge of water management is hence to meet multiple objectives for water use, addressing competition for water, and mitigation of water risks in ways that are equitable, inclusive, and sustainable. Alongside the ecological and economic dimensions of water, there are therefore political dimensions. Societies need water policies and governance that enable negotiation and regulation of how water is managed and allocated, and the trade-offs between different uses, and hence which investments are made in the development and management of water resources and services. Societies also need water-wise policies in domains that affect water availability and use, and exposure and vulnerability to water-related risks.

Financing and investment related to water must weigh complex considerations. With human welfare, different economic sectors and services from ecosystems all dependent on water, investments in water generate returns that extend across society's goals. Water investments protect health and health systems, make agriculture and factories productive and create jobs. If water investments are done well, they have benefits for women's equality and opportunities for youth, and for nature conservation and sustainable development. Now, in a changing climate, they must also build resilience to climate change. The agenda for water and related investments in water management and services—one that addresses the breadth of needs, competing demands, and the diversity of opportunities for water to contribute to sustainable and inclusive economic benefits—must embrace complexity. If successful, financing and investment for water can promote integration of water alongside other societal goals and help to catalyze systemic change and innovation at the heart of sustainable development.

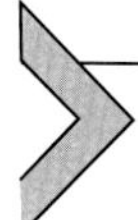

1.2 Priorities: evolving global and national water agendas

Societal goals for water resources were dominated for much of the 20th century by the top-down "hydraulic mission," entwined with development of the state, of using large-scale engineering to "tame nature," harness rivers for hydroelectric power and spread irrigation (Molle et al., 2009). The need to accommodate diverse uses of water and to integrate social and environmental goals, as well as economic, in water policies took hold as the hydraulic mission's toll of inequity and ecosystem degradation emerged after mid-century. The Mar del Plata Action Plan, adopted at the UN Water Conference in 1977, then laid the foundation for the 1992 Dublin Principles and a policy framework at global level for integrated water resources management (IWRM) (GWP, 2000; Smith and Jønch Clausen, 2018). With agreement at the World Summit on Sustainable Development, in Johannesburg in 2002, to develop national IWRM and water efficiency plans (Jønch Clausen, 2004), governments adopted the consensus that, guided by the rubric of IWRM, coordinated management of water should meet multiple social, economic and ecological objectives (GWP, 2000).

By including a dedicated goal on water in the Sustainable Development Goals, adopted in 2015, the global community then went a step further. As SDG 6—"clean water and sanitation for all"—is interlinked with the other 16 SDGs (UN-Water, 2016) societal goals for water are now explicitly systemic. Water runs through the 2030 Agenda, connecting the SDGs, Sendai Framework for Disaster Risk Reduction, the New Urban Agenda and, because of the centrality of water management to climate change adaptation, the Paris Agreement on climate change. Now, in the global agenda on water, it is acknowledged that action on water affects action on other goals, that action by others affects water. This implies that financing and investment for water is systemic too—that investment in water can achieve (or put at risk) outcomes across multiple sectors, and investments by other sectors can achieve (or put at risk) water outcomes.

1.3 Foundations: the investment case for water

Available estimates of future financing requirements for water show that the needs are huge and the gap is wide (WWC & OECD, 2015). Additional investment needed to achieve universal and equitable access to safe and affordable drinking water was estimated at US$ 1.7 trillion for 2015-2030, necessitating a tripling of 2015 expenditure levels (Hutton & Varughese, 2016; Hutton, This volume). Projections for global financing needs for water infrastructure are higher still: US$ 6.7 trillion by 2030 and US$ 22.6 trillion by 2050, with inclusion of financing for development of water resources for irrigation and energy raising these estimates further (OECD, 2018).

The case for such outlays seems clear when weighed against the costs of failure to invest. Global economic losses were estimated in 2015 at US$ 260 billion per year, for example, from inadequate water supply and sanitation, US$ 120 billion just from urban flood damage, and US$ 94 billion from inadequate water management in irrigation (Sadoff et al., 2015). Moreover, the World Bank has estimated that water-related losses in agriculture, health, income and property will reduce GDP growth by 6% in some regions by 2050 (World Bank, 2016).

Despite strong evidence of investment needs, however, the case for investment in water management is often beset by uncertainties. Placing an economic value on water is difficult, and the value proposition for investment in water management can, as a consequence, appear unattractive. Difficulties with valuing water stem from its systemic nature—from water

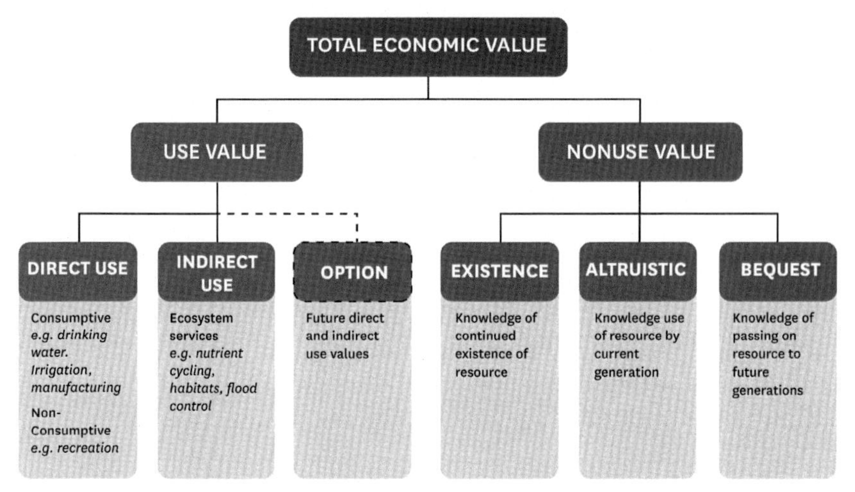

Figure 1.1 Framework for total economic value (adapted from Pascual et al., 2010; Morgan and Orr, 2015).

having multiple uses, and water flowing and cycling between uses (Garrick et al., 2020). The value of water is both dependent on the use to which it is put and interdependent with other uses. In addition, some uses for water provide private benefits, some have benefits for public accounts, and others have benefits that accrue more broadly to society through human well-being (e.g. health, equality, recreation) or ecosystems and their services (Morgan and Orr, 2015). Some water-related values are highest in the short term (e.g. meeting immediate supply needs for drinking or irrigation), some in the medium term (e.g., managing seasonal or annual variability), and some in the long-term (e.g. large-scale storage or ecosystem restoration). As a result, private, government and civil society stakeholders tend to weigh benefits differently when comparing investment options for water management.

In theory, the value of water can be captured using the framework of Total Economic Value (Pascual et al., 2010), as shown in Fig. 1.1. This combines "direct use values" for water in, for example, water supply, agriculture, industry, the energy sector and recreation, with "indirect use values" from, for example, watershed protection, flood control and other regulating or supporting ecosystem services, and "option values" for future direct and indirect uses. Alongside these are "non-use values" associated with cultural, aesthetic or heritage values of water. In practice, however, comparing values for water on this basis is generally not possible. Embedded

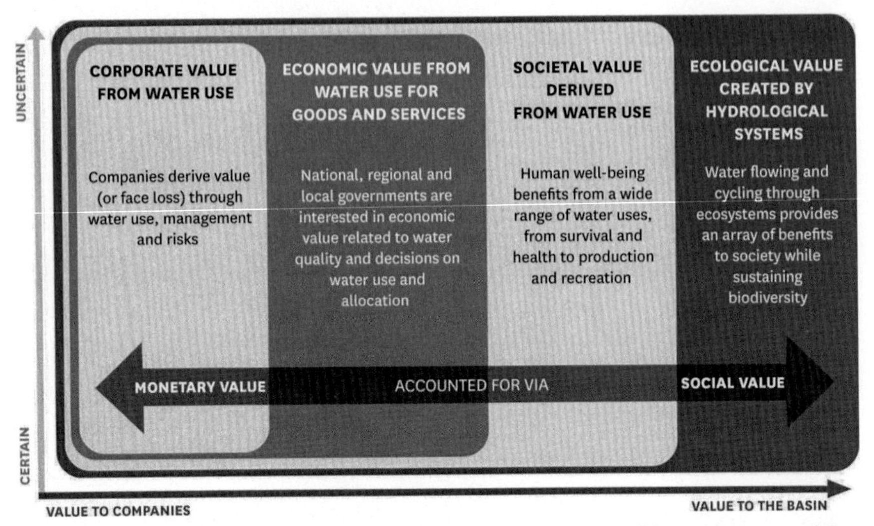

Figure 1.2 Private, public, and societal values for water (adapted from Morgan and Orr, 2015).

within such valuations are large uncertainties and undervaluation of water-related ecosystem services, as well as cultural or heritage values that, while not readily measured, can exceed the other component values combined (Garrick et al., 2017). As the climate changes, in addition, how returns on investment in water are weighed in decision making needs to be adjusted to account for future changes in water regimes and risks (Smith et al., 2019; and see below). These challenges are compounded by variation in values and their uncertainties according to the perceptions of different stakeholders. As summarized in Fig. 1.2, private benefits from direct use values are of highest concern to business, but more weight is given to indirect, social, and ecological values in the public domain or at societal level.

With a framework of values, the case for investment in water could, logically, be based on weighing the changes in benefits that an investment provides—such as better access, more storage or cleaner water—against the costs. Ultimately, however, decisions on water financing are not based on such assessments—at least not solely. Such valuations may help to guide decisions, but at stake are also competing uses from other sectors for available (public) finance, as well as constraints on financing where institutions have low capacity to absorb investment and use it effectively. In addition, if water security and the tradeoffs involved in achieving it are dependent on the perceptions of the parties involved, then political economy shapes how water, and its associated risks, are valued. How preferences for investment are set depends—in addition to conceptions of valuation for water—on power dynamics and access to decision making, and also therefore on stakeholder engagement and collaboration to bridge divides in perceptions of private, public, and societal values for water.

1.4 Transitions: failing assumptions of plenty

Perceptions of needs in water management are changing because pressures on water resources around the world are changing alongside shifts in the water cycle under climate change. As a consequence, flaws in old assumptions about water, held across sectors, are being increasingly exposed. Per capita available renewable freshwater resources declined worldwide by an average of more than 20%, and by 41% in sub-Saharan Africa and 32% in West Asia and North Africa, between 1997 and 2017 (FAO, 2020). Per capita availability has dropped below 1700 m^3, the defined threshold for "water stress," in South Asia, and below 1000 m^3 in West Asia and North Africa, the threshold for "water scarcity." Regional data mask intraregional and subnational disparities, but, in total, some 3.2 billion people worldwide live in areas affected by water stress and high drought frequency, including around 520 million living in severely water-constrained areas in South Asia, and 460 million in East and South-east Asia. In sub-Saharan Africa, per capita availability is declining quickly as the population grows, and 5% or 50 million people live in areas exposed to potentially catastrophic impacts from severe drought on cropland and pastures (FAO, 2020). Exposure of populations to climatic hazards, including floods and droughts, is projected to increase under climate change, and be amplified further by additional warming (Farinosi et al., 2020). The main driver of increases in water scarcity, population growth, is made worse by water pollution making water unavailable for other uses, and higher demands on water resources associated with changing dietary patterns and increasing wealth (FAO, 2020). Old assumptions—whether implicit or stated explicitly—that planning and management of water use for one purpose or one sector could be done is isolation from others are, therefore, untenable today. These were based on assumptions of plenty that are now failing.

There is no shortage of lessons from history and experience showing how failure to manage sharing of water, tradeoffs among sectors, and risks can lead to hardship, economic losses and, ultimately, catastrophe. The story of the Aral Sea in Central Asia is a stark lesson in the damage wrought by failure to recognize that water resources are finite, and therefore to manage water—and invest in water—as a systemic resource that interconnects users across sectors as well as with the natural capital against which the economy is secured. Diversion of the Amu Darya and Syr Darya rivers by Soviet engineers, beginning in the 1960s, for water supply to massive irrigation schemes ultimately led to the destruction of the Aral Sea, which is now just 12% of its original size. With the consequent collapse of the fishing industry, the economy in coastal towns has all but vanished (Micklin, 2010; Djumaboev et al., 2019). In India, for example, where 14 of the 20 largest power utility companies were forced to shut down at least once due to water scarcity in 2013-2016, there are fears that increasing competition for water across the agricultural, domestic and energy sectors will cut future economic growth (Luo et al., 2018). And, in a succession of cities worldwide—most notably São Paolo in 2014, Cape Town in 2018 and Chennai in 2019—growth in demand for water in combination with drought has caused water crises, because of assessments that they were in imminent danger of running out of water (Ahmadi et al., 2020). Planning of water management for one purpose or one sector in isolation from others is hence a very risky proposition.

The need to understand and resolve tradeoffs when managing water and making water investments is not new. It is inherent in IWRM and SDG 6, and has been recognized for decades. Today, as per capita water availability declines and exposure to water scarcity, as well as other water risks as the climate changes, increases, these needs coincide with growing recognition of the imperative for development to respect planetary boundaries and avoid systemic tipping points, including their implications at regional levels (Rockström et al., 2009; Zipper et al., 2020). Completing the long-term process of transition in how decision making on water is made, and how multiple competing needs are accommodated, is now urgent.

While the urgency of integrating decision making on water across sectors and managing tradeoffs in water investments is intensifying, the challenges that must be tackled are also not static, but are expanding. Changes needed in water management are increasingly encompassed within goals for large-scale transformation of social-ecological systems at planetary scale. Meeting the food security needs of a future population of 10 billion people, for example, without transgressing planetary boundaries relating to biodiversity, deforestation, nitrogen pollution and freshwater withdrawals, while adapting to climate change, will demand major changes in how—and where—water resources are exploited. Improving performance in agricultural water management is part of what is needed, but will not be enough by itself, and cannot be approached as a narrowly segmented, sectoral activity. Modeled assessments estimate, on the contrary, that development within planetary boundaries will in addition require a cut in overall water use for irrigation, expansion of irrigation in regions where water is abundant but abandonment in other regions, together with restoration of degraded lands (Gerten et al., 2020). Such grand challenges represent major opportunities, however, for financing of water management and investment to align with, and help to leverage, the large-scale processes of systemic change needed to tackle the climate crisis, food system transformation, and biodiversity crisis, as well as poverty and inequality.

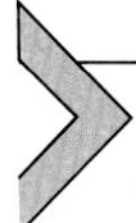

1.5 New imperatives: adapting to climate change and the new systemic mission

Water management in the 21st century is grappling with new imperatives in a context where water is just one of many development, environmental and resource-security issues jostling for the attention of politicians, business leaders and the media. The old "hydraulic mission" is long gone, but now there is a new mission, in which water security must help drive—and is in turn dependent on—systemic solutions for climate change, food and nutrition security, ecological breakdown and social and economic inequality. As a consequence, policy makers are faced with a multitude of priorities, within which it is easy to lose sight of the benefits that investment in water—when done well—can have across sectors and societal goals. There are, without doubt, competing demands on investment—but also potentially new ways to achieve results for water security by aligning financing for water with systemic approaches. With this, however, comes added complexity.

Climate change adaptation is inescapably an entry point for the new systemic mission for water. Water regimes are shifting as the climate changes, and, as a consequence, more frequent drought and exposure to flooding, more frequent and intense storms, and changes in the timing of precipitation and river flows, are among the principal ways that people experience the impacts of climate change. As water-related risks propagate across the economy as a result—putting at risk investments and assets in multiple sectors and locations—incentives for enhanced investment in water security are growing (Alaerts, This volume). Sectors that depend on water availability or are vulnerable to water-related risks should factor water into their investment decisions. Every decision about water management and every water investment is, furthermore, an opportunity to strengthen climate change adaptation, and to direct flows of financing toward building resilience to water and climate risks.

As water regimes shift, they are also becoming more uncertain, with past hydrology an increasingly unreliable means of understanding the future availability or seasonality of water resources or of quantifying water risks. How financing and investment decisions relating to water are made are being reassessed and reconfigured as a result. Water-related risks can no longer be assumed to be constant for long-lived projects, but will change over time as the climate changes and as uncertainty about future water resources increases (Milly et al., 2008). The assumption of a single knowable future climate, which traditional methods for assessing costs and benefits for water investments have relied upon implicitly, is therefore breaking down. As decisions made today on water infrastructure, for example, will have legacies over decades, new approaches are needed for assessing risks and returns for water investments. A range of approaches are emerging for decision making on water investments and financing under uncertainty, as a consequence, based on extending cost-benefit analyses over longer periods of time, and evaluating performance in terms of robustness across a range of possible future water regimes and flexibility to adapt to unexpected events (Maier et al. 2016; Haasnoot et al., 2019; Smith et al., 2019).

Beyond climate change, and in adopting the SDGs, governments, as well as business and civil society, have demonstrated an increasing level of comfort with integration of policy, investment and action across systems. A number of high-profile frameworks have emerged for bringing coherence to catalyzing change in complex systems, and to aligning investment and action at global to local levels. Political and business leaders, and investors from the multilateral, public, and private spheres, are now among the most visible advocates for, for example, building resilience, food system transformation, nature-based solutions and the circular economy. Behind each of these policy goals lies a need for profound—and difficult—changes in the systemic relationships among the economy, development, climate and nature. Management of water resources and of water risks are critical to each of them, and therefore financing for each could—and should—integrate investment in water.

In making the case for investment in water, and for water-wise investments across other policy domains and economic sectors, there is an intensifying need—in addition to weighing conventional options—to consider how water investments can contribute to leveraging systems change across societal goals. In aligning and embedding water investment with priorities around which there is strong or emerging consensus, such as climate change adaptation, societal resilience or nature-based solutions, avenues may open for attracting new and expanded financing for water. Doing so successfully is likely to benefit from application of systems analysis to prioritize the most effective entry points for action (e.g., Smajgl et al., 2016), as well as depend on changing what can be the prevailing mindset among water managers that financing for water-related investments can be satisfied solely from government budgets (see Alaerts, Gietema, both This volume). More broadly, such constructs for systemic change will, in addition, make the same basic demands on water management: integration of multiple goals across sectors and water uses, reduction of water-related risks, social inclusion, multi-level governance that enables negotiation of trade-offs, and innovation and scaling of technologies.

1.6 Way forward: toward a new financing and investment agenda for water

Water security is a multifaceted problem. Water is a finite (though infinitely recyclable) resource, interlaced through economies and ecosystems, and the linkages between them. Improving water security requires management of the availability of water and therefore water flows and storage, control of water pollution, and mitigation of water-related risks—all in ways that are equitable, inclusive, and sustainable. Water security underpins an array of critical societal goals, including poverty reduction, good health, food security, energy security, biodiversity conservation, and resilience to climate change. It can contribute to advances in gender equality, job creation, and peace and security. Conversely, water insecurity puts investments and assets at risk in multiple sectors. There is therefore much at stake for financing and investment related to water—from well-being in individual households and success across economic sectors, to integrity of ecosystems, geo-political balance, and creation of pathways to sustainable futures.

With the systemic nature of water resources, and of delivering water services equitably and sustainably, investment in water must weigh complex considerations. A financing and investment agenda for water can begin with understanding of the value of water, but it cannot end there, because valuation of water is both uncertain and dependent on whose needs and preferences are considered, including what can be sharply contested differences in priorities among private, public and civil society stakeholders. To be useful, therefore, in helping to guide water-related investment, valuing water has to be complemented by an appropriate policy, legal and regulatory framework for water management as well as institutions that enable negotiation of competing demands and tradeoffs among alternative priorities and choices for water services, allocation, infrastructure, and environmental protection.

The case for investing in water can, in principle, be guided by the contributions that water makes to building value across a range of social and economic needs and broader environmental benefits. Today, this case can be increasingly amplified by the benefits of water investment for climate change adaptation and by the contributions that water management can make to large-scale systemic transformation needed on pathways for development within planetary boundaries. To do so, however, will require both overcoming old constraints that have held back financing and investment for water (see also Martini, Gietema, both this Volume) as well as new approaches to how investments are planned and evaluated. The systemic mission for water—thus leveraging systemic benefits from water investments together with benefits *for* water from investment for climate resilience, food systems transformation, nature-based solutions, and circular economy, for example—will require a new convergence of investors, both public and private, as well as of frameworks that guide investment. To confront and navigate this new, more complex landscape of options and opportunities for water financing and investment, investors and institutions, and water managers from across sectors will need help. Key priorities will be capacity development for both investors and recipients and new analytical tools and data to support priority setting and decision making that integrates systemic change and water security.

References

Ahmadi, S.A., Sušnik, J., Veerbeek, W., Zevenbergen, C., 2020. Towards a global day zero? Assessment of current and future water supply and demand in 12 rapidly developing megacities. Sustain. Cities Soc. 61, 102295. https://doi.org/10.1016/j.scs.2020.102295.

Alaerts, G.J., 2019. Financing for water—water for financing: A global review of policy and practice. Sustainability 11, 821. https://doi:10.3390/su11030821.

Djumaboev, K., Anarbekov, O., Holmatov, B., Hamidov, A., Gafurov, Z., Murzaeva, M., Sušnik, J., Maskey, S., Mehmood, H., Smakhtin, V., 2019. Surface Water Resources. In: Xenarios, S., Schmidt-Vogt, D., Qadir, M., Janusz-Pawletta, B., Abdullaev, I. (Eds.), The Aral Sea Basin: Water for Sustainable Development in Central Asia. Routledge, London, pp. 25–38.

FAO, 2020. The state of food and agriculture 2020. Overcoming Water Challenges in Agriculture. United Nations Food and Agriculture Organisation, Rome https://doi.org/10.4060/cb1447en.

Farinosi, F., Dosio, A., Calliari, E., Seliger, R., Alfieri, L., Naumann, G., 2020. Will the Paris Agreement protect us from hydro-meteorological extremes? Environ. Res. Lett. 15, 104037. https://doi.org/10.1088/1748-9326/aba869.

Garrick, D.E., Hall, J.W., Dobson, A., Damania, R., Grafton, R.Q., Hope, R., Hepburn, C., Bark., R., Boltz, F., De Stefano, L., O'Donnell, E., Matthews, N.,

Money, A., 2017. Valuing water for sustainable development. Science 358, 1003–1005. https://doi.org/10.1126/science.aao4942.
Garrick, D.E., Hanemann, M., Hepburn, C., 2020. Rethinking the economics of water: An assessment. Oxford Rev. Economic Pol. 36, 1–23.
Gerten, D., Heck, V., Jägermeyr, J., Bodirsky, B.L., Fetzer., I., Jalava, M., Kummu, M., Lucht, W., Rockström., J., Schaphoff, S., Schellnhuber, H.J., 2020. Feeding ten billion people is possible within four terrestrial boundaries. Nature Sustain. 3, 200–208. https://doi.org/10.1038/s41893-019-0465-1.
Grey, D., Sadoff, C.W., 2007. Sink or swim? Water security for growth and development. Water Policy 9, 545–571.
GWP, 2000. Integrated Water Resources Management. TAC Background Paper No.4. Global Water Partnership, Stockholm.
Haasnoot, M., van Aalst, M., Rozenberg, J., Dominique, K., Matthews, J., Bouwer, L.M., Kind, J., Poff, N.L., 2019. Investments under non-stationarity: economic evaluation of adaptation pathways. Clim. Change 161, 451–463. https://doi.org/10.1007/s10584-019-02409-6.
Hutton, G., Varughese, M., 2016. The Costs of Meeting the 2030 Sustainable Development Goal Targets on Drinking Water, Sanitation, and Hygiene. The World Bank, Washington, DC.
Jønch-Clausen, T., 2004. "...Integrated Water Resources Management (IWRM) and Water Efficiency Plans by 2005." What, Why and How? TAC Background Paper No. 10. Global Water Partnership, Stockholm.
Luo, T., Krishnan, D., Sen, S., 2018. Parched Power: Water Demands, Risks, and Opportunities for India's Power Sector. Working Paper. Washington,DC. World Resources Institute.
Maier, H.R., Guillaume, J.H.A., van Delden, H., Riddell, G.A., Haasnoot, M., Kwakkel, J.H., 2016. An uncertain future, deep uncertainty, scenarios, robustness and adaptation: How do they fit together? Environ. Modell. Softw. 81, 154–164. https://doi.org/10.1016/j.envsoft.2016.03.014.
Micklin, P., 2010. The past, present, and future Aral Sea. Lakes and Reservoirs. Res. Manag. 15, 193–213. https://doi.org/10.1111/j.1440-1770.2010.00437.x.
Milly, P.C.D., Betancourt, J., Falkenmark, M., Hirsch, R.M., Kundzewicz, Z.W., Lettenmaier, D.P., Stouffer, R.J., 2008. Stationarity is dead: Whither water management? Science 319, 5863. https://doi.org/10.1126/science.1151915.
Molle, F., Mollinga, P.P., Wester, P., 2009. Hydraulic bureaucracies and the hydraulic mission: flows of water, flows of power. Water Alternatives 2, 328–349.
Morgan, A.J., Orr, S., 2015. The Value of Water: A Framework for Understanding Water Valuation, Risk and Stewardship. WWF International and International Finance Corporation, Gland/Washington, DC.
OECD, 2018. Financing Water: Investing in Sustainable Growth. OECD Environmental Policy Paper No. 11. Organization for Economic Co-operation and Development, Paris.
Pascual, U., Muradian, R., Brander, L., Gomez-Baggethun, E., Martin-Lopez, M., Verma, M., Armsworth, P., Christie, M., Cornelissen, H., Eppink, F., Farley, J., Loomis, J., Pearson, L., Perrings, C., Polasky, S., 2010. The economics of valuing ecosystem services and biodiversity. In: Kumar, P. (Ed.), The Economics of Ecosystems and Biodiversity Ecological and Economic Foundations. Earthscan, London and Washington, DC.
Ray, P.A., Brown, C.M., 2015. Confronting Climate Uncertainty in Water Resources Planning and Project Design: The Decision Tree Framework. The World Bank, Washington, DC.
Rockström, J., Steffen, W., Noone, K., Persson, A., Chapin, F.S., Lambin, E., Lenton, T.M., Scheffer, M., Folke, C., Schellnhuber, H., Nykvist, B., De Wit, C., Hughes, T., van der Leeuw, S., Rodhe, H., Sorlin, S., Snyder, P.K., Costanza, R., Svedin, U., Falkenmark, M., Karlberg, L., Corell, R.W., Fabry, V.J., Hansen, J., Walker, B., Liverman, D., Richardson, K., Crutzen, P., Foley, J., 2009. Planetary boundaries: exploring the safe operating space for humanity. Ecol. Soc. 14, 32. http://www.ecologyandsociety.org/vol14/iss2/art32/.

Sadoff, C.W., Hall, J.W., Grey, D., Aerts, J.C.J.H., Ait-Kadi, M., Brown, C., Cox, A., Dadson, S., Garrick, D., Kelman, J., McCornick, P., Ringler, C., Rosegrant, M., Whittington, D., Wiberg, D., 2015. Securing Water, Sustaining Growth: Report of the GWP/OECD Task Force on Water Security and Sustainable Growth. University of Oxford, Oxford.

Smajgl, A., Ward, J., Pluschke, L., 2016. The water-energy-food nexus – realising a new paradigm. J. Hydrol. 533, 533–540. http://dx.doi.org/10.1016/j.jhydrol.2015.12.033.

Smith, M., Jønch-Clausen, T., 2018. Revitalizing IWRM for the 2030 Agenda. World Water Council, Marseilles.

Smith, D.M., Matthews, J.H., Bharati, L., Borgomeo, E., McCartney, M., Mauroner, A., Nicol, A., Rodriguez, D., Sadoff, C., Suhardiman, D., Timboe, I., Amarnath, G., Anisha, N., 2019. Adaptation's thirst: accelerating the convergence of water and climate action. Background Paper prepared for the 2019 report of the Global Commission on Adaptation. Global Commission on Adaptation, Rotterdam and Washington, DC.

Tickner, D., Opperman, J.J., Abell, R., Acreman, M., Arthington, A.H., Bunn, S.E., Cooke, S.J., Dalton, J., Darwall, W., Edwards, G., Harrison, I., Hughes, K., Jones, T., Leclère, D., Lynch, A.J., Leonard, P., McClain, M.E., Muruven, D., Olden, J.D., Ormerod, S.J., Robinson, J., Tharme, R.E., Thieme, M., Tockner, K., Wright, M., Young, L., 2020. Bending the curve of global freshwater biodiversity loss: an emergency recovery plan. Bioscience 70 (4), 330–342. https://doi:10.1093/biosci/biaa002.

UNICEF & WHO, 2019. Progress on Household Drinking Water, Sanitation and Hygiene 2000-2017. Special Focus on Inequalities. United Nations Children's Fund and World Health Organization, New York, NY.

UN-Water, 2016. Water and Sanitation Interlinkages Across the 2030 Agenda for Sustainable Development. UN-Water, Geneva.

World Bank, 2016. High and Dry: Climate Change, Water, and the Economy. The World Bank, Washington, DC.

World Economic Forum, 2020. The Global Risks Report 2020. World Economic Forum, Geneva.

WWAP/UN-Water, 2018. The United Nations World Water Development Report 2018: Nature-Based Solutions for Water. UNESCO, Paris.

WWC & OECD, 2015. Water: Fit to Finance? Catalyzing National Growth Through Investment in Water Security. World Water Council, Marseilles.

WWF, 2020. Living Planet Report 2020 – Bending the Curve of Biodiversity Loss. World Wildlife Fund, Gland.

Zipper, S.C., Jaramillo, F., Wang-Erlandsson, L., Cornell, S.E., Gleeson, T., Porkka, M., Häyhä, T., Crépin, A.-S., Fetzer, I., Gerten, D., Hoff, H., Matthews, N., Ricaurte-Villota, C., Kummu, M., Wada, Y., Gordon, L., 2020. Integrating the water planetary boundary with water management from local to global scales. Earth's Future 8 (2), e2019EF001377. https://doi.org/10.1029/2019EF001377.

CHAPTER TWO

Water, physically connected yet institutionally fragmented—Investing in its strategies, asset classes, and organizations

Guy J. Alaerts
IHE-Delft International Institute for Water Education

2.1 Introduction

Treatises on water policy typically apply a perspective which is social ("water is a human right"), technical ("what technology can help us achieve") or economic ("the value of this scarce resource for agriculture, industry, health, etc., or, conversely, of the protection against floods, pollution, and other risk"). Analyses on water financing are scarce and tend to be generic, e.g., forecasting global investment needs, or polemic, such as on privatization. Water impacts upon, and in turn is intensely impacted in a multitude of ways by our living habits, our economy, and the earth's landscape and ecology. Countries possess institutional architectures—the frames encompassing political priorities, policies, regulations, public and private organizations, and other arrangements—to manage water and water services that come with distinct formats through which capital is sought for investments and operations. This chapter will explore how sustainable investments are shaped by the water management strategies, asset classes, and the different organizations in the sector. It will also explore how these strategies that are essentially "productive" in the sense that they aim to expand or improve a service such as water supply, increasingly have to navigate the physical limits and the interconnectedness of the water system and invest in environmental approaches to secure sustained availability of water. The chapter will offer an overview of the asset classes and financing needs in the water subsectors, and the challenge to raise revenue to cover costs, reflect the water's value and invest in a resilient future.

From a global perspective, capital markets have shown a modest appetite in most water investments thus far compared to the industrial and the

Financing Investment in Water Security: Recent Developments and Perspectives.
DOI: https://doi.org/10.1016/B978-0-12-822847-0.00012-0

other infrastructure sectors (notably transport, power, housing, and telecoms) although substantial differences exist between the richer and the developing economies. Distinct investments classes have shown appeal, e.g., when they can be ring-fenced more easily such as hydropower and desalination plants and equipment manufacturing. A mismatch is apparent between the high ambition levels in national and global policies on water development on one side, and the engagement level of the capital markets on the other. While some of this can be attributed to concerns about affordability and restrictions in the global financial system (Martini, This volume), part of the cause lies in specific attributes of water and the water business. Firstly, nations, multilateral institutions (such as the UN), and the water profession have enunciated water (and economic) development policies that set targets guided by ethical, policy, and political considerations, such as the Sustainable Development Goals (SDGs). The financial system, in contrast, is operating on instrumental principles aiming at return on investment, and it responds to business opportunities (although nowadays some investors are also seeking societal or environmental "impact" or aspire to ESG [environmental, social and governance] aims). Secondly, water is nonsubstitutable in many situations.[1] Whereas it is possible to substitute, say, rice for bread, and private for public transportation, no alternative exists for water when it runs out. Hence, water in most situations is perceived as a public good, rarely as a private good, and this can quickly turn political. Thirdly, water is pervasive across all societies as well as ecologies, and it is rife with externalities.[2] This interconnectedness can make individual water use difficult to separate out and ring-fence. Moreover, whereas roads or telecom networks can be added one after the other, water is pre-eminently part of natural cycles, subject to the vagaries of the weather and limited in availability. A stock of water in a river basin can be exhausted, and the more of it is used in household and industry, the more polluted water is generated which must be treated lest it endangers public health or the ecology which receives the

[1] Exceptions exist in specific locations or under specific conditions, e.g., where different water sources are present close to each other, between drinkable tap water and bottled water, and in places where desalination can turn seawater into freshwater or where recycled wastewater can be reused as lower-quality water such as in industrial processes or irrigation. Also, local water can be substituted by import of agricultural products (fodder, soy, meat, etc.) grown on water in another country, i.e. as "virtual water". However, such shifts are often expensive.

[2] An externality is a cost imposed on a third party (a person, community or ecological system) that is caused by a water consumer who enjoys the benefit from that use but is not paying or compensating for that cost. For instance, a city or factory upstream a river that is discharging polluted water renders the river water unfit for use by users downstream; the upstream polluters should either treat their wastewater or compensate the downstream users.

wastewater. Fourthly, notwithstanding this physical interconnectedness water businesses are local and fragmented. Generally, local governments such as municipalities are assigned the responsibility to provide safe or tap water and collect and treat wastewater for the local community; irrigation and flood management are under purview of other dedicated organizations. Finally, and importantly, water is bulky and easily polluted, so keeping it reliably available for household use necessitates extensive infrastructure, secure access to part of the natural water resource, and an administrative capacity to transport it from one place to another or to enhance its quality to make it fit for use. This may turn out very expensive compared to the incomes of many users. Therefore, it remains challenging to control costs well in order to keep the water services expanding and affordable while at the same time safeguarding the coherence and sustainability of the natural hydrological system for the long term. Articulating realistic business cases, thus, requires knowledge of the water system and of how it is managed.

2.2 Water and land: Investing productively, recognizing limits, seeking efficiencies

2.2.1 Water and land are intimately connected, pose limits

Water and land are intimately connected; this shapes water management and its investments. Precipitation is collected on the land surface of a river catchment and the landscape relief directs this water to rivulets, brooks, and rivers, down to the estuary where it flows into the sea. Quite a lot of this precipitation, however, is taken up by plants and trees or percolates into the soil, feeding aquifers. Ninety-seven percent of accessible liquid freshwater is stored as groundwater; only 3% is flowing over the surface. This surface, or river, flow, though, is replenished more or less continuously by rainfall, by snow- or glacier melt, and by the seepage from high groundwater tables, and can be accessed directly. Groundwater, in contrast, usually must be pumped up. Thanks to the filtering effect of the soil groundwater is commonly of drinkable quality and requires little additional treatment, different from surface water which easily gets polluted by human activity. Still, many aquifers are degraded too by anthropogenic or natural pollutants. In western European countries typically 40% to 60% of all tap water is abstracted from aquifers. In wet countries with high groundwater tables the rural population almost exclusively relies on shallow hand pumps, a major tool in combating water-related diseases associated with ingesting polluted water from ponds and rivers. Cities and farmers in arid regions such as Yemen, Libya, Saudi Arabia, and south-west United States rely on very deep groundwater, at ever-rising cost, which often is fossil, collected in aquifers millennia ago and irreplaceable.

Because in many places the abstraction from aquifers occurs faster than their recharge, concern is growing across the globe—from the vast Ogallala aquifer in the mid-west United States, to south-western Europe, India's Rajasthan, and China's north-eastern Hebei-Shandong region—that many local stocks will run dry in the next decades unless drastic measures are taken to return to sustainable abstraction rates by reducing demand and shifting to other sources. To compound this challenge, over-pumping in urbanized areas and for irrigation is now leading to land subsidence, by 10-30 cm per year[3], over vast areas in Indonesia, the Beijing-Shandong basin, Mexico City (where land in some locations has sunk by 12 meters over the past century), the Central Valley in the United States, the Ganges plain in India, the Teheran plain in Iran, etc. Most of this subsidence started in the 1980s-1990s. Now a quarter of the Indonesian capital Jakarta, on the seashore of the Java Sea, has sunk beneath sea level and must be protected with a sea dike. In August 2019, the Indonesian government decided to start relocating its capital to Borneo which would leave many sunk assets sinking and stranded. Elsewhere land is subsiding due to deep drainage to make marshy land more amenable to agriculture, plantations and settlement. This affects notably large parts of Indonesia, Egypt, Vietnam, the US Gulf Coast, the Italian Po delta, and the western half of the Netherlands. Herrera-Garcia et al. (2021) estimate that from 2010 to 2040, this will affect 1.2% to 1.6% of the global land mass and expose a population of 480 to 660 million people and 12% to 16% of global GDP to recurrent and worsening flooding. This phenomenon illustrates starkly how compartmentalized water exploitation—and lack of management—will lead to exhaustion of the groundwater resource while at the same time dramatically increasing flood risk.

The landscape also determines whether water is retained temporarily in marshes and lakes where it nurtures vibrant and diverse ecosystems, or drains into rivers. Too fast drainage can create floods downstream that can have devastating effects on unprotected communities but flood pulses, in turn, are often engines for environmental rejuvenation, such as the annual flood pulse on the lower Mekong which for five months expands the Tonle Sap Lake in Cambodia and the vast plains of the delta boosting productive

[3] To place this in perspective, the sea level rise due to climate change is occurring at 3–4 mm per year.

fisheries and agriculture. Construction of hydropower dams upstream in P.R. China, Vietnam and Laos, and irrigation in Thailand, however, are gradually attenuating this pulse; the annual value of the fisheries at risk is about US$17 billion (Nam et al., 2015). River flows contract and expand seasonally; cities and farmers who wish to protect against flooding must give space to the river.

Riverine flooding brings major economic cost (see further). Climate change will exacerbate this by causing more intense rainfall episodes. The main driver for increased flooding risk, however, are man-made changes in land use that channel rain run-off faster to the river: removal of forest canopies and marshland degrades the natural buffering capacity, and the expansion of "hard surface" such as roofs, parking lots, highways, and paved gardens shunts water into drains and rivers building up into compact flood waves. However, already 75% of the earth's land surface has been significantly altered (and 95% will be by 2050) and 85% of wetlands lost (IPBES, 2019). Water is also a quintessential natural resource upon which all aquatic and terrestrial ecosystems depend. These are by themselves very productive in a narrow economic sense (fish, timber, produce) but the combination of the landscape and vegetation ("green infrastructure") in turn supports a healthy and regenerative hydrologic system. The global "productive value" of ecosystems and biodiversity is estimated at about US$140 trillion (WWF, 2018; OECD, 2019), or 1.5 times global GDP, yet these systems are being destroyed at accelerating rate, also jeopardizing water security. Beside land conversion, water pollution is a main driver of this decline. Finally, at continental scale, vegetative cover evaporates moisture into the atmosphere which turns into rain elsewhere. Moisture evaporating from the western Eurasian continent is responsible for 80% of China's water resources. In South America, the Río de la Plata basin depends on evaporation from the Amazon forest for 70% of its water resources. The main source of rainfall in the Congo basin is moisture evaporated over East Africa; the Congo basin in turn provides rainfall in the Sahel (van der Ent et al., 2010).

Economic activities driving land-use changes, the widespread over-abstraction of surface and groundwater, and their pollution have a significant impact on the water system at local and regional scales. The complex interconnections between the compartments in the water system and their dependence on land raise a fundamental challenge. Most water institutions are assigned to "develop" water to provide services, and only few specifically to manage the water as a limited resource. Moreover, the water system is intensely impacted upon by the policies, investments, and actors in other sectors that are not under control of water authorities. Land and spatial planning, urbanization and industrial development, agriculture, forestry and in many countries even groundwater fall under separate authorities and ownership regimes. The water organizations are expected to manage the water system for human and economic productivity while ensuring that the resource base is sustained, yet these other authorities tend to underestimate the severity of their impacts on the water system and the viability of investments in water-dependent activities (Smith, This volume).

2.2.2 Toward investments that are productive and sustainable

Most water management policies and investments have economically productive aims such as the expansion of a city's tap or safe water provision, of irrigation, or of cooling or process water for industry. Decisions on such investment are often made in isolation without taking account of the broader interconnectedness because it simplifies the individual activities as expertise, asset classes and customer bases differ between subsectors. Some investments and policies may concurrently have environmentally protective aims (e.g., to reduce water use, recycle water, or for safeguarding or rejuvenating ecosystems) but it is becoming critical to strike a balance between productive and protective investment in view of the growing stress on the water system in many locations. Decisions are now associated with pronounced trade-offs between competing uses as well as between immediate consumption and sustainable management. The negative impacts of productive water investments can to a significant extent be mitigated or compensated by additional measures, usually to protect human or ecological interests that otherwise would suffer. This mitigation results from an Environmental (Impact) Assessment (EA) which is mandatory in most countries, though EAs often tend to overlook the systemic context and are sometimes not well enforced. EAs are reactive, responding to an investment initiative; similarly, as the interventions in a water system keep expanding, the options to mitigate such negative impacts on an individual project basis may be running out.

Therefore, governments and multilateral development banks rely increasingly on Sectoral Environmental Assessments which start from a sector-wide analysis anticipating all related demands and pressures on the water system as well as all possible opportunities. This facilitates making up-front choices, rationalizing designs and identifying options that enhance sustainable water use, either by better designs that generate synergies and environmental co-benefits, by explicitly promoting reuse and recycling options, by safeguarding environmentally sensitive water stocks, or by directing water allocation to uses with the highest economic or social values. Since the 1990s integrated water resources management (IWRM) at the scale of a river basin or other significant hydraulically coherent territory, is being used as a pro-active planning instrument. IWRM starts from the recognition that all water is connected through the hydrologic system and land uses, and through its competitive uses. It allows to decide which parts of this system are essential to support strategic functions such as urban or industrial development, or environmental values (e.g., to sustain biodiversity or tourism) and, from there, to identify and cost the options to manage water for maximum aggregate benefit through an allocation mechanism and pricing, and to invest in physical structures for controlling supply and demand. At the same time, the quality of EAs and IWRM rests on the (public) availability of reliable and sufficiently complete data and information which may pose a challenge especially in less developed economies. They may also require access to more sophisticated simulation and forecasting models when longer-term interests are involved. And, finally, follow-up and enforcement of the agreements and plans will depend on local administrative capacity and political prioritization. Nonetheless, financiers are advised to conduct due diligence on the water impact of their investments and the longer-term security of the water allocation arrangements as these affect their sustainability.

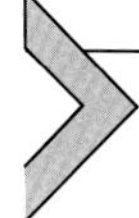

2.3 Water-related assets, operational practices, and institutional architectures

2.3.1 Water supply, wastewater management, and sanitation services for the public

2.3.1.1 The business of water services

The provision of safe and tap water to households and larger institutional customers (typically businesses, small workshops and public agencies, and for firefighting), the collection and treatment of their wastewater, and in general ensuring sanitary conditions in the urbanized area, are in most countries entrusted to local governments, commonly municipalities and counties or their local equivalents. To capture economies of scale Scottish Water, Irish Water, operators in Armenia, Portugal, Romania, and other countries are nowadays organized regionally. In the Netherlands, the over 1000 local tap water utilities before 1940 were encouraged to amalgamate into 52 utilities by 1990, and into 10 large professionally run regional organizations plus two "bulk water" companies by 2012; wastewater and local flood protection are managed now by 21 Water Agencies which are defined by hydraulic region. In France, Belgium, Germany, and other European countries utilities are often managed by for instance *Intercommunales* and *Verbände* (corporations owned by several local governments), provincial governments or other regional cooperative arrangements. In some countries, a dual arrangement exists, like in Croatia, Uganda and Bangladesh, where also a national agency is operating under the auspices of a central government ministry to serve certain, often rural, jurisdictions, and to provide technical and design expertise to local governments. England and Wales in the UK, on the other hand, have aggregated water supply and wastewater treatment in eight water companies in river basin territories that are privately owned and overseen by the Water Service Regulation Authority (Ofwat).

For the tap water service, for which the local clients generally are willing to pay, local governments usually establish corporatized utilities supervised by appointees of the local government. The wastewater management function, on the other hand, is usually handled by a differently styled utility or directly by a department in the local government administration because of technical reasons, and as households and users are less willing to pay a fee covering the costs of this service provision. Where the collected wastewater is not treated, the street gutters and sewers are thought of as part of the public space to be managed by the department that also manages garbage collection or road maintenance. As their economies develop countries start investing also in the treatment of the collected wastewater prior to discharge into the river, which adds significant cost and necessitates new infrastructure and know-how.

Water supply and wastewater management are technically and managerially complex operations and challenging for any local government, with respect to both the treatment and the transportation component. Water supply and wastewater management operations commonly consist of two components: a technical facility to treat the water ("raw" river water into the potable product, and, at the city's outflow side, removing the pollutants from wastewater before discharge into the river), and a sprawling reticulated network for distribution of tap water or collection of wastewater, consisting of pipes, canals, and pumping stations. The networks represent a very large part of the utility's fixed assets. In rich economies these operations are efficient (achieving high purification and supply rates) and reliable, whereas in developing economies technological capabilities are modest yet in many locations reasonably effective, albeit often unable to serve the whole population around the clock.

Water supply systems may require a third component, namely dedicated facilities to draw and convey the "raw water" from distant sources to the treatment plant. Utilities depend on the year-round availability of, and secure rights to good-quality surface or groundwater. In many instances, though, large dammed reservoirs and long-distance conveyance systems are essential to store water or to access alternative sources in order to overcome dry spells. With increasing pressure on the river basins (due to more competitive abstraction, pollution and changes in land use and climate) and on groundwater aquifers water utilities increasingly are being confronted with declining availability while trying to satisfy rising demand. Water utility managers often are not, or not deeply involved in addressing this challenge because in many countries the water resources fall beyond their remit: the utility is typically a local-government business whereas water resources are under the authority of a national ministry or regional agency. In rich economies water utilities are increasingly compelled to lobby for more secure water sources (see further reference to the EU's Water Framework Directive) but in developing and emerging markets water utilities take a junior position. Many large, growing cities are already facing severe limits in their water resources, such as Manila, Jakarta, Chennai, Beijing, Mexico City, Sao Paulo, Lima, and Sana'a, and this penury will grow more precarious in the next few decades (He et al., 2021). In 2018, the city of Cape Town started rationing water supply through public standpipes as its reservoirs in distant mountains, managed by the central government, had run dry after three consecutive dry years (rainfall partly replenished them since). Given that lead times to devise investments, obtain permits, and construct the necessary infrastructure span 10–25 years, it is, therefore, improbable that the SDG 6.1 ("clean water for all by 2030") will be fully achieved and rather probable that after 2030 the service coverage will deteriorate in many locations instead of improve.

2.3.1.2 Performance challenges

Two technical objectives are central to all water supply businesses, regarding their cost-effectiveness and financial health on one hand, and their environmental sustainability on the other. Firstly, utilities aim to minimize their Non-revenue Water (NRW). NRW comprises physical losses, notably leakage from the distribution system, and billing losses (clients in default, undercharging because of poor metering or fraudulent action, etc.). NRW can range from 3% to above 50% depending on soil conditions, age, and quality of pipes and fittings, managerial capacity, regulatory and financial incentives, etc. Only a few countries such as Singapore, Denmark, Switzerland, and the Netherlands manage to keep average losses below 5%. However, investing in NRW reduction below 20% is not always cost-effective as replacement of pipes in a congested city can prove more expensive than revenue to be earned from the saved water. Pro-active asset management and preventive pipe repair and replacement are advisable. Secondly, customers tend to be more willing to pay if presented with a bill reflecting their metered consumption. Installing water meters, however, is costly in investment and in meter reading. Although most utilities in rich economies use meters, many do not: the Amsterdam utility is in the Netherlands the only one not to meter consumption yet average consumption does not deviate from elsewhere. In some countries metering is standard practice but the economic benefit is controversial because of its cost and the low elasticity of domestic water use to price (Leflaive and Hjort, 2020).

Wastewater management has two similar technical challenges. Firstly, treatment works tend to run more efficiently and cheaply with concentrated sewage. Most conventional sewerage systems are "combined" collecting sewage and rainwater from roofs and streets; the rainwater dilutes the sewage and turns it into a lower-value polluted resource. Thus, combined sewerage systems are best replaced by "separate" systems which allow to recycle the rainwater separately. Secondly, at the city's outflow side of the treatment plant, the surface water receiving the discharge will determine the level of treatment required: first-stage only, to remove only debris, flotsam, and silt; second-stage with biological processes to remove most organic pollution; or third-stage, sometimes called polishing, usually to remove nutrients (nitrogen and phosphorus) and/or neutralize pathogens or remove specific micropollutants. Secondary and tertiary treatment are expensive in investment as well as operation. Operational expenses concern technically qualified staff, energy to power aerators and pumps, and treatment and disposal of residual sludge. Whether this cost is warranted depends on the vulnerability of the receiving river or lake but from a sustainability perspective it is desirable that the effluent has a quality comparable to or better than that of the receiving water in order to avoid imposing damage or costs on the downstream users.

Water service systems benefit from large economies of scale notably in the treatment plants, the abstraction of "raw water," the sludge management (of residual matter resulting from the treatment), and in various key services such as laboratory functions, technical design, maintenance, and billing and administration. As water supply utilities critically depend on secure water sources, they need to have a strong technical and political voice to be able to appropriate their share. Many individual operational and managerial functions require expensive expertise as well as financial capability which is well beyond the means of small utilities. This level of complexity also exceeds that of other municipal services such as road construction and maintenance. Therefore, the case is clear for amalgamation of utilities into larger organizations at regional or intermunicipal scale, as described above. This amalgamation, however, is still progressing slowly in many countries, a cause of weak managerial and financial capabilities.

2.3.1.3 Demand management, and circularity

Providing an adequate service to households—and meeting the SDG6 targets—can also be achieved by lowering the demand per household by making water use more efficient. This demand management is not only critical to make do with water stocks that are increasingly constrained, but these measures often bring savings too. Minimum requirements for drinking and cooking for a person are estimated at 3–5 liters per day and minimum water availability in the household is set at 50–100 liters per person and day (UNDP, 2006; UN et al., 2010). In South Africa, the Constitutional Court stated that utilities should provide households with 25 liters per person and day (UN et al., 2010). In European countries household demand is satisfied at 100–150 liters per person and day; in the United States, this rises to 150–250 but now well below the 300–500 that was common until the 1990s. Clearly, much of this water was, and still is, not used efficiently, placing a burden on the resource base and on the utility alike. Demand can be influenced, however, often at a cost below the incremental cost of augmenting supply. Such measures comprise primarily pricing (tariff) instruments, smart appliances (such as low-volume flush toilets, low-flow shower heads, and recycling of moderately polluted "grey" water for garden watering and car washing), "nudging" and education. Rainwater can be used for washing and cleaning purposes. Also, other measures can be contemplated, such as incentives in the arid US states for the conversion of lawn gardens into gardens with desert vegetation, and restrictions on golf courses and new housing development.

Both in water supply and wastewater treatment important opportunities exist to recover marketable resources, such as calcium, catalytic iron, and struvite (phosphate) pellets (from water softening, iron removal, and tertiary sewage treatment, respectively). The organic sludge from biological ("second stage") sewage treatment can be anaerobically digested and converted into biogas which covers a large portion of the plant's electricity and heat demand, or it can be applied as fertilizer in agriculture if phytosanitary concerns are addressed. Effluent from wastewater treatment plants is valuable too and is often used for agricultural irrigation or as process water in industry; in Los Angeles and in Tel Aviv, the purified effluent is used to recharge groundwater aquifers. In Europe, the United States, and Japan several treatment plants have combined above measures with solar panels and wind energy and have achieved carbon-neutrality and energetic self-sufficiency. In Stockholm and in Amsterdam *aquathermy* is piloted to recover heat from the city's sewage in winter, and cooling from groundwater in the summer, lowering the city's carbon footprint.

2.3.1.4 Assets

The assets of water supply and wastewater treatment systems are typically composed of civil-engineering items such as tanks, reservoirs, filter beds, and pipes, and, for a significant portion, of electro-mechanical equipment, such as pumps, blowers, aerators, valves, meters, mixers and presses. The civil engineering assets tend to be of a long-term "sunk" nature with depreciation periods of 20–40 years (though some infrastructure can technically last longer still) and relatively low annual operation, maintenance and replacement costs (unless in soft soils or within congested urban areas). The electro-mechanical equipment, on the other hand, is intensive in operation, maintenance and replacement with typical depreciation periods of 15 years and even shorter for certain electronic components. Depending on for instance age and design of the plant, and the local water resources security (that determines the cost to bring the "raw water" to the treatment plant, or of effluent discharge), the operational and the capital annual expenditures each make up about half of the total cost of the water treatment. In arid regions, notably the Middle East, and some Caribbean and Canary Islands, seawater (or brackish groundwater) is desalinated to produce drinkable water or high-quality industrial process water (like in Chennai); such high-performance plants carry high energy and operational costs. In developing and emerging markets which find import of specialized electro-mechanical equipment expensive and where land can be cheap, designs are often more traditional relying more on civil-engineering-based technologies (such as large pre-settling basins for water supply, and land-extensive lagoons for sewage treatment) which lower the operational expenditures.

For many investors, such assets are less appealing: they are "sunk," immovable and hard to ring-fence (with the exception of some high-tech distinct components such as desalination plants) (Baker, This volume), they do not offer easily accessible collateral (with collateral then to be sought from the securitization of tariff payments from customers), they are sensitive to political interference and to disruptions in the natural water system, and they are slow in generating revenue. On the other hand, the monopolistic nature, the likely growth of demand for water, and the opportunity to capture gains from growing efficiency over time can offer advantage.

2.3.1.5 Public and private organizations

The influential International Conference on Water and the Environment, Dublin, ICWE (1992) stated that people have a basic right to access to water and sanitation but that water must be managed as an economic good. This dual public–private nature of water is also reflected in the institutional architectures in place to manage the water supply and wastewater systems. In the majority of cases, water supply and wastewater management organizations are incorporated with local governments owning the shares. A minority of organizations are part of the local government's own administrative structure. In a few other countries the private sector's technical and financial capabilities are engaged to carry out large parts of the management. As mentioned before, England and Wales are the only jurisdictions that sold their drinking water supply and wastewater management in 1989 to firms specialized in water services. In France, about 60% of water services takes place through a menu of management contracts, leases, and concessions to private firms which carry out the operation and management of significant parts of the system, and, in the case of concessions, also part of the investment.[4] Other contractual forms comprise build-operate-transfer and its variants in which the management, financing and risks are distributed among the public and private partners (see also Gietema, This volume). In China (Wu, This volume), Senegal, Côte d'Ivoire, the Philippines, Colombia, the Czech Republic, Romania, and several other countries, various forms of public–private partnerships are arguably effective, but applied often only in selected cities. In China, this applies more to wastewater management than to tap water supply because, it is argued, water supply is "strategic." In many other countries, on the other hand, such forms of equity participation are not allowed (like in the Netherlands) or politically unpalatable, sometimes after earlier failed initiatives (like in Bolivia). In several cities in Germany

[4] For the agglomeration of Paris, in 1984 a mixed-ownership company, *Eau de Paris*, was established for water supply treatment, owned by the City and two private firms; the distribution and metering in two city halves was delegated under management contract to these two firms. In 2008, the City retook full ownership of *Eau de Paris*, buying out the private partners. The wastewater collection and treatment have always remained in public hands through its Syndicat inter-départemental pour l'assainissement de l'agglomération parisienne (Barraqué, 2012).

and some other countries the water supply, wastewater management and city cleaning are brought together with other municipal tasks such as public transportation and gas distribution in so-called *Stadtwerke* (City Corporations) allowing pooling of resources and some degree of cross-subsidization, with water and energy typically subsidizing public transportation. Water services are, in principle, a viable business proposition in rich countries as most households can afford and are willing to pay the full tariff covering operation, maintenance, and capital expenditures. Major international firms active in this field are Veolia Water, Suez Environnement (both French, and expected to merge by 2021), United Utilities (UK), Aguas de Barcelona (Spain), Ayala Corp. (Philippines), Bechtel (US), and Mitsubishi (Japan), with the first three possessing the largest track record. The ownership structure of these firms is of an increasingly global signature and their shares are traded. At the same time, in several countries, such as China, Russia, Colombia, the Philippines and Uganda, the number of larger and smaller specialized firms operating only at national scale is growing.

Water utilities, even where allowed to be managed as autonomous businesses, remain subject to close regulation. This pertains to their environmental impacts, the service quality (performance criteria) and tariffs. Depending on the country this can be through a generic Consumer Protection and/or a Competition Authority, a Food and Drugs Administration, or a regulatory office dedicated to the water service industry. Regulators typically seek to impose service delivery standards such as minimum periods of uninterrupted service provision, NRW reduction, and response times after complaints, while keeping tariffs in check. On the financial side, they may specify thresholds for tariff categories, e.g., imposing a formula for tariff-setting and revision; capping the financial liabilities a utility may incur or the return-on-investment it may generate; or approving investment plans.

Smith (This volume), Hutton (This volume), and Rosenstock and Leflaive (This volume) show that the financing needs of the water utilities are very large, but that both public and private financiers perceive risks as significant and the sector as complex. In developing and emerging economies the public budget often takes up this responsibility, sometimes in conjunction with Multilateral Development Banks which at the same time bring sectoral expertise and, often, support for a reform program. In richer economies and in countries where robust economic regulation is in place the sector has become mature and reliable, and financing has become mostly commercial also in cases where water utilities are held in public hands. The utilities here are able to formulate commercially and environmentally viable proposals that meet financiers' requirements, and they are able to provide track records and comfort to investors as well as much-prized long-term stability. The utilities are credit-worthy and solvable thanks to good governance, professional accounting, effective management of the water resources that secures their sources, and an adequate willingness of their customers to pay the bill. Thus, for example, in the United States the EPA assists in attracting and structuring commercial finance for local water services (Gebhardt et al., This volume) and in the Netherlands the Netherlands Water Bank acts as intermediary between water utilities and financial markets; both enjoy high credit ratings.

On the other hand, in other countries, also inside the European Union, challenges remain and in many cases the utilities still lack the economies of scale and specialist competence that large utilities offer, as well as sound profitability. Commercial financing is slower to develop in these situations but various new instruments are being launched such as dedicated Special Purpose Vehicles and blue, green and climate bonds (e.g., Hydro-bonds in Italy to help finance smaller utilities in the middle and south of the country) with adjusted risk-return expectations.

2.3.1.6 Water supply and sanitation in rural and peri-urban areas

Water supply and sanitation in rural areas and in urban fringes such as shantytowns and *favelas* in emerging and developing economies pose different sets of challenges. Low incomes (and sometimes, the absence of a monetized economy) render imposition of realistic tariffs problematic. In small rural towns the long distance between dwellings makes piped water distribution expensive, whereas in urban fringes the congestion, poorly planned urban lay-outs, legal impediments as well as neighborhood insecurity often preclude good infrastructure. Similarly, the collection and disposal of wastewater and other drainage are precarious, often insufficient, and thus risking to pollute the water supply. A variety of low-cost technologies are available for water supply, nonetheless, such as shallow and deeper hand-pumps, dug wells, protected ponds, rainwater harvesters, and simple piped systems without or with only basic treatment in locations where solar or diesel energy can help overcome chronically unreliable, or absent public power supply. For sanitation and wastewater management, single- and double-pit latrines, septic tanks and lagoons are cost-effective; in many locations also communal latrines and centralized fecal matter collection have been traditionally accepted. However, acceptance of such toilet facilities requires them to be well designed, well maintained and safe for use, especially for women.

These facilities can be managed by a group of households, a township, Community-based Organization, NGO or cooperative. In some countries a national or regional public authority supports the establishment of water kiosks or public stand-posts where the local population can fill up jerry-cans with water against a set price. Financing of these operations primarily occurs from public budgets and support from (international) NGOs. However, increasingly micro-credit systems help providing local finance for water-related investments by households or neighborhoods, such as these run for decades already by BRAC in Bangladesh and other countries, and Water.org across Africa and Asia. Credits amount to typically US$10 to 1000, and repayment rates are reported to exceed 90%. Clearly, such operations rest on local expertise and come with high transaction costs, but seem financially viable and can scale up thanks to capital injection from impact and philanthropic investors. In China and India, sanitation has also attracted private entrepreneurs, such as Sulabh International (India), providing reliable services to urban lower-income neighborhoods against payment of small fees.

2.3.2 Industrial water

Few industrial activities can exist without water for their processes or for cooling. Beverage industries, steel mills, microchip manufacturing, and leather tanneries alike consume copious quantities of water which they discharge again as polluted wastewater. Agri-industries such as sugar, cotton, potato, coffee, palm oil, and other mills use water to rinse, soak, ferment and transport the beans and tubers. Power generation depends heavily on large flows of cooling water, and in the case of hydropower on steady river flow or dammed reservoirs. The array of technologies to produce the process water of appropriate quality, and at the other end to treat wastewater effluents, is large and diverse, well proven, and often innovative. Similarly, as water intake increasingly comes at a cost where the opportunity cost of using water and the cost of pollution are reflected in well-designed and enforced abstraction and pollution charges, the industry is seeking to reduce its net "water footprint," e.g., by recycling water, or abandoning processes that are too water-intensive or generate wastewater that is too expensive to treat. Also for these purposes technologies are available in the market. Corporates finance these investments in the same way they finance their production processes, drawing capital from the markets or their own balance sheet.

Industry is facing particular challenges as water use in stressed river basins is turning more competitive and water availability less secure, and as regulations and penalties are ever-tightening for discharges of wastewater in rivers. Across all countries factories have been forced to temporarily shut down during dry spells. In the Netherlands, power plants along the Rhine river have had to suspend operations during several summer weeks in the past decade when high water temperatures caused by low summer water flow were pushing the water intake temperatures at downstream water treatment plants above 25°C, the maximum allowable under EU Directive. Similar events have been occurring in the eastern and western United States such as along the Brazos and other Texan rivers (Fowler, 2011). For India, Gassert (2017) and WRI (2018) reviewed the vulnerability of over 400 power plants and found that more than a third which depend on freshwater for cooling are located in areas of high or extremely high water stress. These plants have, on average, a 21% lower utilization rate than their counterparts located in low or medium water stress regions. Fourteen of India's 20 largest thermal utilities experienced at least one shutdown due to water shortages between 2013 and 2016, costing the companies US$1.4 billion. Thus, it is likely that due to growing competition for water and climate and other changes the long-term viability of many investments in industrial and power plants risks becoming compromised. This is a concern for their financiers, when these physical risks turn into financial risks.

2.3.3 Flood and drainage management

Protection against floods, and the drainage of excess water from inundated or marshy land call for a distinct class of water management interventions. As observed earlier, flood pulses in rivers have also important positive effects on ecologies, and they adduct sediment which builds and strengthens deltas. Floods originate from very high river discharge caused by high precipitation, from land-use changes that canalize land run-off straight into the river, or from very local cloudbursts which local drainage systems cannot cope with. Coastal flooding by storm surges is occurring more frequently the past decades as ocean water is warming up. Yet, an important root cause of flooding is humans settling natural floodplains along rivers and coasts. Past flood defenses consisted only of high dikes, hemming in the river, but often unable to manage the pressures at very high water level. Critically, when upstream communities construct dikes to let the river wave flow past quickly, they may be creating larger problems downstream as under natural conditions the river wave would have been attenuated by expanding into plains upstream where it could also have recharged the groundwater.

Measures to manage floods include foremost proper land use planning avoiding human settlement in floodplains, and giving enough room to the river. Dikes remain central to flood protection; they are usually built of dirt, with an impermeable clay layer or membrane. Dikes necessitate regular monitoring, repair, and upgrading, but generally the operational expenditures are modest (at 1%–3% of construction cost per annum). To enhance the hydraulic discharge capacity of the river, dredging is used to deepen the main channel in the bed and remove sandbars and other obstructions. Groynes (spur dikes perpendicular to the river bank) and regulating weirs also help moderate flow rates and maintain adequate depth in the river's central channel. Retention reservoirs and reserved overflow areas can store part of the flood wave; such areas can be landscaped and managed to support biodiversity, enhance ecology, as well as tourism (angling, biking, kayaking, etc.), creating significant secondary benefits. These "soft" nature-based solutions, sometimes called *green infrastructure,* tend to work with, not against the natural systems, and add intrinsically to overall sustainability. Within urbanized areas, local rainwater must be drained via the sewer system, or excess water is to be temporarily stored in underground road tunnels (Kuala Lumpur) or in squares and parks constructed below street level (e.g., Copenhagen, Rotterdam). Also here, smart designs integrating flood management with other urban functions, can achieve multiple benefits. In many cities, derelict riverfront real estate such as old factories, warehouses, and docks are being redeveloped and this creates the opportunity to, at the same time, reduce flooding hazard or potential damage. Finally, rather than just "protect" against floods, it often pays off to take an integrated approach and invest in "preventive" measures, notably those which relate to land use, such as retaining water upstream, reafforesting catchments, and reducing hard surface. For instance, in Flanders (Belgium) building regulations now restrict pavement in gardens and seek to hydraulically disconnect rainwater falling on roofs from the sewer system. Critically, land use and spatial planning, and building permits must take full account of flood hazard and potential damage.

Well-designed flood management brings large economic and financial gains (see further), partly from avoided damage and, to an increasing extent, from cobenefits that are to be identified through integrated approaches in landscapes and cityscapes. However, the cost to mitigate very rare events may not warrant the expenditure of additional infrastructure and insurance policies may then be cost-effective to cover the residual risk.

Financing of these investments can take place at various levels, applying different instruments. In or near cities, local governments, often jointly with a developer, can attract commercial financing to be repaid from the proceeds of the sale of land and/or higher local municipal property tax revenues. Larger-scale flood management, along long stretches of rivers and coasts, typically is financed by the national or state general budget, however, often associated with commercial capital furnished by banks or through bonds dedicated to that specific investment. Increasingly, new arrangements are being developed to seek more flexibility and efficiencies. For example, one of the key components in the Netherlands' sea defense system, the Closure Dike (*Afsluitdijk*) which closes off the central IJssel Lake from the sea, is being upgraded and expanded by a consortium that includes private financiers (Gietema, This volume); repayment on this investment is from the general budget as flood protection commonly has no distinct fee-paying customers and is perceived as a public task. Several river management operations such as dredging and port maintenance which depend heavily on annual maintenance works offer the opportunity to establish contracts that cover not only the works but also long-term follow-up river maintenance. Such arrangements recognize that the high recurrent costs can significantly influence the basic design choice of the operation, e.g., between dikes which are expensive to construct but cheaper in maintenance, and channel dredging which has a low up-front cost but requires repetitive operations. Flood management investments increasingly are undertaken by river basin organizations, which are typically (semi-)autonomous entities or authorities which manage river infrastructure such as reservoirs, barrages, resorts and wastewater treatment facilities and which possess a customer base to generate revenue. Such entities are potentially also able to raise commercial capital.

2.3.4 Agricultural water

Agricultural water is by far the largest water appropriator: on average in the world, it is responsible for 69% of all water abstraction though this varies from 92% in South and Central Asia to about 20% in Europe. About 6% of all spending on agriculture is water-related (AQUASTAT, 2020). Irrigated agriculture covers 275 million hectares or 20% of cultivated land and accounts for 40% of global food production (WWAP, 2021) underscoring the importance of water adduction. In most instances this water is used to irrigate agricultural land to enhance crop growth (notably of rice, cotton, corn, soybeans, alfalfa, berries, nuts, sugar beets and sugar cane, and various vegetables and fruits) both with respect to their quantity and quality. Irrigation helps grow produce under optimal conditions so it can meet high specifications and supply markets reliably; such produce can catch a premium price. Cattle watering can locally be a determining element. However, perhaps an equally large area supports agriculture by virtue of draining residual irrigation water or, more impactful, reclamation and deep drainage of marshy lowlands for plantations.[5] Both irrigation and drainage lead to significant change in the regional water system. The livestock's manure, on the other hand, is a major source of pollution, but also land run-off to the surface and ground water can raise problems as it carries large silt loads, nutrients, pesticides and herbicides.

Large-scale investment in public irrigation has taken place across the world up to the 1990s. Given often very low use efficiency and local limits to water availability, new public irrigation development all but halted in the 1990s; henceforth, more modest investments targeted system modernization and techniques to yield "more crop per drop". However, much commercial agriculture and forestry on vast plantations is owned by multinational conglomerates such as Cargill, Bunge and United Fruit (US), New Hope Liuhe (P.R. China), Bakrie Group (Indonesia), Asia Pulp & Paper (Singapore), Sime Darby (Malaysia), Unilever (UK), Nestlé (Switzerland), BRF (Brazil), etc., whose operations have a critical impact on the local water systems through water abstraction, deep drainage and general conversion of natural land into agricultural estates. Such corporates are highly specialized and well able to attract commercial capital at global scale.

Like for water supply, irrigation water is typically abstracted from a reservoir on a river or from groundwater, conveyed by canals, pipes, and pumping stations over significant distance, and then flows or is sprayed over the arable land, or is delivered by tubes under the top soil in the root zone in small targeted quantities by drip irrigation. More recent innovations further enhance the water use efficiency, for example by introducing new cultivars that are more drought-resistant, or by small meteorological stations and soil moisture sensors that allow to synchronize the water delivery with the growth phase of the plants and the actual and forecasted weather conditions.

The economics of irrigation are complex as farm-gate prices depend heavily on production costs, weather conditions, pests and diseases, labor availability, and markets and consumers' dietary preferences, which all are

[5] A current grave concern relates to low-lying forested coastal peatlands of which nearly half lies in Indonesia. Here, currently nearly 5 million ha have been deforested and converted into large-scale plantations (palm oil, and *Acacia* for pulp/paper) by draining the marshland. This process causes the land to settle by 1–3 meters per decade; now about a quarter of this land already lies under the high-water sea level, expanding to nearly half by 2050. In this process peat is oxidized releasing vast quantities of CO_2, making the country the second or third largest carbon emitter in the world (Hooijer et al., 2012).

more or less variable. Many crops and livestock production require significant up-front investments with revenue generation only after several years. Different from household water users, agricultural water users are in essence entrepreneurs entertaining different investing and cropping strategies. They are diverse posing vastly different financing positions: from very poor rural smallholders—often subsistence or part-time farmers tilling a quarter of a hectare—to wealthy conglomerates. In richer and poorer economies alike, regulations and subsidies influence produce prices and farmer incomes. In developing economies policies aim at rural development and to safeguard a minimum farmer income. In many developing and emerging economies, they may also aim at national food security. Policies in richer economies rather aim for competitiveness and approach farms as commercially viable businesses. Here, farms are at competitive scale, say, 50 hectares and upward, depending on the crop, and upward of 250 cattle head; they have access to functional markets, technology, logistics as well as financing (including subsidies and insurance) and often seek multi-year supply contracts with wholesalers, supermarkets (for produce) and specialized processing industries (for sugar, potatoes, seed oils, cotton, rubber, meat, etc.). Investments in irrigation, water management, and drainage are part of the overall farm business. Still, in these countries, agriculture and rural development are becoming increasingly intertwined with nature and landscape management seeking sustainability. The current EU's European Agriculture Fund for Rural Development which subsidizes farm investment, incentivizes more sustainable management of landscapes, waterscapes, and soils by farmers.

The above four objectives—raising farmer income, food security, competitiveness, and business viability—often prove incompatible and require deep trade-offs, new technologies and practices, and regulation. Large agricultural estates contribute to food security at national or global levels and offer the robust business cases financiers are seeking but are among the driving forces for expansive land subsidence, and deteriorating ecologies, biodiversity, and water security. Smaller ones contribute to the livelihood of rural communities but are less likely to be connected to global value chains.

In richer economies, farmers often associate in Irrigation Districts or similar organizations to defend their interests when it comes to water access, and to run and maintain the water infrastructure. In developing and emerging economies, in contrast, public investment in irrigation is often a tool in a broader social policy program. Farmers are encouraged to join the local Water User Association (WUA) or similar cooperative arrangement to distribute the irrigation water in their system, maintain the canal infrastructure, collect an irrigation fee to pay (partially) for this—often with labor or produce—and represent the farmers in meetings with the water authority. Typically, such fees are very modest and WUAs are unable to make significant investments in infrastructure. Notwithstanding, such organizations are crucial to secure sustainable operation of the system and take on essential, basic maintenance tasks. On the other hand, in a few countries such as China, Vietnam, Jordan, Mexico, and Egypt comparatively large technical incorporated organizations dedicated to regional irrigation management or the supply of bulk irrigation water to farmers could potentially evolve into well-run utilities able to own and develop assets and raise capital; however, experiences are still patchy as the managerial autonomy and capability of these organizations are still modest. Commercial investment in such distorted and often politicized markets is tenuous. In other places river basin organizations (or equivalent institutions) operating with a broader remit and owning, for example, multipurpose reservoirs and canal systems, are better positioned to raise capital for and manage the large off-farm infrastructure that supplies irrigation water, while collecting fees from larger farms or WUAs. Such institutional set-up is relatively common in richer economies such as in the Murray-Darling Basin (Australia), the Tennessee Valley Authority (USA), the Canal de Provence (France), most of the basin management organizations (*Confederaciónes Hidrográficas*) in Spain, as well as the North African countries.

Irrigation fees can be based on the acreage of the tilled land (in case of flood or furrow irrigation from canals) or, preferably, on the metered abstraction from a canal or standpipe. In many irrigated areas in richer economies farmers can use an internet-based site to book and pay for a water quantity and time slot from a specified offtake point which is equipped with a remotely controlled valve to dispense the water. Such pumped systems are well-controlled offering confidence to users and operators. But irrigation management brings three economic and financial challenges that need to be addressed to secure the system's sustainability and effectiveness. Firstly, withdrawals from gravity-fed open canal (or river) systems generally cannot be well monitored, and a common challenge is that upstream farmers may tend to withdraw more than their agreed share which depresses the willingness of tail-end farmers to financially cooperate. Secondly, in the western United States and Chile water users have title to their water rights and this often leads to, from an economic point, suboptimal water use. Although the rights are tradeable, in some cases in the United States, the "first in time, first in right" arrangements provide disproportionately secure rights to long-held entitlements compared to new entrants. For example, farmers with such rights in California may grow alfalfa for fodder, a commodity of low economic value, as their water comes at little cost. Lastly, farmers typically are not compelled to draw water from the system and, thus, can avoid payment, for instance when sufficient rain arrives for their crop, or if they opt for minimizing their expenditures, instead of maximizing their production. In this case, the irrigation system risks becoming financially unsustainable.

2.3.5 River basin management organizations

The above analysis has illustrated that the different water "compartments" are physically connected but that their management is carried out by separate institutions. Sustainable water management in a rapidly expanding global economy in many instances benefits from an approach that integrates the compartments of the water cycle with the water supply and water demand dynamics, and which takes into account the mutual dependence of water, land, and nature. Because these physical, economic, and social interrelationships are strongest within a river catchment or river basin, the Dublin Principles (ICWE, 1992) and professional circles advocate that water be managed at the scale of the hydrographically coherent river basin. The EU Water Framework Directive (EC, 2000) imposes such framework in the European Union for the purpose of planning investments for water quality management. OECD (2016, 2021) acknowledges them as good practices, while emphasizing the benefit of, and challenges related to, combining and, as much as possible, aligning water management practices and tools at different geographical scales. For planning purposes, IWRM provides a practical planning framework at the (sub-) basin's scale to balance the different competitive uses of water, notably that by nature, to identify trade-offs, synergies and economies of scale, and to prioritize investment and management measures that foster the sustainability and integrity of the resource while creating co-benefits.[6] Such analyses can apply simulation

[6] Relevant examples are the major expansion of water supply for New York (USA) and Amsterdam (The Netherlands) in the 1990s where the utilities achieved major investment savings by purchasing large tracts of land in catchments and buying out any potentially polluting activity (the upper Delaware, and coastal dune and polder areas, respectively) thus safeguarding very clean water resources, instead of conventionally scaling up only the filtration treatment facilities (Hu, 2018; Waternet, 2021). This investment in "green infrastructure" led to significant net cost savings, higher storage and supply reliability, and creation of valuable nature and recreation facilities. Similarly, Poland's flood management project on the Odra river captured co-benefits: near the port of Szczecin a deteriorated wetland is being restored to enhance wildlife and adapted as "green infrastructure" to buffer flood waves; river barrages are equipped with fish ladders to rejuvenate fish stocks; and Wroclaw city used the rehabilitation of

models to forecast the benefits and disadvantages of different investment scenarios on energy generation, agriculture, fisheries, job creation, and nature values (e.g., on the Mekong [Alaerts and Makin, 2007], and the Zambezi, [World Bank, 2010]).

The operational functions are generally better addressed by organizations that benefit from specialized sectoral skills, local political and administrative anchoring, and knowledge of the client base. In an increasing number of situations it does make economic and managerial sense to combine also operational functions in a river basin management organization (RBO). Even without these, the RBO adds value already by providing a formalized collaboration and negotiation platform and shared database to the basin stakeholders. "Asset-heavy" operational RBOs are typically those that own reservoirs for hydropower, shipping infrastructure (locks, ports, etc.), irrigation, flood management infrastructure, and city water supply, such as, for example, the Tennessee Valley Authority (US) that was established in the 1930s to generate hydropower and combat floods but evolved to become essentially a power utility; the Damodar Valley Authority (India) that manages reservoirs to supply water to industry and agriculture and controls water pollution; the *Ruhrverband* and *Wahnbachtalsperrenverband* (Germany) that operate series of reservoirs on tributaries of the Rhine and piped conveyance systems for city water supply and ecological functions; the *Confederaciónes Hidrográficas* (Spain) that control the river infrastructure; the 21 Water Agencies (the Netherlands) that collect and treat all wastewater, manage water quality and protect from regional floods; and the *Organisation pour la mise en valeur du fleuve Sénégal* (Senegal, Mauretania, Guinea and Mali) and the Niger Basin Authority (nine north-western African countries) that own and manage dammed reservoirs, and river infrastructure and boat landing facilities. They operate the hydraulic system but also engage with agriculture, freshwater fisheries, and ecological restoration, and address floods and droughts. China's River Basin or River Conservancy Commissions for the Changjiang (Yangtze), Huanghe (Yellow River), Huaihe and others fall into this category (although called Commissions, in communist-system terminology they are implementation field offices of the Ministry of Water Resources). "Asset-light" RBOs are, for instance, the nine French *Agences de l'eau* which convene all stakeholders of the basin and assist in co-financing water-related investments of municipalities,

embankments to create riverine boulevards and restore the riverside historic Castle with city and private funds, enhancing the city's cultural appeal (World Bank, 2015a).

and the international Rhine, Mekong, Sava, and Nile Basin Commissions and the International Commission on the Aral Sea that do not own assets but convene the stakeholders, prepare basin-wide plans, assist in resolving conflicts and manage key information. In Indonesia two *Perum Jasa Tirtas* solely operate and maintain the five main rivers' infrastructure the cost of which they cover with the revenue from selling bulk water to large customers; investment, and asset ownership, is the remit of local River Agencies that are departments of the Ministry of Public Works. The Murray-Darling Basin Commission (Australia) is an independent body that synchronizes the water management and investment plans of the three riparian states, and it manages the trading platform for the water use rights that water users can bid for. De Bièvre and Coronel (This volume) describe corporatized catchment management and restoration organizations that are particularly successful in Latin America with public and private shareholders.

Among typical tasks of RBOs are the apportionment of water flow to different bulk users, and the forecasting of the impacts of growing demand, and change in land use and climate on the river system. As water demand will inexorably keep rising and relative availability will be decreasing due to these changes, water use rights and their trading will become more prominent instruments to manage access to water and provide security and comfort to investors. The entities of the "asset-heavy" RBOs are generally well managed and familiar with investments, and can attract commercial capital; the RBOs of the second category may not attract capital themselves but are nonetheless key for developing consensus on what constitutes productive yet sustainable investments in the basin. Repayment of financing occurs from revenues from selling bulk water and electrical power to large industrial, agricultural, and water supply customers, as well as from access and concession fees to lake and riverfront areas. Thus, these organizations tend, in principle, to possess revenue-earning power.

2.4 Value, cost, and price—sometimes aligned, often not

2.4.1 Value, cost, and price

A systemic feature of water is the often large discrepancy between its value, cost, and price, which is distinctive in comparison to transportation and power, for instance (Fig. 2.1). Typically, the value of water can be high, but

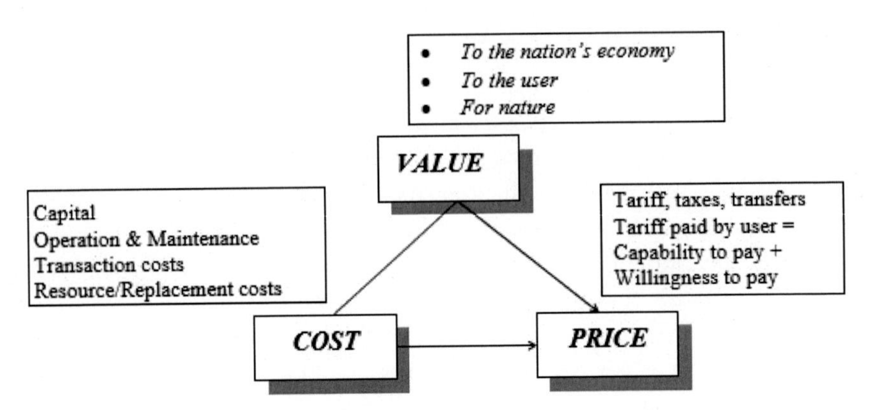

Figure 2.1 Value, cost, and price (tariff) of water and water services.

its price can be low (or nil); both value and price can be disconnected from the cost of providing water services or mitigating water-related risks. Smith, and Leflaive et al. (both This volume) describe water as an input to many processes which lend it an intrinsic economic, social, environmental, or other *value* in achieving a given purpose. Safe drinking water and sanitation are particularly valuable to achieve public health goals. As water is quintessential to human and environmental activity a basic available amount carries a very high economic and social value, but the value of incremental water will vary between alternative use scenarios. With regard to economic value, Schellekens et al. (2018) estimate that for the European Union, for example, 24% of employment and 26% of Gross Added Value (or €3.4 trillion) depend directly on secure water. While such value can be demonstrated, decisions affecting water are taken by how citizens and politicians *reveal* their effective valuation. Often the scarcity and fragility of water are not appreciated for lack of knowledge and it is a well-known phenomenon that people who first suffered from flooding are quick to forget the lesson to prepare for and avoid the next flooding event. Even when people profess that water is valuable to them they may resist to translate this into commensurate action and refuse to take action to save water or pay for it; this revealed "effective" value can be significantly lower than the economic value. The *cost* refers to the expenditure necessary to satisfy the demand for water from households, industry, and nature, which is usually considerable; it tends per unit of delivered amenity to be much higher than in the transport, power and telecoms industry, and is set to rise with declining relative water availability and quality.[7] An important element is the resource cost (replacement cost) that can rise prohibitively high, for example when fossil groundwater gets exhausted. The *price,* then, is what the water user actually will have to pay as fee, tariff or tax for this service. Where the user cannot be distinguished or beneficiaries are diffuse, e.g., for flood protection or environmental values, or where poor communities need support, a proxy such as the state or third party may pay the bill. Economic orthodoxy holds that polluters should pay the environmental costs. Users or beneficiaries are best placed to pay the resource costs, i.e., the price signaling the degree of water scarcity. Water users pay to recover the financial cost covering the operational (recurrent) and capital expenditures. However, from a social perspective (partial) tax- and subsidy-based payments are legitimate to achieve more equity: they are meant to either cover the public-good dimension of the service (part of the public-health benefits of sanitation, for instance) or to address equity and affordability concerns. Also from a financier's perspective the origin of the payment is not necessarily relevant.

The (revealed) value sets the price for many commodities as it reflects what the user is willing to pay. For many public services without a profit motive and for which no substitute is available the cost typically determines the price. In rich countries abundant in water richer households can easily pay the full cost but poorer ones may not. Where water services are very expensive even richer households may be unable to pay the full cost. Typically, policy, then, will aim to recover at least the recurrent costs and shift the capital provision and debt service to the State.

2.4.2 The value of water (in)security

Lately, many studies have made attempts to quantify the value of water in economic and financial terms as benefit or as damage avoided, sometimes

[7] The EU's Water Framework Directive refers to the financial, environmental and resource costs of water. The financial costs cover the capital and operational expenditures to deliver the service. The Directive defines environmental costs as the costs of damage that water uses impose on the environment and ecosystems and those who use the environment (e.g., a reduction in the ecological quality of aquatic ecosystems or the salinization and degradation of productive soils). Resource costs are defined as the costs of foregone opportunities that other users suffer due to the depletion of the resource beyond its natural rate of recharge or recovery (e.g., due to over-abstraction of groundwater) (EC, 2021). In principle, all costs are to be recovered by a combination of revenues from user tariffs, charges, taxes and transfers (the latter in the case where an outside financier provides subsidies). Tariffs apply to the provision of a service (typically supply of safe water, and collection and treatment of wastewater). Abstraction charges can be designed to capture the opportunity cost of using the water resource. Pollution charges are best designed to cover the cost of pollution (environmental cost).

through proxies such as Disability-adjusted Life Years or lives lost. Drought is deemed to be the deadliest physical hazard with at present an estimated 3.6 billion people (nearly half the global population) living in areas that are water-scarce for at least one month per year. This population could increase to some 4.8–5.7 billion by 2050 (i.e., 55%–65% of the world population) with over 40% of the world population living in water-scarce river basins under severe water stress, up from about 16% in 2010 (OECD, 2012; Sadoff et al., 2015; WWAP, 2018; Smith, This volume). Poor water and sanitation services are estimated to cost emerging and developing economies about US$260 billion per year due to poor health, pollution, and inefficiencies in households and industries. This amounts to, on average, 1.5% of their Gross Domestic Product (GDP), but may, in specific countries, rise to as high as 10% (World Bank, 2016a; World Bank/UNICEF, 2017). OECD (2012) estimates that the number of people per year affected (including, killed) by water-related disasters (of which two-thirds can be attributed to floods) will rise from 100–200 million in 2010 to 1.2 billion in 2050, or almost 20% of the world population.

Significant riverine flood episodes in Pakistan, Bangladesh, Thailand, and other emerging economies since 2000 are reckoned to have cost these countries 1 to 4% of GDP (Sadoff et al., 2015). In the coastal and deltaic areas, the damage may rise even more steeply. Hallegatte et al. (2014) identify growing populations and asset values, land subsidence (due to over-abstraction of groundwater) and the changing climate as key drivers of coastal vulnerability. They estimate average global flood losses in the 136 largest coastal cities in 2005 at US$6 billion per year, increasing to US$52 billion by 2050 assuming that only the projected socio-economic changes will take place. When including the effect of climate change and subsidence, unacceptable losses of US$1 trillion or more per year would be incurred. Even if adaptation investments in flood protection would maintain constant flood probability, subsidence and sea-level rise will increase global flood losses to US$60–63 billion per year in 2050. Ligtvoet et al. (2018) estimate that, currently, flooding affects about 106 million people annually and causes the largest economic damage, at US$31 billion per annum. In 2050, the economic value of the assets at risk of floods is expected to be around US$45 trillion, or 350% higher than in 2010, partly because of higher flood frequency, but mostly driven by the growing value of the assets at risk. While the *number* of affected populations and assets is much higher in the emerging and developing economies, the financial and economic *value* of the assets at risk is much higher in the rich economies. This is borne out by, for example, the financial costs from hurricanes in the US, resulting primarily from flooding damage: Katrina near New Orleans (2005) US$161 billion; Harvey (2017) US$125 billion and Maria (2017) US$90 billion, both on the East Coast; and Sandy (2012) in New Jersey and New York US$71 billion. Only a minor part of this financial damage is reimbursed by insurers (NOAA, 2018). The Harvey and Irma episodes depressed the fourth-quarter GDP growth forecast for the United States by 0.8 percentage points (Ciolli, 2018). World Bank (2016b) for many regions in the world forecasts a decline in GDP of up to 6% by 2050 caused by water-related losses in agriculture, health, income, and property, with some regions in the world facing sustained negative growth. The above studies calculate economic cost-benefit ratios of 3-5 for water supply and sanitation expansion and 5-9 for investments in resilience of water resources. The foregoing forecasts assume continuation of current or moderately enhanced policies and sector performance.

2.4.3 Financial assets values grow vulnerable to systemic water-related risk

While good water management adds value, poor management detracts value. The above discussion shows that not only human livelihoods and aquatic ecologies but also large assemblies of financial assets—factories, power plants, supply and distribution chains, real estate, etc.—are exposed to risks created by poor water management practices and by weak pro-activity regarding large-scale changes in land use and climate. For the US, the National Climate Assessment (USGCRP, 2018) confirms that, without proper adaptation, the specific climate change-related phenomena will cause significant damage to infrastructure, property and to the production, supply chains, and trade of US businesses, inside and outside the country. It cautions that most risk assessments do not yet reflect compound extremes (co-occurrence of multiple events, as happened with the 2012 hurricane Sandy when the sea surge concurred with extreme rainfall) and the risk of cascading infrastructure failure. Without proper action, aggregate impact by 2100 could lower GDP by up to 10%.

Physical productive assets, real estate, and other investables are turning vulnerable in the next two decades to quickly evolving water-related extremes exacerbated by climate change, for which we are not well prepared and for which little alternative exists. This poses new risks for individual firms but, because of its systemic nature, also for the global financial stability. Financial consultancies, such as Moody's Global Solutions,

Acclimatise, Aquantix AI Group and S&P Global Trucost have started assessing the "climate resilience"—mostly related to water—of investment classes linking physical risks (i.e., the "natural" risks, but as mitigated under current protection arrangements) with the viability of firms and investment portfolios. Mazzacurati et al. (2017) assessed the climate vulnerability of the CAC40—the benchmark 40 most significant firms listed on the Paris stock exchange—and of Asian banking, food, and power stocks, diversified to Supply Chain, Market, Operation, and Location Risks. Such scoring helps better understand the nature of the vulnerability and the need for adaptation strategies. Similar exposure analyses have also been piloted on real estate in the US. Moody's now also assesses exposure and preparedness of sovereigns to climate-related stress; in January 2020 it issued credit ratings of countries particularly exposed to climate (water) risks and retained the (existing high) rating of Japan and the Netherlands as these countries had adequate countermeasures in place, but was critical of Vietnam, the Bahamas and Egypt for inadequate preparedness (Moody's, 2020).

Many future water-related hazards will differ from past events that were considered "disasters" because they will be affecting regions and sectors in structural ways and can be forecasted with simulation models with increasing accuracy. Also, the connectedness of local and global economies and the steep rise in the investment value of all asset classes multiply the risk. The traditional fashion of diluting net risk in one location by balancing it out against the performance in many unaffected other locations, may no longer be always valid. Therefore, the Financial Stability Board and the Bank of International Settlements (Basel) commissioned a Task Force on Climate-related Financial Disclosures (TCFD, 2017); the European Union is drafting regulations to make climate- (and water-)related risk to the financial sector quantifiable and manageable (EC, 2019) and will issue in 2022 a taxonomy of green and climate finance. The TCFD concludes that risks related to climate change are one of the most significant and perhaps most misunderstood risks that organizations and investors face today, with impacts both in the short and long term. Thus, the value at risk to the global stock of manageable assets would range from US $4.2 to 43 trillion between now and 2100 through weaker growth and lower asset returns. The Task Force recommends that firms and financial institutions disclose publicly in their mandatory annual filings both their exposure to the risk and their response, around four core thematic areas of their operations: their governance related to the risk, their strategy including how the risk also creates commercial opportunity, their risk management, and the metrics and operational targets applied to manage the risk and opportunities. As per 2020, these recommendations are being subscribed to by a small but growing part of the financial sector.

2.4.4 Large financing needs but a challenging business case

Several efforts have been made to estimate the sector's capital requirements applying different scopes and methodologies, as reviewed by Winpenny et al. (2016) and Hutton and Varughese (2016), and, across all sectors, McKinsey (2016). Hutton, and Rosenstock and Leflaive (both This volume) offer updates for developing and emerging economies and for the EU, respectively. Given the high economic value of water, the high cost-benefit ratios of investment, and the high objective financing needs, one would expect that readiness to invest has grown proportionally. However, this is not the case across the board. Broadly, the "needs" in the richer economies can be largely, though not everywhere fully, met across the different water asset classes with a substantial and growing role for commercial financing.[8] Notwithstanding, capital outlays likely need to increase significantly in the next two decades and innovative arrangements will be in demand to invest in the circular economy and the climate transition, and achieve more robust environmental sustainability. In the emerging and developing economies, on the other hand, typically only 20%–50% of the calculated needs are being met. Here, the bulk of commercial financing for infrastructure goes to power, roads and telecoms. For both Asian and Latin American developing countries, about 91% of the water finance is public against about 13% in telecoms, 52% in power and 70% in transportation (World Bank, 2018).

Moving forward, this raises two considerations. Firstly, as Martini (This volume) points out public financing capabilities will become more constrained in the next decades, which implies a higher reliance on private financing. But secondly, obstacles of institutional and affordability nature seem to obstruct the flow of private capital to "needed" water investments, and financing instruments, regulations and institutional capacity are not yet attuned to cater to this market.[9]

[8] However, the need to replace decaying infrastructure in the developed economies remains insufficiently recognized at national and utility levels (for the EU, see Rosenstock and Leflaive [This volume]). Associated investments are huge: many assets need to be renewed, upgraded, and adjusted to shifting conditions such as higher social expectations in terms of water security and environmental quality, and the changing climate.

[9] Taxation also plays a role, as in many countries tap water services are taxed at commercial rates even where the government considers water service a task of public interest. Many resource recovery initiatives, usually commercially precarious, are burdened by taxation. In other words, if a water utility discharges a residual sludge as waste this is not taxed but imposes an economic cost on nature, but if

A first, central challenge with water services is the combination of *cost* and ensuing *price (tariff)* which tend to be high compared to the current capability and willingness to pay of the customers. Richer households in western Europe, Japan, the United States and in well-to-do city quarters around the world can afford the price but middle- and lower-income households, and farmers where it concerns irrigation, in the emerging and developing economies may be constrained in their ability to pay. In addition, when able, they may not be willing to pay, for instance, because the service is of poor quality or unreliable, or the utility perceived as poorly run. The monopoly position of water service providers and local vested interests may reduce the incentive of management to become more effective. Subsidies can be helpful under certain conditions but risk turning into political patronage. Strong governance arrangements, managerial capability, and accountability to the customers and citizens, then, prove essential to eventually render water services more performing and affordable, and to strengthen the business case for investment.

The managerial, financial, and environmental performance of water utilities is being monitored systematically (albeit on voluntary basis) in many countries as a tool to support peer learning and enhance performance, e.g., through the International Benchmarking Network for Water and Sanitation Utilities (IBNet) which monitors Key Performance Indicators of nearly 3000 utilities mostly in emerging markets (World Bank, 2021) and the European Benchmarking Initiative (https://www.waterbenchmark.org) (Eureau, 2015). World Bank (2015b) assessed the Operating Cost Coverage ratio for representative samples of tap water and wastewater utilities in 15 Central European countries (Fig. 2.2) illustrating the challenge in emerging economies. The ratio measures net billed sales over operating expenses; utilities should have a ratio above 1 to be financially self-sufficient for O&M, and above 1.4 to include forward-looking investment capability and creditworthiness. A majority of the utilities manage to recover their operating expenses from their own revenues, though on average, Hungary and Ukraine do not meet this basic requirement (nor comply with the EU requirement on cost recovery) due to strong central tariff control. Only Austrian utilities are, on average, creditworthy. The situation is not positive, also considering that utilities in a number of these countries fail to collect

the utility invests to convert the residual in a recyclable resource and sells it to recover part of the cost, it is liable to taxation by the government. Thus, Ministries of Finance and tax regulations are critical to achieve government objectives of sustainable development.

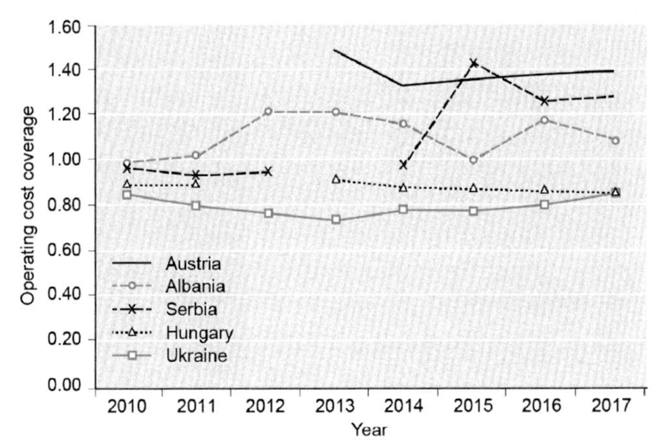

Figure 2.2 Country-average operating cost coverage ratios of a sample of water and wastewater utilities in 15 countries in Central Europe (World Bank, 2015b and data from www.iawd.at/eng/danube-toolbox/benchmarking/danubis-org-pi-database/). Not shown, with ratio 1.0–1.2: Czech Republic, Romania, Bulgaria, Slovakia. Not shown, with ratio 0.8–1.0: Bosnia and Herzegovina, Kosovo, North Macedonia, Slovenia.

a significant share of billed revenues and that provisions are seldom made for accounts receivable write-offs. The long-term trend regarding operating costs coverage shows weak improvement (with Serbia performing better) pointing to structurally weak managerial capacity.

In an in-depth study of 690 utilities across the world, World Bank (2017) found that only 15% were generating enough revenue to cover their O&M costs and earn a basic cash surplus. However, the study showed that this ratio could be raised to 77% by simple yet realistic management improvements: by collecting all bills, lowering nonstaff expenditures by 15%, and by reducing NRW (leakage) with 10%. Importantly, this could be achieved without raising tariffs or discharging staff. This study underscores the broader observation, also mentioned in the context above of micro-credit arrangements in poor rural settings, that households are generally quite willing and able to afford basic water service—provided the service is reliable and adequate. Thus, although affordability is a concern, weak managerial and institutional capacity in many instances is a greater challenge. Where affordability is an issue adroit technical, financial, and social measures such as subsidies, pro-poor tariffs, and fee waivers targeting the poorest households can often mitigate it. The managerial constraints are also a root cause of the shortage of financing-ready investment proposals, weak creditworthiness, and the too-small operational scale of many utilities. Thus, financiers are expressing concern that in many places the capacity to absorb finance is the main challenge rather than the access to finance (e.g., ADB, 2017; European Court of Auditors, 2018; OECD, 2018). In conclusion, the weak readiness to get access to access to finance is the result of the (often low) value attributed to water, the cost of service provision and (often poor) institutional readinesss.

2.5 Conclusions

The progressing deterioration and exhaustion of land and water resources combined with irrevocably growing expectations from civil society will likely raise the demonstrated and revealed values of water, the investment needs and the willingness to pay a realistic price for secure water. Until recently plentiful, or at least adequate, availability of water was taken for granted in the majority of countries, and water did not figure prominently on national agendas as a financing challenge, or as a challenge to the viability of financial assets. The water sector, however, is now entering a fast transition over the coming two decades shifting between two paradigms. The conventional paradigm assumed that all demand—by households, agriculture, industry, etc.—could be satisfied and that any negative impacts of water appropriation could be mitigated through add-on measures. The new paradigm accepts that the water system has inherent constraints, that it is interconnected across its compartments, and that it is deeply linked to land and ecosystems. These linkages are becoming very visible in the form of over-exploitation, land subsidence, and extensive and progressing degradation of ecosystems and biodiversity, with in addition the consequences of climate change. Policies, operational practices, and infrastructure are needed to enhance on one hand the security and quality of the water service—essential for robust business cases—and on the other the sustained integrity and functioning of the natural systems—the precondition for a future that is sustainable, i.e., where the natural water system will still be able to continue providing these services in a way that is in balance with the demand for it. In many situations, notably in developing and emerging economies, finance will remain required for conventional "productive" investments such as for expansion of water supply or sewerage networks and treatment plants. But in a rapidly growing number of instances investment will be warranted to increase the efficiency of existing water use, protect the land and ecological systems of catchments, reduce the "water footprint" of an economic activity, establish a "circular economy," and arrive at more efficient water allocation and easing of water stress. Integrating approaches in water management and land management over shorter and longer time horizons should be part of any investment planning. Moreover, as many sectors outside the water sector have disproportionally large and often negative impacts on the water system, financiers should conduct their due diligence and consider to amend or discontinue the financing of activities detrimental to water security irrespective of the short-time business case, because of reputational and material longer-term risks.

While the physical water system is highly interconnected, the institutional architectures tend to be fragmented and focused on local stakeholders and local priorities. Policy and management, the vulnerability to water-related insecurity, and the scenarios for financing differ significantly between the water supply and wastewater sector, the irrigation sector, the flood management sector, etc. Increasingly, water-related investments are embedded in urban and industrial development programs and in agricultural ventures; therefore, the impact of each investment on the overall water system needs to be quantified and taken into account.

The financial markets have an important role supporting the development and sustainable management of water. Proper water management and the provision of water services, however, rely on heavy investment in infrastructure, on getting payment for their operation and maintenance, and on the development of a worldwide knowledge base. Going beyond financing water services per se, financial organizations and markets have an important role to play to ensure that finance in other sectors (e.g., agriculture, land use, or urban development) does not expand exposure and vulnerability to water risks and, thus, build future liabilities. Failing to do so is likely to be consequential for the sector, with potential spillover effects across value chains and the financial system. Public financing will remain essential in many situations but is likely to become constrained, opening opportunities for private investors. A strategic use of public finance to crowd in (and not crowd out) private finance is a promising way forward.

In the richer economies the water organizations are already effective in partnering with commercial financiers and mobilizing private finance. In emerging and developing economies these organizations often have still too small a scale and modest performance to offer appealing financing opportunities. Blended finance arrangements that combine commercial and "impact" or public investment are showing promise for crowding in risk-averse capital. Still, global and country average figures obscure the important fact that in each place some water organizations are already better performing and are able to offer good business cases. Across both richer and developing economies distinct differences occur; wealthier city quarters and industrial estates can be credit-worthy and should be encouraged to demonstrate the way forward to their weaker peers. While many local water utilities in Brazil struggle to balance their books and present too high a risk for commercial financiers, the São Paolo State Water and Sanitation Company SABESP is well able to raise capital in the international markets. Such institutions should become active role models and support peer-learning. Both the water sector and the financial sector stand to gain from expanding and diversifying their instruments and strengthening their capability.

References

ADB, 2017. Meeting Asia's Infrastructure Needs. Asian Development Bank, Manila.

Alaerts, G., Makin, I., 2007. Mekong Water Resources Assistance Strategy. The World Bank, Washington, DC, and Asian Development Bank, Manila.

Barraqué, B., 2012. Return of drinking water supply in Paris to public control. Water Policy 14, 903–914.

AQUASTAT, 2020. UN Food and Agriculture Organisation (FAO), Rome. http://www.fao.org/aquastat/en/overview/methodology/water-use (accessed October 20, 2020).

Ciolli, J., 2018. Hurricanes Irma and Harvey Cause Goldman Sachs to slash its outlook for the US economy. http://www.businessinsider.com/goldman-sachs-hurricane-irma-harvey-gdp-impact-2017-9 (accessed September 11, 2018).

EC, 2000. European Union Water Framework Directive. European Commission, Brussels https://ec.europa.eu/environment/water/water-framework/index_en.html.

EC, 2019. Report on Climate-Related Disclosures; Technical Expert Group on Sustainable Finance. European Commission, Brussels.

EC, 2021. https://ec.europa.eu/environment/water/water-framework/economics/pdf/Information_Sheet_ECO1_Cost_Recovery.pdf.

European Court of Auditors, 2018. Commission's and Member States' Actions in the Last Years of the 2007–2013 Programmes Tackled Low Absorption but Had Insufficient Focus on Results; Report 17. European Court of Auditors, Brussels.

Eureau, 2015. How benchmarking is used in the water sector. https://www.eureau.org/resources/position-papers/135-benchmarking-october2015/file (accessed May 10, 2021).

Fowler T., 2011. More power plant woes likely if Texas drought drags into winter. Fuel Fix, http://fuelfix.com/blog/2011/08/24/more-power-plantwoes-likely-if-texas-drought-drags-into-winter/ (accessed 20 January 20, 2021).

Gassert, F., Landis, M., Luck, M., Reig, P., Shiao, T., 2017. Aqueduct Global Maps 2.1. Working Paper. World Resources Institute, Washington, DC.

Hallegatte, S., Green, C., Nicholls, R.J., Corfee-Morlot, J., 2014. Future flood losses in major coastal cities. Nat. Clim. Chang. Lett. 3, 802.

He, C.Y., Liu, Z.F., Wu, J.G, Pan, X.H., Fang, Z.H., Li, J.W., Bryan, B.A., 2021. Future global urban water scarcity and potential solutions. Nature Communications 2021 (12), 4667–4677.

Herrera-García, G., Ezquerro, P., Tomás, R., Béjar-Pizarro, M., López-Vinielles, J., Rossi, M., Mateos, R.M., Carreón-Freyre, D., Lambert, J., Teatini, P., Cabral-Cano, E., Erkens, G., Galloway, D., Hung, W.-C., Kakar, N., Sneed, M., Tosi, L., Wang, H.M., Ye, S.J., 2021. Mapping the global threat of land subsidence. Science 6524, 34–36.

Hooijer, H., Page, S., Jauhiainen, J., Lee, W.A., Lu, X.X., Idris, A., Anshari, G., 2012. Subsidence and carbon loss in drained tropical peatlands. Biogeosciences 9, 1053–1071.
Hu, W., 2018. A billion-dollar investment in New York's water. The New York Times, (accessed January 10, 2021).
Hutton, G., Varughese, M., 2016. The Costs of Meeting the Sustainable Development Goal Targets on Drinking Water, Sanitation and Hygiene. Technical Paper 10317. The World Bank, Washington, DC.
ICWE, 1992. Proceedings International Conference on Water and the Environment. Dublin. United Nations Environmental Programme, World Meteorological Organisation. January 1992.
IPBES, 2019. Summary for policymakers of the global assessment report on biodiversity and ecosystem services of the Intergovernmental Science-Policy Platform on Biodiversity and Ecosystem Services. Intergovernmental Science-Policy Platform on Biodiversity and Ecosystem Services, Bonn S. Díaz, J. Settele, E. S. Brondízio E.S., H. T. Ngo, M. Guèze, J. Agard, A. Arneth, P. Balvanera, K. A. Brauman, S. H. M. Butchart, K. M. A. Chan, L. A. Garibaldi, K. Ichii, J. Liu, S. M. Subramanian, G. F. Midgley, P. Miloslavich, Z. Molnár, D. Obura, A. Pfaff, S. Polasky, A. Purvis, J. Razzaque, B. Reyers, R. Roy Chowdhury, Y. J. Shin, I. J. Visseren-Hamakers, K. J. Willis, Zayas, C. N. (Eds.). Intergovernmental Science-Policy Platform on Biodiversity and Ecosystem Services, Bonn.
Leflaive, X., Hjort, M., 2020. Addressing the social consequences of tariffs for water supply and sanitation. OECD Environment Working Papers, 166. Organisation for Economic Co-operation and Development, Paris https://doi.org/10.1787/afede7d6-en.
Ligtvoet, W., 2018. The Geography of Future Water Challenges; PBL Netherlands Environmental Assessment Agency, The Hague.
Mazzacurati, E., Vargas Mallard, D., Turner, J., Steinberg, N., Shaw, C., 2017. Measuring Physical Climate Risk in Equity Portfolios. Deutsche Asset Management Global Research Group, Frankfurt; New York, NY.
McKinsey, 2016. Financing Change: How to Mobilize Private Sector Financing for Sustainable Infrastructure. McKinsey & Partners, New York, NY.
Nam, S., Phommakone, S., Vuthy, L., Samphawamana, T., Son, N.H., Khumsri, M., Bun, N.P., Sovanara, K., Degen, P., Starr, P., 2015. Economic value of lower mekong fisheries. Mekong River Secretariat, Catch Culture 21 (3), 4–7.
Moody's, 2020. http://moodys.com/viewresearchdocs,aspx?docidPBC_1175883 (accessed November 30, 2020)
NOAA, 2018. National Oceanic and Atmospheric Administration. https://coast.noaa.gov/states/fast-facts/hurricane-costs.html (accessed October 12, 2018).
OECD, 2012. OECD Environmental Outlook to 2050. Organisation for Economic Co-operation and Development, Paris.
OECD, 2016. OECD council recommendation on water. https://www.oecd.org/water/recommendation/#d.en.431326 (accessed June 12, 2021).
OECD, 2018. Green Finance Forum. Organisation for Economic Co-Operation and Development, Paris www.oecd.org/cgfi/2018-forum-documents.htm.
OECD, 2019. Biodiversity: finance and the economic and business case for action. In: Report prepared for the G7 Environment Ministers' Meeting. Organisation for Economic Co-operation and Development, Paris.
OECD, 2021. Toolkit for Water Policies and Governance: Converging Towards the OECD Council Recommendation on Water. OECD, Paris https://doi.org/10.1787/ed1a7936-en.
Sadoff, C.W., Hall, J.W., Grey, D., Aerts, J.C., Ait-Kadi, M., Brown, C., Cox, A., Dadson, S., Garrick, D., Kelman, J., et al., 2015. Securing Water, Sustaining Growth: Report of the GWP/OECD Task Force on Water Security and Sustainable Growth. University of Oxford, Oxford.

Schellekens, J., Heidecke, L., Nguyen, N., Spit, W., 2018. The Economic Value of Water - Water as a Key Resource for Economic Growth in the EU. Deliverable to Task A2 of the BLUE2 project "Study on EU integrated policy assessment for the freshwater and marine environment, on the economic benefits of EU water policy and on the costs of its non- implementation". Directorate-General for the Environment. European Commission, Brussels.

TCFD, 2017. Final Report—Recommendations of the Task Force on Climate-Related Financial Disclosures. Bank for International Settlements, Financial Stability Board, Basel.

UN, OHCHR, UN-Habitat, WHO, 2010. The Right to Water. Fact Sheet No. 35. United Nations, New York, NY.

UNDP, 2006. Human Development Report 2006. Beyond Scarcity: Power, poverty and the global water crisis. United Nations Development Programme, New York, NY.

USGCRP, 2018. Impacts, Risks, and Adaptation in the United States: Fourth National Climate Assessment. In: Reidmiller, D.R., Avery, C.W., Easterling, D.R., Kunkel, K.E., Lewis, K.L.M., Maycock, T.K., Stewart, B.C. (Eds.). Report-in-Brief, II. U.S. Global Change Research Program, Washington, DC.

van der Ent, R.J., Savenije, H., Schaefli, B., Steele-Dunne, S.C., 2010. Origin and fate of atmospheric moisture over continents. Water Resour. Res. 46, W09525. doi:10.1029/2010WR009127.

Waternet, 2021. https://www.waternet.nl/en/ (accessed January 10, 2021).

Winpenny, J., Trémolet, S., Cardone, R., 2016. Aid Flows to the Water Sector: Overview and Recommendations. The World Bank, Washington, DC.

World Bank, 2010. The Zambezi River: A Multi-sector Investment Opportunities Scenario Analysis. The World Bank, Washington, DC.

World Bank, 2015a. Project Appraisal Document—Odra and Vistula Flood Management Project. The World Bank, Washington, DC.

World Bank, 2015b. Water and Wastewater Services in the Danube Region. The World Bank, Washington, DC.

World Bank, 2016a. Water Overview. The World Bank, Washington, DC.

World Bank, 2016b. High and Dry: Climate Change, Water and the Economy. The World Bank, Washington, DC.

World Bank, 2017. Easing the Transition to Commercial Finance for Sustainable Water and Sanitation. The World Bank, Washington, DC.

World Bank, 2018. Private participation in infrastructure database: https://ppi.worldbank.org/ (accessed September 11, 2018).

World Bank, 2021. International benchmarking network for water supply and sanitation Utilities IBNet. https://tariffs.ib-net.org/ (accessed January 10, 2021).

World Bank/UNICEF, 2017. Sanitation and Water for All: Priority Actions for Sector Financing. The World Bank/United Nations Children's Fund, Washington, DC/New York, NY.

WRI, 2018. World Resources Institute. https://www.wri.org/blog/2018/01/40-indiasthermal-power-plants-are-water-scarce-areas-threatening-shutdowns (accessed November 12, 2018).

WWF, 2018. In: Grooten, M., Almond, R.E.A. (Eds.), Living Planet Report - 2018: Aiming Higher. World Wildlife Fund, Gland.

WWAP, 2018. World Water Development Report 2018: Nature-Based Solutions for Water. United Nations World Water Assessment Programme, UNESCO, Paris.

WWAP, 2021. Facts and Figures. United Nations World Water Assessment Programme http://www.unesco.org/new/en/natural-sciences/environment/water/wwap/facts-and-figures/all-facts-wwdr3/fact-24-irrigated-land/#:~:text=Rainfed%20agriculture%20covers%2080%25%20of,40%25%20of%20global%20food%20production.

CHAPTER THREE

Financial structuring: key tool for water sector investments

Hein Gietema
CSC Strategy & Finance, Rotterdam, The Netherlands

3.1 Introduction

No sustainable development goal (SDG) is left untouched by water, and climate change speaks to us through its impact on and by water. The world is feeling the threat of too little, too much, or too polluted water. In market terms, we are dealing with a growing demand, finite supply and no substitutes for water.

For sound economic reasons, our focus has traditionally been on the public-good aspects of water and their impact on society.[1] The common prominent role of government in water management arises from the common-good and monopolistic nature of many water management activities and services. Water security safeguards broad societal benefits such as public health, and economic growth. This may have however given rise to the impression that private investment and management are not needed or even desired, as they are associated with privatization or profit-seeking behavior. The harsh reality today is that water and sanitation remain far from being available to all and that we, therefore, still need to keep investing heavily in pipes and dikes to enjoy "God's free gift" in our homes and to keep our feet dry.

The public sector cannot nor should shoulder this task alone: the expanding investment requirements are exceeding the capacity of public budgets (see i.a. Martini, This volume), and more effective regulation and better governance are opening up new roles for the private sector. The water sector needs private-sector investments and skills at scale to be able to deliver on the many SDGs that depend on water. The SDG 6 goal to "*ensure availability and sustainable management of water and sanitation for all*" implies two important activities, namely construction and operational services, for which the private sector is able and willing to take part of the responsibility.

[1] The economic case for investments in water services and water security has been well made, for instance by Sadoff et al. (2015).

Financing Investment in Water Security: Recent Developments and Perspectives.
DOI: https://doi.org/10.1016/B978-0-12-822847-0.00005-3

Private investments, however, can only enter the sector if bankable initiatives (i.e., projects, companies/businesses, and funds) are available. Bankable is defined here as fully developed investment opportunities offering an adequate return on investment, with a transparent risk-return profile and features acceptable to the financier. We acknowledge that in practice bankability varies per type of asset (e.g., company, fund, and greenfield/brownfield project), financier (government, venture capitalist, development finance institution, commercial bank, etc.), and phase in the project lifecycle (development, construction, and operation phase). This chapter will argue that financial structuring is an essential tool to enhance the viability and sustainability of infrastructure investments in the water sector, and to turn economically valuable investment propositions into bankable operations, and, most importantly, provide the sustainable water services we all seek.

Water as an investment opportunity has hardly been on the agenda of investors, although water risks and how they affect investors' portfolios get growing attention.[2] Both perspectives, water as an opportunity versus water as a threat, are relevant from the investor's perspective. The latter may surface as part of the regular due diligence for new investments or triggered by portfolio assessments. The former necessitates more proactivity and, given the nascent state of the market, requires some sort of financial structuring. Not many investors are willing or capable to bring in these structuring efforts and related risk capital, certainly not if risks are perceived as high and with the comparatively low returns associated with the water sector. Only by improving the risk-return perspective and aligning the interests among stakeholders, can the private sector be enrolled.

This chapter is organized around the role of financial structuring in the subsequent phases of investment. The section *Financial Structuring* introduces the basic concept and its constituent parts. It discusses finance versus funding, risk-return perspectives, and financial instruments. The section *The Project Cycle* introduces the phases and risks along the project cycle. The subsection *Development*, the phase for conceptualizing and structuring, discusses the need for reducing risk for progressing projects. If done well, this leads to *Contractual Close* and *Financial Close*, which is the formal end of the development phase, committing and incentivizing relevant stakeholders contractually and financially. In the *Construction* subsection, distinct characteristics of the water

[2] The Global Risks Report 2021 (World Economic Forum, 2021) ranks the likelihood and impact of environmental threats that are closely related to water (e.g., from extreme weather, climate action failure, biodiversity loss, etc.) among the top risks to the global economy, as it did in the past five assessments.

subsectors (safety, security, and production) are introduced. The *Operation* section closes the project cycle; it is the most important phase as all efforts in the phases before converge here to achieve the ultimate objective of sustainable service delivery. The *Comparison with other sectors* section explores the differences with other infrastructure sectors and draws lessons from these.

3.2 Financial structuring

The purpose of financial structuring is to bring about an investment opportunity—the asset itself and its appropriate governance—with the objective to generate sustainable services able to serve the needs of both beneficiaries and investors. Financial structuring is an integral part of the development phase within the project cycle. Although the wider enabling environment is an important consideration, here the focus is on the transaction level, taking everything else basically as given. Financial structuring aims to get the risk-return balance of the proposed transaction acceptable from the investor's perspective: (1) providing investors the necessary insights to make that assessment, (2) addressing the roles and interests of, and the possibilities created by the relevant stakeholders, (3) helping design the appropriate governance, the finance blend and the contractual structure, and (4) working on the project's scope and features to help minimize risks and enhance the societal and financial returns.

It is important to first distinguish finance and funding. Finance concerns the moneys provided to an enterprise or its projects with the expectation to be repaid including an extra return to compensate for risks and opportunity costs incurred by the financier. Funding is the actual recurrent repayment of installments drawn from the revenues generated by the investment; alternatively, such funding could also be through a grant or subsidy payment from an outside party or the national budget. Only if funding is secured can finance be arranged. Within the water sector funding typically takes the form of the so-called three T's: tariff, transfer, and taxes. Policymakers sometimes mistake finance for funding; however, a private financier may be willing to finance a project but obviously is not the one to fund it. In the end all infrastructure-based services must be paid for by either users (through tariffs), domestic tax-payers (through taxes) or the (inter)national community[3] (through transfers).

[3] Notably philanthropies, foreign tax-payers in the case of official development assistance, and the so-called "structural funds" within the European Union.

A second important note concerns the variety of risk-return perspectives among investors. Finance providers each have different expertise and investment mandates concerning risk appetite, return expectations, and preferences on project features. Pension funds can do "big and boring" projects and seek value in the form of low but steady financial returns that are aligned with their long-term liability pattern of paying pensions. Impact investors may want to invest small tickets in innovations whilst expecting high financial returns in the short term. Donors may be rather risk neutral but keen on societal return like employment, climate resilience, public health, etc. As private investments also generate these nonfinancial returns, and certainly so in the water sector, there is much opportunity for financial structuring to align interests and blend finance.[4]

Investors will carefully consider the features of a possible investment. Features like sector, country, currency, ownership, greenfield versus brownfield project, a project versus a corporate investment, ticket size, transaction cost, tenor, collateral, presence of sovereign guarantee, etc., are all critical determinants of risks and returns. For example, the water supply and sanitation sector may be less attractive for many financiers as in general it has governance concerns with politicians influencing tariff-setting and interfering in contract procurement and daily operations. Risks of novel greenfield projects may be perceived as higher than brownfield corporate transactions: the former may face construction risks and may depend on an unproven, probably single-revenue stream, while the latter has more diversified revenues and a track record. Size, collateral, currency, and tenor may be other relevant features that may increase or mitigate risks for investors. Water infrastructure often takes the form of a bulky investment of which the derived services are paid for in local currency over a long period of time whilst practical collateral is lacking. As long-term domestic finance is usually not available, again financial structuring is needed to explore how to match project requirements and blended finance options including currency hedging.

Different finance instruments are associated with different risks, require different financial returns and, therefore, the blend determines the overall capital cost[5] (see also Martini, in this volume). Table 3.1 offers a

[4] Blended finance is the strategic use of development finance for the mobilization of additional finance toward sustainable development in developing countries (OECD, 2018).

[5] The Weighted Average Cost of Capital (WACC) gives the overall cost based on the costs of its constituents, mostly equity (dividend expectations) and debt (interest rate). The internal rate of return

Table 3.1 Financial instruments for (blended finance) used over the project cycle.

	Development	Construction	Operation
Public (societal return objective)	**1.** Grant **2.** Budget	**4.** Budget **5.** First loss **6.** Output-based aid	**12.** Viability gap funding **13.** Insurance (export credit)
Private (financial return objective)	**3.** Equity	**7.** Equity **8.** Junior/subordinated debt **9.** Construction debt **10.** Deferred payments **11.** Performance bonds	**14.** Equity **15.** Senior debt **16.** Bonds **17.** Currency hedging **18.** Insurance **19.** Refinancing

1. Nonrepayable funds deployed to public and/or private entities for specific public purposes.
2. Estimated expenditures for project *plan,* often part of administrative budget planning process or program.
3. Ownership and value to shareholders (when eventually all debt is paid off), the return is dependent on profitability. Equity is a prerequisite to (eventually) attract debt. It is also costlier than debt. Equity has different characteristics in, and may possibly "exit" from each of the project phases. For the development phase it entails a hands-on strategic perspective often targeting the highest equity returns so as to compensate the high development risks. For the ensuing phases equity may take a more hands-off financial orientation but still targeting higher returns because riskier than debt. It is helpful to distinguish "strategic" and "financial" equity from the project proponents who wish to expand their business operations, and third-party investors and financiers that seek investment opportunities, respectively. Also "sweat" equity is relevant, valuing the inputs of resourceful individuals, often the initiators, with crucial knowledge and skills but limited financial means.
4. Estimated expenditures for project execution/capital expenditures (CAPEX) often part of administrative budget planning process or program.
5. Credit enhancement tool providing first level of support bearing bulk of financial risks (i.e., lower revenue than expected) and lowering the overall investment amount that requires (commercial) financial returns.
6. Link delivery of assets (and ensuing services) to targeted performance-related subsidies (see also "result-based financing").
7. See 3.
8. Debt is an obligation that requires one party, the debtor, to repay principal and interest to another party, the creditor. *Junior debt* has lower priority of repayment than, and is subordinated to, *senior debt* in case of default. The instrument may also have equity characteristics, i.e., sharing in the upside/profitability of a project.
9. Debt incurred for the construction period with objective to be refinanced on better terms after construction is completed.
10. Postponed payment to contractor which would imply a lesser need for (senior) debt.
11. Issued to one party (client) of a contract as a guarantee against the failure of the other party (contractor) to meet obligations specified in the contract.
12. Enhances financial feasibility and service delivery at an affordable tariff for the community.
13. Guarantee of compensation for specified loss or damage in return for payment of a specified premium. Some (public) schemes protect exporters against foreign customer default due to political or commercial risks.
14. See 3.
15. See 8. Senior debt is prioritized for repayment in case of bankruptcy or default, it therefore has the lowest risk and typically carries lower interest rates.
16. A fixed-income (debt) instrument issued by either government, large corporates or for large projects. Corporate bonds are sometimes listed on stock exchanges.[6]
17. Reduces the effects of currency fluctuations on the value of xthe investment. Especially important for projects that have local currency revenues and hard currency debt obligations (currency mismatch).
18. Promise that something is of specified quality, content, benefit, etc., or that it will perform satisfactorily for a given length of time.
19. Replacing an existing loan or other financing instrument with a new loan that pays off the debt of the first one. The new loan should ideally have better terms or features. It involves a re-evaluation of the project (risks). Refinancing can also ztake place in the preceding phases.

(IRR) of a project should be higher than the WACC to justify the investment. Grants or subsidies can help to enhance net returns or to lower risks or the WACC; the latter may be preferable.

[6] Martini (This volume) highlights the importance of bonds and capital market as "banks have largely moved to an "originate and distribute" business model."

nonexhaustive list of public and private financial instruments in use across the project development cycle. These public and private instruments have a predominant societal or financial return objective. Each instrument has its specific risk-return profile. Risks and associated expected financial returns are lowered when a project is maturing and/or thanks to appropriate blending with public support instruments. Financial instruments in one phase can replace or add to instruments deployed in earlier phases (see further).

Efficient financial structures can be created by blending finance sources. It remains challenging, though, to determine how much (costlier) equity is needed to attract (cheaper) debt. And from a public sector perspective, scalability is key; how much of scarce grant resources is needed to attract private equity and debt that is in principle available in abundance. Financial structuring remains tailor-made until a track record of replicable transactions becomes available.

Grants or subsidies play an essential role in many water initiatives. To be well structured, grants should be results-based and not open-ended, and bring comfort to the stakeholders involved.[7] To the grant providers, such as national governments, donors, and philanthropies, they should bring societal impact. For the customers, they should lead to better services at an appropriate and affordable price. In the form of Viability Gap Funding they may help as an intermittent subsidy to overcome temporary inability to pay. Importantly, grants can help to reduce the risk for private investors which is often needed given an oft-skewed risk-return profile in the water sector that cannot compete with alternative attractive investment opportunities in other sectors available to them. The availability of well-structured grant instruments can provide the private sector with just the level of comfort they need to step into a challenging market or initiative.[8]

The potential effectiveness of leveraging private investors by derisking investment with targeted grants may be evident; less public investment is required to achieve the societal objectives. In addition, these societal

[7] Subsidies can become open-ended and create dependency, their fiscal impact can explode, they can undermine financial discipline and blur accountability. Furthermore, they tend to postpone much-needed reform. Subsidies should be contingent on performance improvement and a last resort after costs have been minimized through competition, regulation, appropriate technology and service standards, or public enterprise reform (World Bank, 2001).

[8] For example, the Dutch government provided €50 million of callable first loss to TCX (The Currency Exchange), an innovative fund that enables local currency finance by absorbing currency variations. This subordinated capital triggered multilateral finance institutions, development banks and private banks to invest €300 million. Ten years later TCX facilitated over €1.5 billion worth of transactions in 37 exotic currencies. Importantly, the €50 million first-loss tranche is still intact.

objectives may be more sustainably served as private investments at project level can bring in the financial and operational discipline the sector requires: books do not close after commencement as repayments expect that the water services keep properly functioning. As much of the private financing for the water sector is presently absorbed by the government, and as its servicing is not directly related to the proper functioning of water systems this financial and operational disciplining effect is largely lost (see also Box 3.1).

BOX 3.1 The two faces of concessional finance

The water sector in developing countries is traditionally (grant) financed by the national government. Whilst Multilateral Development Banks (MDBs) such as the World Bank, European Investment Bank (EIB) and the African Development Bank can also do private project lending, for the water sector they typically provide long-term hard-currency sovereign concessional loans to the national Ministry of Finance or Treasury. These sovereign loans proceeds are on-lent or granted to local water agencies or water service providers. While generally welcome, sovereign loans also bring unintended effects that distort incentives.

First, they impose less of a disciplining effect on water usage or utility operations compared to commercial finance. Although MDBs follow stringent procedures to assess the feasibility and societal impact of programs before committing capital, once the capital has been deployed follow-up may turn less rigorous. The MDB's primary concern regarding loan repayment indeed lies with the country's Treasury and less so with the agreed water service operations. Failing maintenance or misappropriation of budgets impact the water services to the people but it does not affect repayment to the MDB.

Second, the loans have a long maturity with pay-back periods of 25-40 years, which makes the debt service economical on an annual basis for the borrower country. But this often results in a "reversed maturity" mismatch, with the loans having a longer lifetime than the assets being financed. This contrasts with the common maturity mismatch where the economic lifetime of assets exceeds the tenors commonly available in commercial domestic financial markets. This reversed mismatch implies that a country is still indebted over assets that have passed their economic use. Assets in developing countries are often not well maintained as financial incentives or budgets are lacking, which tends to further shorten their useful economic lifetime. This unintentionally adds to the country's indebtedness without a matching productive base.

Third, these loans are in hard currency whilst the services are being paid for in local currency as user fee or tax. The country's Treasury thus bears the risk of a devaluation of the domestic currency, which is, also given the long tenor, a substantial risk.

(*continued*)

BOX 3.1 The two faces of concessional finance — cont'd

Ways forward?

By their charter MDBs generally lend to national Ministries of Finance or Treasuries, and they assist them by strengthening national policies, governance and the nation's implementation capacity. Typically, MDBs will commit to programs that generate public-good value, or are innovative in the country and trigger reform and a learning curve; these programs bring higher risk and transaction costs than private financiers would accept. MDBs are also engaging in subsovereign lending with (public and private) creditors closer to local operations and their cash flow, and with less recourse to the domestic tax-payer base. The MDBs' private project lending and structuring capabilities may gradually become more relevant for the water sector. Certain promising contractual arrangements may improve productive use and maintenance of the assets over their economic lifetime (see Box 3.2 on DBFM contracts). Also, MDBs offer lending windows for the corporate sector and for guarantee systems.

BOX 3.2 DBFM contracts for nonrevenue generating water infrastructure assets[9]

The Netherlands successfully contracted out a series of locks and flood barriers to the private sector concerning billions worth of assets, among which the Sealock IJmuiden and the *Afsluitdijk* (Sea Closure Dike), situated on important waterways having crucial water management, navigation and flood defense functions.

The nearly 90-year-old *Afsluitdijk* is one of low-lying Netherlands' key sea defenses. The Dutch government has embarked on a future-proofing project to strengthen the iconic 32 kilometer (20 mile) long structure. With climate change bringing more powerful storms and rising sea levels, the dike is getting a major makeover. The *Levvel* consortium, consisting of Dutch contractors and a financial adviser providing the equity, and pension funds providing debt, was awarded a DBFM (Design-Build-Finance-Maintain) contract and will be paid *availability payments*[10] by the government for making a functional dike available over the 25-year contract period. Work is expected to continue until 2023. Despite this progress, it is also transpiring that the number of large contractors that possess the requisite technical and financial capability, and thus the resilience and appetite to absorb the risks, may prove limited.

Ingredients for success were:

- A credit-worthy government willing and able to pay *availability payments* against performance over the concession period;

[9] Gleijm, 2015; and https://www.rebelgroup.com/nl/projecten/winnende-plan-versterking-afsluitdijk/

[10] Payments made over the lifetime of a contract in return of the private party making the infrastructure available. As a matter of risk allocation the government should only pay for the asset as long as it is available (so, never before construction is completed) and to the extent that it is available, while considering "exemptions" such as planned maintenance periods or certain events out of the control of the private

- Comfort with DBFM type contracts and processes developed in other infrastructure sectors, by both public and private parties;
- Well-structured procurement process for instance using Competitive Dialogue, a method used in mature PPP markets to detail contract scope and capture innovations by bidding parties whilst preserving competition;
- Experienced transaction advisers with appropriate financial structuring skills, on both sides.

Lessons for emerging countries?

- As maintenance of water infrastructure assets is often not budgeted for and neglected, DBFM structures provide a workable alternative for sustainably delivering the desired infrastructure services to beneficiaries. The DBFM contract may add to efficiency (less bureaucracy, direct incentives) and effectiveness (no appropriation) of water services.
- Governments can take leadership and also support crucial infrastructure without an associated revenue stream. Understandably, the present focus is on PPPs for revenue-generating assets (toll roads, power, health, etc.) to relieve the fiscal burden (see also i.a. Martini, This volume). Value capturing may be possible, for instance by combining dikes with toll-roads and/or real estate development nearby. In practice, for instance in an earlier scenario in Indonesia to protect its capital Jakarta from sinking below sea level by constructing a sea dike, the scope may become overly complex and prone to corruption accusations delaying interventions—Indonesia now decided to move its capital, not protect it.
- Multilateral and regional development banks can offer support by enhancing governments' creditworthiness. Loan and guarantee programs could be used to comfort private consortia that availability payments can be paid during the concession period. By supporting the use of these types of contracts the banks are also assured that proceeds have been used effectively and according to agreements and objectives.

Experience with the relevant DBFM fundamentals is widely available and can be procured and/or acquired by governments and/or multilateral institutions. Governments may also need advice to address accounting rules and procedures allowing for longer-term commitments.

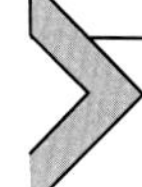

3.3 The project cycle

3.3.1 Introduction

A project cycle typically comprises three main phases: development, construction, and operation (Fig. 3.1).[11] Most of the financial structuring takes

partner. The availability risks are ultimately operational risks, intrinsically manageable and dependent on the performance and management capabilities of the private partner (APMG, 2016).

[11] The development phase of project investment may have various sub-phases requiring conceptualization, (pre-)feasibility studies, permits and (government) approvals, team set-up, legal structuring, commercial

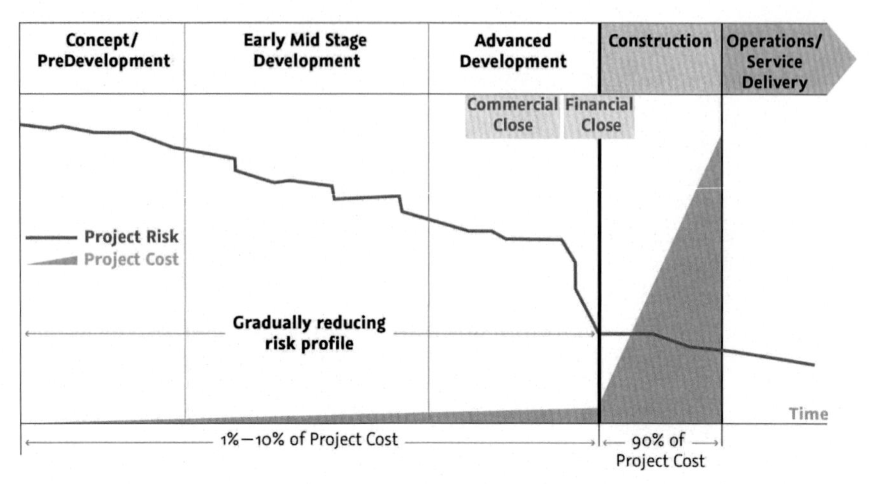

Figure 3.1 The project cycle, risk, and cost by phase (adapted from InfraCo/KIFFWA [2020]).

place in the development phase. Understanding, mitigating or distributing/allocating, and pricing risk is done in the development phase of the project. During development, the many unknowns on scope, feasibility, structure, etc. are unraveled with the objective to gradually reduce the overall risk—and capture the return—of the project, or, in case this proves impossible, abort the project. Up to a point, (private) investors may feel comfortable enough to step in, finance the construction phase, and stay on during the operation phase if it is not being refinanced. If a transaction is well-structured in the development phase—i.e., the technical, financial, and societal risks are properly understood, distributed/allocated, and priced—it will bring the desired long-term sustainable water services. This is the terrain of the financial engineer, and derisking may be the prime objective.

Fig. 3.1 shows the overall risks decrease over the lifetime of a project. The practitioner literature on infrastructure financing consistently emphasizes the importance of derisking projects in order to make them bankable but infrastructure risk can be an amorphous concept. A taxonomy developed by OECD (2015) also uses the aforementioned project lifecycle approach,

contracts negotiations, fund raising, etc. Project and corporate finance (as applicable to businesses) follow different paths, however. The financial close thus works out differently when financing individual projects or businesses. Project finance often applies to distinct greenfield projects that are geographically defined with a single revenue stream and a financial close followed by commencement of operations. Corporate finance typically applies to businesses that are expanding (sometimes beyond geographic boundaries) with multiple and gradually growing revenue streams and often requiring multiple rounds of financing in line with the evolution of the business.

adding a termination phase, and lists the main political, macroeconomic, and technical risks arising during the respective phases (Table 3.2).

The political and regulatory environment has impact via, e.g., the permitting process which can lead to rising costs in each project phase, or even to stalling the project. It further impacts the project contract negotiation processes, its contents (tariff, taxation, currency convertibility, etc.), and its enforceability (social acceptance, collateral, tariff regulation, etc.). The macroeconomic and business risks impact the project financials via financing availability (liquidity, local currency, hedging, etc.) and pricing (inflation, perceived risk, exchange rate, etc.). The counterparty default risk, the risk associated with a party (client) not meeting its obligations, may also fit this category. The technical risks of implementation are typically under better control of the developer/initiator. It comprises the governance and management of the project, the technology in use, construction timing and costs, and the quality of the assets and/or services. Force majeure comprises the risks that are outside anyone's control (Acts of God). To truly comprehend these risks both the likelihood and the impact of the risks need to be assessed.

To progress along the project lifecycle, the financial instruments in use (Table 3.1) need to match the associated risks and returns of each phase. The high risks and uncertain returns in the development phase require government budgets or strong public support for private developers (sponsors/initiators) providing high-risk capital (equity) and effort ("sweat equity"). Debt type instruments may only become relevant for the construction phase although many banks are wary of construction risks, notably cost and time overruns. They may come in only in combination with additional instruments such as equity, first loss, possible targeted subsidies, deferred payments by contractors, and performance bonds providing enough financial comfort. Capital market instruments like bonds provided by (domestic) pension funds and insurance companies may enter in the operational phase once a project turns "big and boring" and/or when refinancing is required.

With help from financial structuring, blending, and (contractual) innovations it is possible to decrease and/or allocate risks within and over project phases. Privately managed publicly funded entities like KIFFWA[12]

[12] The Kenya Innovative Finance Facility for Water (http://www.kiffwa.com) is a co-developer of commercial water interventions (projects, companies and funds) in Kenya. KIFFWA aligns with and de-risks private lead developers by providing capital and financial structuring expertise. The objective is to drive interventions to reach financial close and mobilize private finance in the sector. KIFFWA is only compensated if financial close has been reached. It is operational since 2017, has a pipeline of

Table 3.2 Risks linked to infrastructure assets over the project cycle (adapted from OECD, 2015).

Risk categories	Development phase	Construction phase	Operation phase	Termination phase
Political and regulatory	Environmental review	Cancellation of permits	Change in tariff regulation	Contract duration
	Rise in preconstruction costs	Contract renegotiation		Decommissioning
				Asset transfer
			Currency convertibility	
		Change in taxation		
		Social acceptance		
		Change in regulatory or legal environment		
		Enforceability of contracts, collateral, security		
Macro-economic and business	Prefunding		Default of counterparty	
			Refinancing risk	
	Financing availability		Liquidity	
			Volatility of demand/commercial risk	
		Inflation		
		Real interest rates		
		Exchange rate fluctuation		

(*continued on next page*)

Table 3.2 Risks linked to infrastructure assets over the project cycle (adapted from OECD, 2015)—cont'd

Risk categories	Development phase	Construction phase	Operation phase	Termination phase
Technical		Project governance and management		Termination value different (from expected)
		Environmental		
	Project feasibility	Construction delays and costs overruns	Qualitative deficit of the physical structure or service	
		Technology and obsolescence		
			Force majeure	

(van Oppenraaij et al., This volume) support private developers in Kenya in structuring bankable water projects to reach financial close. Innovative organizations such as Climate Investor Two[13], a developer-investor of water projects, provide support over the project phases by offering separate funds for the development phase (grant-funded) and for the execution phase (blended fund). Integrated contracts like Design, Build, Finance and Maintain (DBFM) incentivize the private sector to achieve lifecycle optimization, to avail agreed quality services over the asset lifetime, and to allow capital market players to participate in a greater part of the project cycle, even for nonrevenue assets (Box 3.2).

3.3.2 Development phase

Project conceptualization and scoping start with the right narrative—by the right initiator or sponsor—that convinces people that the project is relevant and will succeed in bringing the desired benefits and impact. The basic risk is that during the structuring in the development phase the necessary trust among relevant stakeholders, including the sponsor himself, does not get, or loses momentum. The feasibility risks depend on technical, financial, and societal aspects that are assessed in design, feasibility and environmental, social and governance (ESG) studies. These studies typically address items such as the regulatory and legal environment, permitting, tariffs, taxation, development budget, financing availability, technology, market risk, environmental review, currencies, etc. (Fig. 3.1 and Table 3.2).

Project development in general is a nonlinear risky time-consuming process, it can take easily more than three to five years, with many puzzles to solve. Water infrastructure projects may even come with bigger puzzles due to numerous regulations, for instance due to considerations on public health and the monopolistic nature of water services. Although the project development phase may require relatively modest budgets at typically below 10% of the overall capital investment (CAPEX), development budgets can easily be lost as many projects do not make it to financial close.

Potentially bankable projects may not get developed because of the risks developers perceive and/or are not willing to take. Projects also fail if they are

17 projects representing €250 million CAPEX and first financial closes are expected in 2021. KIFFWA was set up with a €10 million grant capital injection.

[13] Climate Fund Managers (www.climatefundmanagers.com) managing Climate Investor One (energy) recently added Climate Investor Two (water) to provide financing solutions for infrastructure developers in the renewable energy, water, sanitation and ocean sectors worldwide. The fund structures investments across the entire project lifecycle using blended finance structuring to attract public and private capital.

not properly structured because the developer may lack certain experience or skills; many possess more technical knowledge than financial or business-case structuring expertise. The nascent state of the market, the sector, and business-case complexity, the level of origination and structuring effort required and the risk-return expectations may, without some government support, further deter investors from deploying risk capital (development equity) for developing the projects. As mentioned, project development is the riskiest phase of the project cycle.

BOX 3.3 The role of Procurement

Within the project development phase and certainly when transitioning to successive phases entailing large capital expenditures, transparent procurement procedures bringing value for money should be observed. Especially if public budgets are involved, which is often the case, prevailing procurement rules need to be applied. Early engagement of the private sector, in the development phase of public projects, may help solve challenges and add innovation. Unsolicited proposals for public sector projects, i.e., initiatives from the private sector without an explicit request, may help governments' planning process (identify and prioritize projects), expand the range of potential solutions, assess the preliminary feasibility and indicate market interest. Many projects, however, that originate as unsolicited proposals experience challenges, including diversion of public resources away from the strategic plans of the government, providing poor value for money, and leading to patronage and lack of transparency, particularly in developing countries. Alternatives for unsolicited proposals are also available. For a broad infrastructure challenge, e.g. a water demand in a city, government bodies could elicit the private sector's ideas. Procurement processes can be structured using output specifications rather than prescribing technical detail information, providing a wider scope for innovation. The process could have multiple stages allowing interaction between Employer and interested parties on project scope, technical solutions and structure, to ensure alignment of needs and deliverables (see also Competitive Dialogue in Box 3.2) (World Bank, 2018).

Projects can be developed by governmental bodies and/or by private parties. Like with any other project, those developed or procured by governments carry their own risks (Box 3.3). Projects may fail as processes within bureaucracies can be slow and politically sensitive, especially when private sector participation is involved. Tender documentation and directives may eventually not match what is on offer in the market. Public sector officers and private entrepreneurs may not speak the same language, and transaction advisers, one type of financial engineers, may be needed (see selected examples in Part 3, and Baker, This volume). It is generally advisable to involve prospective financial partners early on in the process as financing parameters may partly determine the shape of the investment, and to raise confidence and commitment on the financier's side.

The decisive deliverables of the project development phase are the financial close and contractual close, formally committing the (private) investors and financiers of the CAPEX of a project, and the engineering procurement and construction and possibly the operation and maintenance (O&M) contractors, respectively.'

3.3.3 Commercial and contractual close

Various contractual arrangements are used to allocate the responsibilities and risks for water infrastructure service delivery between the public and the private sector. Each intervention needs financial and contractual structuring to address the possibilities and limitations of all stakeholders, the water subsector, and the intervention itself. Table 3.3 categorizes and allocates the main aspects of contractual structuring, i.e., ownership, finance, and operational responsibilities, within any infrastructure project.

As stated, the private sector's perspective is basically a risk-return assessment. These risks increase substantially going to the right side of Table 3.3, moving from management contracts to more integrated contracts to privatization. With the former, the private party's reputation (and possible bonus) is at stake, but with the latter also large investment amounts and future revenues.

Contracts such as DBFM integrate the various aspects and phases into a single long-term contract. Being responsible for the main aspects over the project cycle, the private consortium comprising advisers, contractors, and financiers, is incentivized to seek lifecycle optimization; design innovations may pay off during operation, lowering cost, or adding value. It also mitigates risks of transitions between phases. Despite higher capital costs that reflect the private financier's risk-return assessment, DBFM contracts have been shown to deliver added value (better costs and time performance and incorporating private-sector innovations)(Verwey and van Meerkerk, 2021). A prime gain for emerging economies may be the higher certainty level for stakeholders, including possible donors, local government and particularly end-users, that the water service is actually delivered and secured for the long term. The contractual obligation of government, possibly backed by an MDB, to provide a stream of funding over the contract period helps to avoid peaks in the public budget appropriation and the possibility of infrastructure getting neglected over time (see also Box 3.2).

Table 3.3 Contractual forms and their features.

	I	II	III	IV	V
Ownership	Public	Public	Public	Public/ private	Private
Finance	Public	Public	Private	Public/ private	Private
Operation	Public	Private	Private	Public/ private	Private
Contract type	*SLA* *Corporatization*	*Management contracts* *Lease/ affermage*	*DBFM* *BOT*	*BOOT* *PPP*	*Privatization*

• SLA: the service-level agreement defines the level of service expected by a customer from a supplier, laying out the metrics by which that service is measured, and the remedies or penalties, if any, should the agreed-on service levels not be achieved.
• Corporatization: the process of transforming state assets, government agencies, or municipal organizations into corporations. The result of corporatization is the creation of state-owned corporations (or corporations at other government levels, such as municipally owned corporations) where the government retains a majority ownership.
• Management contracts: a contractual arrangement under which operational control is awarded to a separate operator that performs the necessary managerial services in return for a fee. A management contract can involve a wide range of functions including management of personnel, accounting, marketing services and training.
• Lease/*affermage*: similar to management contracts but the operator tends to bear greater operational risk, the operator now has to deduct its fee from market revenues instead of receiving it from the awarding authority. Like with management contracts, the awarding authority bears the finance/investment risks (see e.g., www.ppp.worldbank.org).
• DBFM (Design Build Finance Maintain): a contractual arrangement whereby a private organization/consortium is awarded responsibility for the design, building, financing and maintenance of a project. Similar to a BOT contract (below) but unlike BOT for certain DBFM cases the market risks remain with the awarding authority, for instance with availability payment structure (see also Box 3.2).
• BOT (Build Operate Transfer): a project delivery model whereby a private organization is granted by the public sector partner (awarding authority) to finance and build a (large) project. The private organization subsequently maintains and operates the project assets for a set period of time during which they can draw user fees. During the concession period the public partner owns the project assets and the private organization accepts most of the risks. Thereafter control of the project is transferred to the public sector partner.
• BOOT (Build Own Operate Transfer): similar to BOT but in BOOT contracts the assets are owned by the private partner during the concession period.
• PPP (Public Private Partnership): no standard, internationally accepted definition exists of a PPP. The term is used to describe a wide range of types of agreements between public and private sector entities, and different countries have adopted different definitions as their PPP programs has evolved (see e.g. www.ppp.worldbank.org).
• Privatization: occurs when a government-owned business, operation, or property becomes owned by a private, nongovernment party.

3.3.4 Management and operations

Accountable management—private or at least at arm's length from political maneuvering—is the most important element to deliver the required water services. Especially within the infrastructure domains where the chances of public interference are high the government's political and social goals may at times conflict with the business orientation of the operator. The objective of financial structuring is to address possibly diverting interests, allocate risks and returns for infrastructure service delivery upfront to the party best able to manage them whilst ensuring that the operator is properly incentivized to deliver the required services.

3.3.5 Ownership

As shown, furnishing private finance differs from privatization. Private ownership of water infrastructure assets usually is a delicate subject with specific examples few and far between; governments have been frequently charged by critics claiming that the "crown jewels" or "matters of strategic interest" are not to be sold to "profit-seekers." Still, it is certainly true that the natural monopoly position which comes with certain types of water infrastructure requires strong countervailing forces from the public sector, including transparent procurement (competition *for* the market), contract structuring, and independent regulatory agencies (see also Box 3.3 on unsolicited proposals).

3.3.6 Finance and investments

For the private investor, and in most instances for public financiers as well, the return should be worth the risk. In financial terms: the project's Internal (Financial) Rate of Return (IRR) should be higher than the cost of capital (for instance expressed as WACC [Footnote 5]). A firm's cost of capital is a blend of own capital (equity) and loans, adjusted for the project risks. All these elements tend to lead to higher costs in emerging economies: (1) loans—if available at all—command higher interest rates, (2) financiers want more of the private operators' own capital at risk, and (3) operators and financiers want a higher return on their capital. Currency mismatches (revenues in local currency and debt payments in hard currency) may add to the risks. Well-structured subsidies can be crucial, especially to shield the poor. Also, opportunities may exist for value capturing; for example, financiers may desire to also participate in associated or downstream activities that the water project unlocks, such as agricultural, industrial or real estate developments.

3.3.7 Financial close

Once proper structuring in the development phase has mitigated and distributed/allocated project risks, third-party investors, and financiers can step in. Many financiers are unwilling or less equipped to participate in the more risky and complex development phase, and may prefer even to participate far into the operation phase when assets can be refinanced or sold through secondary markets or securitizations. As mentioned earlier, the financial and contractual close are the formal end of the development process. The developer may choose to stay in as he may become also the contractor of the assets[14] and/or eventually their operator. Financiers likely require the developer to stay in for at least a period after the commercial operations date, until all start-up problems have been solved. This may, however, raise possible liquidity issues for the developer: he has been investing in the project development for several years and is now requested to wait longer for his payment. At this stage, the project has now locked in third-party financiers; with relatively modest development budgets the large CAPEX investments have now been successfully secured.

3.3.8 Construction

When the assets are being built, the bulk of the finance is deployed. Much of the focus historically has been on this phase of the project cycle as it is tangible, provides opportunities to cut ribbons, and exude optimism about the future. The ensuing operation phase, although factually the most important one as it is where the actual services are delivered, does not have the same appeal. Maintenance habitually receives less attention. As debt service over time depends on providing quality water services, private finance helps to focus the attention beyond the construction phase.

Although the overall project risk has been substantially lowered in the development phase, the construction phase brings specific risks too (Table 3.2). The risks of construction delays and cost overruns are the most prominent. To enhance control over these risks financiers may only want to deploy funds against milestone payments ("certificates"), have budget contingencies, engage a lender's technical advisor, stipulate the sponsor to provide additional own equity, or have the contractor provide guarantees (as "performance bonds" or insurance). Many financiers do not feel comfortable

[14] A contractor's involvement in both development and construction phases may raise transparency and value-for-money concerns (see also Box 3.3). Contractors can be powerful financial allies in developing a project, and by providing equity, possibly in the form of deferred payments, also for construction. The resulting conflicts of interest, however, making profit out of the contract versus out of the co-ownership of the assets, needs to be appropriately managed on behalf of the other owners and the end-clients. Possible ways to handle these conflicts are Open Book Bidding, whereby the contractor reveals his calculations and mark-ups, or using the Swiss Challenge Method, whereby other contractors provide a bid which the involved contractor is allowed to match or overrun with, say, up to 10%.

with construction risks and may only enter the operation phase once services and cash-flow are emerging.

3.3.9 Operations

The efforts during the development and construction phases should lead to a smooth operation phase delivering the required water infrastructure and services. Still, the operational phase may experience start-up problems due to technical glitches and issues concerning contractual roles and responsibilities, and, given the long economic life of water sector investments, some foreseen and unforeseen risks may appear over time (Table 3.2).

Tariff (indexation) risk (also called off-take risk) is prominent in water service operations. Many utilities, for instance, in the emerging but even in higher-income economies, do not charge tariffs reflecting the full cost of the service. They act so for a wide variety of external reasons including having a low-income customer base, inadequate tariff setting procedures, policy considerations, or political interference. The utilities' own operations, however, may also discourage regulators from allowing tariff increases: many public utilities in emerging economies deliver substandard services with below-par water quality and availability and high water and administrative losses (with nonrevenue water loss exceeding 50%). Whilst most utilities in developed economies are financed via private means (e.g., bonds) these challenges in emerging economies impact on the debt servicing capacity and shy away private financiers.

A second main risk for all types of water investments is the currency mismatch when water assets are financed in hard currency but paid back ("funded") in local currency. The ensuing currency risk typically lies with the enterprise or project, or with the Ministry of Finance (Box 3.1) and can be material in case of currency devaluation. The debt obligation to international financiers can turn unsustainable in local terms. Unless long-term local currency financing is available this currency risk can be hedged against at a cost by third parties that specifically cater to impact investments, such as TCX (which is backed by 22 multilateral and bilateral development finance institutions, and microfinance investment vehicles), GuarantCo (backed by the governments of Australia, France, UK, Switzerland, Sweden, and The Netherlands) and MIGA (the guarantee facility of the World Bank Group).

Strong project management capabilities, good governance, and a conducive political and regulatory environment are crucial considerations for financiers. Is management capable of delivering the (debt) services, are they properly incentivized and capable of navigating the risks? And, is governance robust and transparent? Governance can be sensitive to political risks, especially if the public sector is represented in the governing board of the enterprise and/or the regulatory environment is unpredictable.

3.4 Comparison with other sectors

Other infrastructure sectors, notably the power, transport, and telecom sectors, have benefited from an influx of private-sector finance and management and subsequently enjoyed increasing and improved services. For water, the public sector is more dominant and in many instances assets are owned, managed, and financed by public means. Water is generally also more regulated and politically sensitive and perhaps also more fragmented and complex from a business modeling and financial structuring perspective (see also Alaerts, This volume).

The energy sector was first to seek private participation notably for power generation, leaving transmission and distribution networks mostly in the hands of the public sector arguably because these are more immovable, strategic and prone to monopoly. These privately financed independent power producers typically negotiate a power purchase agreement with the state power company implying they have only one customer. The relatively straightforward, tested process and contractual structure concern the energy price[15], contract period, and credit quality of the (public) off-taker[16]. The financier has to satisfy himself with the quality of the private contractor's ability to deliver, which is implicitly also to the benefit of the public off-taker and his clients. Because of the shared risk of nonperformance, a crucial part of the technical due diligence is done by the financier himself. The downstream market risk, i.e., how much energy is sold for what price, remains with the public sector. The market can nowadays propose and develop projects relatively fast as soon as the off-taker indicates interest. Desalination has similar features, when users (a municipality) commit to procure a certain volume of water at a set tariff over a period; process and financing models are tested and standardized, minimizing transaction costs. The creation of comparable markets for, e.g., domestic water supply is however more complicated. Water creates a more natural monopoly as it is, unlike energy, relatively expensive to source and to transport. Many water

[15] With possible (pass-through) indexation and elements like Feed-in-Tariff for renewable energy.

[16] The credit quality of the off-taker and the currency risk often raise concern. Although most contracts are agreed in US dollar terms, the (tax) revenue base for the off-taker is usually in local currency.

supply projects also take the form of add-on projects to existing, often old, and poorly maintained distribution networks, and delivering directly to the many individual end-users instead of to one centralized off-taker.

The transport and roads sector counts a wide variety of clients and markets, with both revenue (e.g., ports, airports, toll roads) and nonrevenue generating (e.g., non-toll roads) assets. Toll roads have clear, ring-fenced, and sustained revenue streams able to service debt and provide dividend to the concessionaire who may own the assets for the duration of the concession period. The market risk, concerning the number of cars passing at a given price, may be borne or shared with government that can guarantee a minimum revenue to the concessionaire. For nonrevenue generating transport assets, the DBFM model with availability payments has proven very relevant (Box 3.2). Road funds, to finance road maintenance and funded by levies on fuel and managed at arm's length from politics, may be another model that could be relevant for water.

The telecoms sector has fully matured from a private finance and operations perspective. Once a license has been awarded, private companies can maneuver relatively independently. Entrepreneurship combined with (equity) capital, well-priced finance provided by vendors often backed by their Export Credit Agency or ExIm bank, use of prepaid cards (instead of subscriptions) securing up-front cashflows, and a receptive market have led to a successful arrangement. The market here is willing and able to pay for good services. Pricing is determined by competition. The telecoms markets have consolidated with respect to market share while new internet- based services (banking, music streaming, location services, etc.) are being added to the core activity. Prepaid metering as a way to secure up-front payment is now also well established in urban water supply as well as for sprinkler and drip irrigation. Increasingly, smart meters are using ICT to provide reading, invoicing, and revenue collection services that add to the efficiency and transparency that facilitate engagement of the private sector.

3.5 Conclusions

Financial structuring is a tool to help bridge the water sector financing gap. The need for investments in the broad water sector in advanced and emerging countries is growing and the private sector has to step up its engagement in order to meet the SDG targets and provide sustained services across all water sectors. The good news is that private sector is increasingly ready to do so.

The investment challenge, however, is not primarily the availability of finance but the shortage of bankable projects. Fully developed investment opportunities are required to engage the private sector, and financial structuring is the way to bring them around. During project development, the financial engineer helps structuring the transaction to fit the risk-return profile of investors, whilst observing broader societal interests.

Risk, a rather amorphous concept, as well as financial and/or societal return are key elements. Political and regulatory, macro-economic and business, and technical implementation risks can manifest themselves along the project cycle. Derisking entails understanding, mitigating and allocating and pricing the likelihood and impact of these risks. Derisking rarely eliminates all risk but it allocates it across a range of public and private partners; cost-effective derisking tends to allocate selected risks to the parties best equipped to deal with them. Governments and international finance institutions can further support this by providing financing instruments to blend, derisk it and encourage private-sector finance, whilst not replacing it or competing with it. Assuming more operational risk exposure, i.e., taking the actual delivery of the desired water service for their debt servicing instead of leaning solely on the tax-payers base, would require and bring a better understanding of the sector. As large segments of the water sector are associated with high risks, low financial returns and high societal returns, public support and reform are opportune and, until these segments have matured, required.

Distinctive financial instruments and contractual incentives have proven helpful in the development, construction, and operation of water assets. These include among others the first examples of privately managed, publicly funded project developments; integrated contracts for nonrevenue generating waterworks; subordinated tranches from government to leverage private investments in assets; and targeted grant instruments underwriting sustainable operations. Financial sustainability, the key to sustained O&M and service delivery, requires committed payments (i.e., funding): for revenue-generating assets, the market-derived revenues possibly supplemented with targeted subsidies; and for nonrevenue generating assets, public payments, or availability payments. Thus, the ability to provide long-term sustainable water services is addressed and to be resolved in the development phase, with help from the financial engineer. Financial structuring contributes to making the water sector more palatable for financiers. The level of financial structuring effort can only be reduced when we can demonstrate investors robust delivery models, build a track record of transactions, and hence show market standards and discipline.

References

APMG, 2016. The APMG public-private certification guide. https://ppp-certification.com/pppguide (accessed January 2021).

Gleijm, A., 2015. Unlocking the value of private sector for water management infrastructures. Water Governance 05/2015, 35–39.

InfraCo/KIFFWA, 2020. Annual Report 2019. Kenya Innovative Finance Facility for Water, The Hague.

OECD, 2015. Infrastructure Financing Instruments and Incentives. Organization for Economic Co-operation and Development, Paris.

OECD, 2018. Blended Finance Principles, for Unlocking Commercial Finance for the Sustainable Development Goals. DAC, Organization for Economic Co-operation and Development, Paris.

Sadoff, C.W., Hall, J.W., Grey, D., Aerts, J.C.J.H., Ait-Kadi, M., Brown, C., Cox, A., Dadson, S., Garrick, D., Kelman, J., McCornick, P., Ringler, C., Rosegrant, M., Whittington, D., Wiberg, D., 2015. Securing Water, Sustaining Growth: Report of the GWP/OECD Task Force on Water Security and Sustainable Growth. University of Oxford, Oxford.

Verweij, S., van Meerkerk, I., 2021. Do public–private partnerships achieve better time and cost performance than regular contracts? Public Money & Management 41 (4), 286–295.

World Bank, 2001. The World Development Report: Attacking Poverty. The World Bank, Washington DC.

World Bank, 2018. Policy Guidelines for Managing Unsolicited Proposals in Infrastructure Projects. The World Bank, Washington, DC.

World Economic Forum, 2021. The Global Risks Report, 16th Edition. http://wef.ch/risks2021 (accessed March 2021).

CHAPTER FOUR

Financing instruments and the ecology of the financial system

Mireille J. Martini
OECD Environment Directorate[1]

4.1 Introduction

This chapter presents a vision of the global financial ecosystem and its implications on the financing of water-related activities. This vision is that of the author, gained through a three-decade-long experience working internationally for various financial institutions.[2] This chapter discusses some key features of the financial system that are particularly influential on the way water activities can access finance. The chapter is intended for readers without a finance background, wishing to understand better why finance is available for some projects and not for others, or how finance professionals are looking at water projects. Most issues raised in this chapter are complex; the present format does not allow for the level of detail that would be necessary to present a fully informed view. In particular, it will not deal with a specific case or country studies (see Parts 2 and 3 of this Volume). Rather, it presents a few key concepts and basics of financial structuring, with the aim to provide foundations for the readers' understanding of finance topics. The objective has been to make complex matters clearer and simpler, at the risk of oversimplification. A bibliography is provided for further reading.

4.2 The ecology of the financial system: macroaspects

4.2.1 Emerging economies lack access to global currencies

4.2.1.1 The Bretton Woods monetary order and the transfer risk

A key aspect of today's financial system is its division in two largely heterogeneous systems. The monetary world is split between two kinds of currencies. Those who are freely tradeable between themselves form "global financial markets." They are a minority: 19 currencies out of a total of

[1] At OECD at the time of writing, after 2021 with the Climate Bonds Initiative.
[2] Nothing in this chapter is to be construed as a statement by the OECD.

Financing Investment in Water Security: Recent Developments and Perspectives.
DOI: https://doi.org/10.1016/B978-0-12-822847-0.00004-1

some 190 currencies worldwide. The other currencies are nontransferrable: they cannot be held or traded outside their country of issuance. Freely transferrable currencies are notably the US Dollar, the Yen, the Euro, and the Swiss Franc. Nontransferrable currencies are for instance the Indian Rupee, the Russian Ruble, and the Brazilian Real.[3] This state of things results from the monetary organization that was agreed at the Bretton Woods conference in the United States in 1944. The International Monetary Fund and the World Bank are managing this system.

It is not possible to hold a Rupee-denominated account in for instance France, nor for a French company to hold an investment denominated in Rupee. Let us assume that an Indian business wants to acquire water equipment from a French company to develop a water project in India. The French company cannot be paid in Rupee. Therefore, the ability of the Indian business to purchase the equipment will depend not just on its wealth in Rupee, but also on its capacity to change these for Euros in India to pay the French supplier in Euros. The Indian business will ask Euros from its local bank, which will, in turn, ask them from the Central Bank of India. As the Rupee is the sovereign currency of India its Central Bank can print as many as it wants. The Indian Central Bank has only three ways to obtain Euros: by exporting goods or services to a Euro-paying country, by borrowing Euros from the World Bank, or on international capital markets. Loans will need to be repaid however, so borrowing ultimately also depends on the capacity of India to generate Euros via exports. In case India does not have Euros in hand at the time when the Indian businessman brings his Rupees for change to the bank, the businessman will not be able to purchase the French equipment. This risk is called the transfer risk. It is different from the foreign exchange risk which is the risk of variation of the exchange rate between the Euro and the Rupee. Various commercial platforms offer coverage of the exchange risk for nearly all currencies (albeit of course at a sometimes considerable cost). But the transfer risk cannot be covered, except as described below. The transfer risk is not a theoretical risk. Many emerging countries have not developed export capabilities and are thus not able to borrow in hard currency for the same reason that they cannot generate hard currency proceeds to repay the loans. The consequence is that these countries cannot access modern technology sold in global currencies, for water projects, or for anything else. This goes not only for instant purchase

[3] China uses a dual monetary system. Its domestic currency, the yuan, is not transferrable. Its currency for external exchanges, the renminbi, is fully transferrable and its exchange rate is tied to the US Dollar.

of machinery, but also for long-term purchase commitments, for instance of running yearly expenses for chemicals in the case of a water treatment plant. This transfer risk is a major obstacle to investment by the cash-rich developed countries in emerging economies. It is also a major obstacle to technology transfer and the achievement of the global development agenda. This is well known and debated in international policy circles and there have been various calls for the revamping of the international monetary order sometimes dubbed a "new Bretton Woods" (Morse, 1983). Meanwhile, only limited solutions are available to cover the transfer risk and those solutions impose a certain structure to investments from global currency economies to the other part of the world, as described below.

4.2.1.2 Covering the transfer risk: export credit agencies and multilateral development banks

The transfer risk can be covered for a premium by Export Credit Agencies (ECA). The French company selling the equipment in India in the above example can get paid in Euros in France upon presentation of the payment in Rupee by a bank in India. However, the ECA, in this case, is in fact lending the Euros to the Central Bank of India and will expect repayment. Some countries may encourage their domestic exports by taking a degree of risk on their foreign counterparts, but generally, any emerging countries not acceptable for World Bank or capital market loans are also excluded from ECA coverage.

Multilateral Development Banks (MDBs), for investment operations, are structurally equipped to cover the transfer risk for the operations they finance. They enjoy the so-called Preferred Creditor Status with the World Bank and other MDBs (IFC, n.d.). This means that when for example an MDB finances a project in Turkey, if Turkey cannot repay because it lacks global currency, then the MDB can include that amount into the restructuring negotiations with Turkey led by the World Bank. The crucial point is that the MDB is covered for the amount it lends, but can also give to its colenders or investors, whether public or private, the benefit of this coverage. This is done via a technical arrangement called a "syndicated A/B loan." For instance, a water treatment plant project in Turkey would require a loan for €400 million. The MDB, operating as leader of the syndicate, offers a fraction—perhaps €50 million—as an A loan, and invites other global banks for 350 million as a B loan. The so-called "B lenders" have no transfer risk with this structure as the MDB will carry the whole €400 million to the negotiation table on their behalf. This Preferred Creditor Status is the technical cornerstone of the lead role of MDBs to attract global currency investment to emerging economies. This structure is the most commonly used to finance water projects in emerging countries in global currencies, or for emerging countries to access technology transfers in the water sector. Similarly, MDBs avoid working with countries that are already at the global debt restructuring table or that risk a transfer problem soon. The vast majority of emerging economies lack access to global currencies; therefore, for the water sector like for other sectors, they cannot easily benefit from technology transfer. Going further, those emerging countries may be tempted to attract priority projects that will generate global currency via exports, and water projects, usually targeting domestic markets, do not present this feature.

4.2.2 In advanced economies, capital markets have overridden lending finance

The global financial system of fully transferrable currencies has undergone drastic changes over the past five decades. These changes take again their origins in the Bretton Woods monetary rules, this time with the end of the gold standard in 1973 that created free and constant fluctuation of exchange rates between freely tradeable currencies. Before 1973, currencies were not fluctuating freely, as they were all pegged to the US Dollar, itself pegged to a fixed amount of gold ("the gold standard"). This in turn created a new business for banks, "financial derivatives," i.e., the trading of optional contracts to cover those risks. A typical option contract is the possibility to buy a currency, say the US Dollar, for a fixed amount of, say, Euros, over a period of time, say, six months. This option is giving a protection against a change in the exchange rate. The new bank activity of trading option contracts (or "derivative products") was more rewarding but also more risky than simple credit. In 1980, a global regulation was imposed on banks, the prudential system,[4] in order to shield the banking system from this increased level of risk. The prudential system is still in force today, and it imposes banks to set aside part of their equity capital as a reserve against risky lending. At the same time, credit was deregulated in most advanced economies: where there had been obligations for banks to lend to certain sectors, and in certain amounts, they became free to lend as they wished.

The combination of prudential regulation and credit deregulation caused a complete change in the role of banks. Banks have largely moved to an

[4] The present version of this regulation is called Basel 3, as it is issued by the Bank for International Settlements, the club of worldwide central banks, based in Basel in Switzerland.

"originate to distribute" business model: they structure financial products (such as a bond issuance for a water utility) and then sell them to capital markets (e.g., institutional investors, asset managers, pension funds). Lending to businesses has become a marginal activity for most large banking groups in advanced economies. Trading has become a core business of banks. And the vast majority (nearing 60% across advanced economies [Jorda et al., 2016]) of their lending concerns mortgages, which bear the intrinsic guarantee of repossession of the asset in case of default on the loan. As an illustration, the BNP Paribas banking group, one of the largest in Europe, had assets worth €1,920 bn as of the end 2019 (BNP Paribas Group, 2019). Out of this, only 44% or €851 bn were loans. A nearly equivalent amount (€800 bn) was held in cash and trading instruments. Besides, the share of assets with a maturity above five years was just 18%.

As a result, businesses in advanced economies have had to increasingly resort to capital markets to finance themselves by issuing bonds, rather than going to banks for credit. In the United States, some 60% of corporate funding comes through capital markets. Capital markets are less influential in Europe but are still the key source of funding for larger corporations (Schildbach, 2013). Capital market bonds are held by institutional investors, asset managers and pension funds. They are less often held by banks, because banks would need to set aside capital for the bonds if they held them, under the prudential regulations. Bonds and loans are two forms of debt that can have exactly the same cost, and both involve repayment of capital and interest. The essential difference is that a loan is a contract and is not tradeable. If the loan is long-term, say, 7 years, it is a long-term relationship between the company and the bank. The lending bank wants the company alive for the whole tenor of the loan. There is some flexibility: a phone call can arrange for the postponement of a payment, because a loan is a private contract between parties.

A bond is not a private contract, it is a market instrument, most usually traded on a stock exchange. Bonds are made to be traded. Technically, the borrower never knows who exactly holds his bonds, because they are sold and bought continuously. Even if the bond has a ten-year tenor, the lender to the company can lend for two days and on-send to another bond investor. Of course, some institutional investors like pension funds will enjoy holding long-term water assets in their portfolios. But still, those investors are not committed in any way for the long term: they will be sensitive to a potential fall in the value of those bonds if the water project or companies run into trouble, and may sell anytime. No relationship exists between the parties: the borrower and "the market." The market entertains no interest in the long-term future of the company: it is made of individual investors who seek profit on their bonds. No flexibility is offered in case of payment difficulty: if interest or capital is not paid on the due date, the market immediately advises all investors that the bond is in default, and procedures apply. Capital markets put short-term pressures on borrowers to repay their debt, because they are short-term in nature, and because they do not offer the long-term relationship and flexibility in case of difficulties of private contracts.

4.2.3 As a result, long-term credit is less available for water projects

Like other sectors, water utilities and water projects in advanced economies have therefore had to resort increasingly to capital markets for long-term finance. The situation is more nuanced in non-OECD countries, with a number of Asian countries, including China, having credit guidance policies that maintain bank credit availability for certain sectors. As argued above, this is putting a constant (market) pressure on them to deliver stable results. With small margins, the water business is generally not cash-rich. If an unexpected event triggers the need for an urgent investment, for instant the need to carry out emergency repairs on a reservoir following a flood, and the water company does not have cash at hand, it may want to borrow urgently. A credit line with a bank may usually be temporary and quickly extended. Raising a bond on the markets takes longer, and may not be an option if markets are depressed at that moment. Besides, bond markets are fluctuating globally, with the cost of debt highly sensitive to market events, much more so than the price of credit, so that unexpected needs for borrowing may be detrimentally expensive.

4.2.4 Since the 2008 financial crisis, debt is cheap, but capital is expensive

The 2008 financial crisis has triggered a structural change in global financial markets with deep consequences on where and how finance is available. In order to avoid a collapse of the global system, advanced economies' Central Banks had to resort to so-called accommodating monetary policies, or quantitative easing (QE). Central Banks printed money massively and injected this money into the banking system in order to avoid banks defaulting and to restore inter-bank lending. The crash created a serious economic crisis crunching the finances of the States. QE not only rescued banks, but also enabled States to borrow massively at a cheap rate even when their credit quality was deteriorating, as Central Banks would purchase State bonds from banks with QE.[5] Central Bank's holding of bonds has been expanding to amounts never seen before in modern history. This monetary policy is keeping interest rates and inflation at a low level but is also creating major uncertainty. A financial crash would occur the minute the policy stops because the price of borrowing would soar for the many insolvent borrowers that have been kept afloat by Central Bank liquidity, including some States, banks, and corporates. Moreover, this prolonged period of artificially low rates has distorted markets. Borrowers have been able to generate returns by investing cheap debt into profitable businesses, such as real estate. As a result, assets and equity prices have been largely inflated. In the past ten years, the MSCI world index of global equity values has tripled (MSCI, n.d.). In an environment with low inflation and soaring equity prices, which has been that of advanced economies over the past decade, investments in the real economy are not attractive unless they can match market equity returns. This is particularly true for water businesses and all public service infrastructure services generally which generate low margins. The corporate sector generally has been paying high dividends because shareholders chose to invest their money on soaring "paper" markets, boosted by QE, rather than in a low-return real economy. In the EU, gross fixed capital formation (investments in business assets) grew barely 3% during the period 2009–2019 (European Investment Bank, 2020). Over that same period, the Paris stock exchange index CAC 40 nearly doubled (Macrotrends, n.d.).

In these distorted financial markets, raising long-term equity capital to finance a water (or any other long-term infrastructure) asset is challenging. Money flows to cheap Central Bank backed debt, or to the high returns of risky leveraged investments. Investors will look for long-term risk guarantees from States, but States themselves are in dire financial situations. They depend on financial capital markets, which will put short-term pressures on them to secure their lending.

The global financial ecosystem is not healthy. Debt has polluted the financial ecosystem (Turner, 2016, pp.190-194). Imbalances in trade and in land use have physical and financial counterparts that are actually deeply interlinked. Addressing the physical imbalances that plague human development cannot be done without addressing the financial imbalances that plague human economies and societies.

[5] In several advanced economies, the EU in particular, direct lending by the Central Bank to the State is prohibited by the Central Bank's mandate.

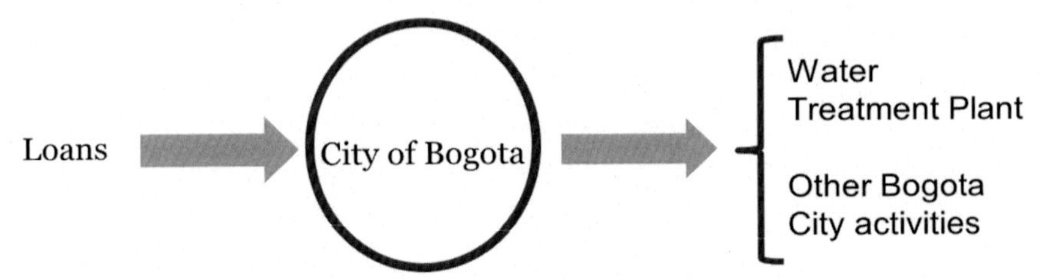

Figure 4.1 Case 1: public financing; lenders finance the city budget.

4.3 The ecology of the financial system: microaspects

Against this backdrop, whether in an advanced or in an emerging country, there are broadly three types of financing structures that can be used for a water project. Each structure has different stakeholders, different risk and return profiles, and different financing channels, creating constraints and opportunities. For the sake of illustration, let us imagine that a given city, say, Bogota in Colombia, has decided on building a new water treatment plant to supply the city with drinkable water. Below, three different financing structures are outlined to finance the plant: public, corporate or project finance.

4.3.1 Public finance

The City of Bogota may borrow to finance the plant. The lenders will lend to the City, and will be repaid out of the general revenues of the City (Fig. 4.1). Such revenues include local taxes, contributions from the Central Government, proceeds from sales and financial income, etc. In such a structure, the lenders have no connection to the plant. Their credit risk is on the City. The lenders are indifferent to whatever charges the City collects or not on water usage, or to the investment or running costs of the plant, as long as they remain a minor part of the City's overall budget.

This financial structure is usually referred to as "public finance," because the borrower is a public-sector entity.[6] Lenders can be themselves public sector entities, like the State or the Region, or a national or MDB. But they can also be private-sector lenders: the City can borrow from capital markets Fig. 4.2 Case 2: corporate financing. (by issuing municipal bonds), or get a loan from a private bank. Again, the credit decision is entirely based on the strength of the City's finances on the planned time span for repaying the loan. If the City's credit risk is not acceptable, lenders may accept the guarantee

[6] In different contexts "public" may mean "listed on a stock exchange" (with shares traded publicly), this meaning does not apply here.

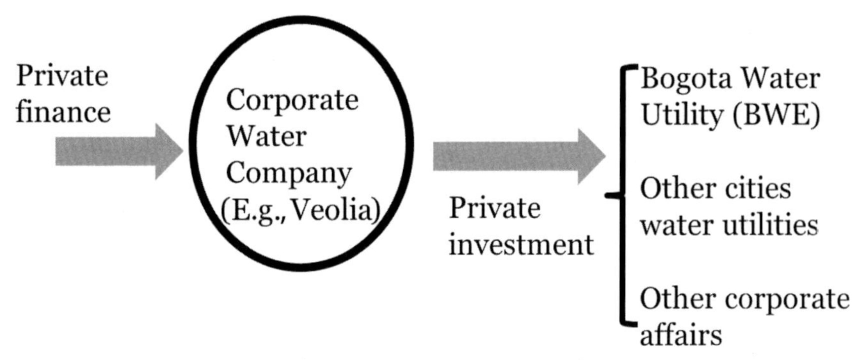

Figure 4.2 Case 2: corporate financing.

of another entity such as the Region or Fig. 4.1 Case 1: public financing; lenders finance the City budget. The Central Government—provided, of course, that their financial strength is better.

4.3.2 Corporate finance

The City of Bogota can also create a corporate company called "City of Bogota Water" (Fig. 4.2). This company may have shareholders from the public and/or private sectors; the distinctive feature of this structure is that the water treatment plant is owned by this corporate company, and not by the City itself. Typically the corporate company will own all other water treatment plans operating in the City. Its shareholders may be the City, the State, contractors, groups specialized in operating such plants or running water distribution. It can also be listed on a stock exchange and have institutional investors as shareholders such as insurance companies, pension funds, and asset managers. The new water treatment plant will be financed on the balance sheet of this corporate structure. It can finance the new plant by issuing shares, by issuing debt, by a mix of both, and by its own available cash.

In this structure, the risk of financiers is that of the corporate entity. Typically, the corporate entity will receive income from the sale of the water, either to the City, or directly to the people and businesses who consume water. Therefore, the financiers will have a direct link to the plant and its performance in this structure. How strong this link is depends on how important the plant is in the business of the corporate.

4.3.3 Project finance

The third option is project finance (Fig. 4.3). This is a structure in which the new water treatment plant is located into a private corporate company

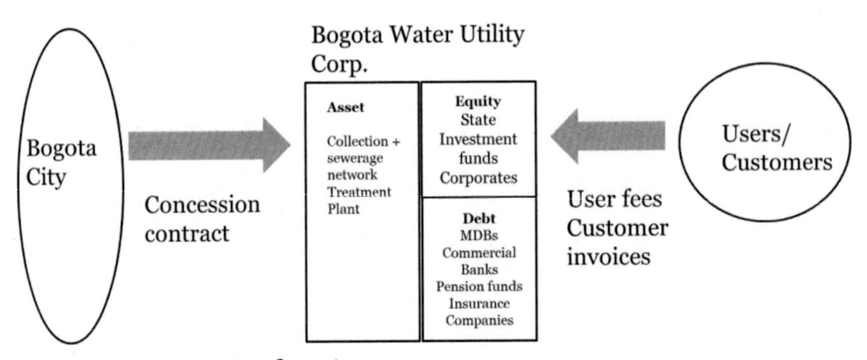

Figure 4.3 Case 3: project financing.

of which it is the sole asset, and that is created for this purpose only. Such a company is usually called a special purpose vehicle (SPV). They key word in project finance is the ring-fencing in a single-purpose SPV structure. The financiers in such a structure are only taking the risk of the plant, not the risk of any other asset or business. They can access all revenues of the plant. This structure is complex and costly but presents investors and lenders with unique possibilities to tailor the risk and return of the project as they best like. The present chapter provides a quick outlook, while Gietema (This Volume) provides a more detailed analysis.

This type of structure is sometimes called a public–private partnership (PPP) (see Gietema, this Volume). Although there are some precise legal definitions of PPPs in certain countries, generally PPP and Project Finance structure are synonymous. This type of structure is very often used for ventures in emerging economies, indeed it is the only structure that allows private global capital markets to invest in emerging markets. MDBs can take a share of the debt to the SPV (the "A Loan") and cover the transfer risk for the B Lenders, as outlined below. Usually, the MDB will structure the operation, i.e., negotiate the creation of the vehicle and the various contracts around it with the parties involved. These parties are the City or Country that tenders the contracts for the building, operation, and transfer of the water treatment plant. "Transfer" means that at the end of the contract the water treatment plant returns in full ownership to one of the parties, generally the tendering party. This transfer clause may have a financial value. Other parties are the civil works contractor that will build the plant and the water treatment operator that will run it. Frequently, the civil works builder and the operator create a consortium with a leading bank (sometimes the MDB) that commits to fund the project if it wins the tender.

Apart from the key advantage of permitting access to MDB coverage for the transfer risk, the structure has other advantages. Because it is created *ad hoc* for the project, the risks and returns can be tailored in a customized manner to serve the risk-return requirements of all parties. First, the financing can be located off-shore from the emerging country, in a global-currency jurisdiction, ruled by global finance laws where enforcement and seizing monies in case of necessity is comparatively easy and safe. Second, the revenues can be paid from the most desired party. For instance, the risk of not collecting water payments from users may not be taken by the SPV if it does not wish to; instead, the City or Country can commit to compensate for missing payments, or even to pay in full an agreed income to the SPV, independently of the water consumed, invoiced or paid. In this case, the water is free for users, but they may pay for it as part as their global tax bill to the City or Country. Besides, part of the revenue may be directed to an escrow account that will hold and maintain enough cash over the project duration to provide a guarantee to financiers for a period of repayment (e.g., 6 months). Third, the shareholding of the SPV is usually split in a manner that results in none of the parties having to fully integrate the structure in their accounts. Therefore, the SPV and its risks do not weigh on any parent company balance sheet.

In order to achieve that, the shareholding is often split between three parties having each around one-third of equity. Typically, the conceding party (State or City) has such a stake and may waive its rights to dividends, so that the equity investment becomes more attractive to private shareholders. A second typical shareholder is the civil works contractor and/or the water plant operator. This party, therefore, gets out of the project not only its revenue from being an equipment and/or service supplier, but also its shareholder remuneration. A third typical kind of shareholder is an investment fund specializing in infrastructure development.

In project finance, shareholders play the key roles because they structure the deal at the time they build the answer to the tender, in case there is a bidding process. They also furnish the initial money; all equity is paid up before debt is called. And they give guarantees to the lending parties for the risks lenders usually refuse to take, i.e., construction risks and volume/price risk on revenues. Volumes and prices are guaranteed in the contracts by the conceding party and the water plant operator. Those guarantees do not come for free: the shareholders will only consider making such investment if the revenue expectation for them, in the form of dividends and increase in the value of equity, covers the risks taken, including the risk involved in giving the guarantees. A rough idea of the yearly rate of return for private shareholders in a project finance water venture is 10%–15% per annum in real terms.

Another key feature of the project finance structure is the long tenors it enables for debt. Global capital markets can offer loans for longer durations than local capital markets. This is because the investor market includes pension funds and life insurance companies, which need to balance their long-term exposure with long-term assets. In many emerging economies, pension and retirement systems are still in their infancy, leading to shorter maturities in capital markets. For example, the longer maturity available in China is 10 years, while some road infrastructure projects in France are financed with 60-years bonds. In the case of water projects, lending to a water treatment plant financed using a project finance structure can be 15 to 30 years long, while the local markets in public or corporate finance may not go beyond 5 or 10 years.

This is of crucial importance to the very existence of the project, particularly for water. The off-shore international structure of SPVs allows the costs of the investment in the single asset owned and financed by the SPV (in our case, the water treatment plant) to be depreciated over the duration of the debt of the SPV. This means that the longer the debt finance, the lower the yearly running cost of the plant. This is particularly important for a public service activity with low margins like water supply. If the water treatment plant needed to be depreciated over a shorter period, the costs of the water (to the conceding party or to end-users) would be less affordable.

Debt providers to SPVs can be banks, asset owners such as insurance companies and pension funds, looking for long-term assets. The role played by banks in such structures is not as lenders, but as financial advisors. In order to answer the bid and create the SPV with its myriad of risk-sharing contractual arrangements, the shareholders of the SPV employ a bank as financial advisor. The bank's fees for this work are in the range of 3% to 5% of the whole amount of finance. This quite often exceeds the amount of interests received, especially if the advisor is lending only a minor share of the total finance.

The long tenor of project finance debt, though making yearly costs more affordable, still has a flip side. While the ratio of debt to total project cost varies a lot, an average of 80% of debt is probably a good order of magnitude. Debt is therefore important. It bears fees at inception (including structuring bank fees as noted before) and then an interest rate which is often higher at the beginning of the project, when perceived risk is higher. The interest rate, even though it can appear as modest[7], runs over the whole life of the project. Therefore, when considering the whole allocation of project earnings to project expenditures over time, and adding dividends

[7] An indication of 2% to 5% can be provided to give an idea, but this will vary a lot between projects.

and bank and legal fees to interest charges, overall financing costs in such structure often amount to roughly half of the project's earnings.[8] Another way to formulate this is to say that half the added-value created by such projects goes to the financial system.

A frequent theme for discussion around project finance is the notion of the cost of capital. There is a theoretical background to this notion that cannot be covered in these short notes.[9] Practically speaking, let us consider a project with 20% equity and 80% debt. Equity yields 15% per annum and debt costs 4% per annum. The average cost of the capital invested in this project is 6.2%. For this project to create economic value, the assets of the project (the water treatment plant) therefore needs to yield more than 6.2% per year (all yields are expressed in real terms here, i.e., after taking inflation into account). This is a substantial amount, and means that somebody, the users or the conceding party, needs to pay a lot for the water. The gap between the substantial cost of capital, caused by the perceived risks in the project, and the impossibility to generate high returns from public service assets, is one of the main reasons why projects do not materialize. The easiest way to lower the cost of capital is to get a patient, state-like investor to invest a large share of equity without expecting dividends. However, we have seen above that under the current monetary policies adopted since 2008, financial investments are generally very profitable (in the range of more than 10% per annum). Such an investment choice is therefore mostly for development-oriented institutions, which are limited in size and risk-taking ability. Besides, as we have seen above, States themselves are often heavily indebted and cannot invest a lot in projects because they lack cash (sometimes, they cannot even tender the project for this reason). This weak financial situation makes any guarantees that they could provide to banks to try and lower the price of debt less valuable (or costlier). Therefore, it is not always possible to lower the cost of capital of a given project so that the price of the good or service provided is affordable to its prospective customers or users.

Economists have also been questioning the high financial cost of project finance structures. From an economic point of view, public service projects like water treatment plants can be justified by socio-economic benefits that are not priced in the sale value of water. For instance, water users are healthier, and water bodies downstream are less polluted by harmful substances. Such benefits however can be estimated. If the cost of finance

[8] Again, this is a broad average that will vary according to particular situations.

[9] Capital Asset Pricing Model (CAPM).

is too high, they offset socio-economic benefits, and the overall socio-economic benefits disappear. The United Kingdom has been using a lot of project financing structures between 1980 and 2000 to develop its economy under the Project Finance Initiative (PFI). UK economists then pointed to the high costs of the project financing structures and developed a "value for money" objective for public policy, which basically comes down to limiting the financing costs with a public policy objective in mind.

In this respect, it is important to have in mind that project finance structures have mostly been used historically to develop projects in emerging economies, because they enable the coverage of the transfer risk via MDBs as outlined above. In advanced economies, infrastructure was historically developed using public (state) investment more than private investment, and bank lending more than capital markets (Monet, 2018). It may also be noted that public investment and bank lending were massively used to develop infrastructure in emerging (now emerged) Asian economies such as China, South Korea, and postwar Japan (Werner, 2003). Of course, the postwar context was very different from the present time. However, in view of the failure in the past several decades to finance the need for infrastructure, both in advanced and emerging economies, it may be worth questioning the generally accepted view that private capital markets are best suited to finance infrastructure because public finance is scarce. This "generally accepted view" is an outcome of the generally accepted economic theory, in particular the assumption in most economic models that finance is neutral, and the efficient-market hypothesis in financial theory (Turner, 2016, p 38-42). Economy and finance, as academic disciplines, may need further research and new approaches.

4.4 Constraints on the financing of water investments in advanced economies

The main constraints to financing water investments are briefly recapitulated below starting with advanced economies with access to global financial markets.

4.4.1 Borrowing capacity of some sovereign borrowers

As outlined above, sovereign and quasi-sovereign[10] borrowing are now difficult in most advanced economies. Of course, the global and severe

[10] Cities, regions, local authorities backed by a State guarantee.

economic crisis triggered by the Covid-19 pandemic is constraining public balance sheets even more. In several countries, direct monetary finance is not allowed, so, States need to borrow from private global capital markets. Central Banks do print massive amounts of money, but they are not allowed by their mandates to lend this money directly to the State. Central Banks purchase securities from commercial banks and States need to borrow on private capital markets. Even though debt ceilings have been released capital markets are still controlling, as they should, that States will have the capacity to repay, under the threat of downgrading of the sovereign debt rating, which would make further borrowing costlier. Scarcity is particularly felt in Europe and Japan, but less so in the United States because the role of dollar as reserve currency allows the United States to run a higher budget deficit than other countries. Because States are the lenders of last resort in the financial system, the financial weakness of States trickles down on subsovereign borrowers that cannot obtain strong State backing. Locally, some cities or regions are still cash-rich and can enter into significant projects (one example is the City of Paris in France with its Grand Paris mass transportation project and its repossessing of the water supply system, see also Alaerts, This volume). However, they are the exception more than the rule at a time when the stock of existing infrastructure, particularly water assets, is aging, and significant expenditure needs may arise as climate change starts to manifest (Smith, and Alaerts, This volume).

4.4.2 Water utilities have limited access to long-term finance

On the corporate side, as outlined above, the general situation is that only States in advanced economies can borrow for long tenors (if they can borrow) on capital markets. Therefore, corporates are able to borrow for tenors inferior to 10 years which makes it difficult to finance water assets on a long enough time frame.

4.4.3 Uncertainty and high financial returns increase the cost of capital for project finance

The cost of capital remains high for project finance structures (also see Gietema, this Volume), and it is difficult to narrow the gap with pricing constrained by economic depression and scarce public budgets. The financial weakness of States is adding to the uncertainty perceived and priced by private lenders and investors. The availability of high returns in other segments of financial markets than real infrastructure investments is pushing equity investors' requirements for returns on the upside.

4.5 Constraints on the financing of water investments in emerging economies

Emerging economies with no automatic access to global financial markets are also facing difficulties around financing water investments, but due to other causes.

4.5.1 Scarce access to hard currency limits technology access

As outlined above, many emerging countries cannot borrow anymore in hard currencies. For those who still can, a low rating in hard currency borrowing makes the cost of such borrowing quite high. As a result, those countries cannot acquire technologies from the developed countries, including for developing modern water projects. Therefore, most projects in emerging economies involving advanced technologies will require transfer risk coverage in the form of export credit or multilateral bank coverage. This limits the possibility of emerging countries to equip themselves, as well as the possibility for advanced economies to export their technologies and their capital in potential growth areas. Only a handful of emerging countries, that managed to develop exports in hard currencies, can attract global capital in water projects. Examples of such emerging economies include Chile, Mexico, and Thailand (Tradingeconomics, n.d.).[11]

4.5.2 Lack of long-term funding increases the cost of domestic projects

Countries without access to global currency funding can develop projects domestically, using technology developed locally. One frequent issue for infrastructure development, including water, is the lack of availability of long maturities on local financial markets (also see Gietema, this Volume). The solution is often to finance longer-term assets by rolling over debt, for instance borrowing three times on a five-year maturity for a fifteen years project. This adds major uncertainty to the project at inception, and therefore

[11] A bond is considered "investment grade" if its credit rating is BBB- or higher (BBB, BB, B, or in the A category) in the Fitch ratings and Standard and Poor scales. A "noninvestment" grade rating signals a high probability of default, and therefore a high cost of the debt.

increases the cost. This is a particular issue for low-margin projects, such as water investments.

4.5.3 Water is not always prioritized in public infrastructure spending plans

The lack of prioritization of water projects in governments' agendas poses a difficulty frequently encountered in emerging economies. One of the reasons is that water projects are unlikely to generate hard currency proceeds. As a result, when these governments decide which infrastructure projects to tender, they may favor extracting, processing, or manufacturing sectors that offer the possibility of hard currency exports (also see Gietema [This Volume] for a sector comparison in project finance). These difficulties and others hamper the financing for the fast attainment of the 6th United Nations Sustainable Development Goal "Ensure access to clean drinking water and sanitation for all." Against this backdrop, however, a new "sustainable finance" sector has emerged in finance, which could offer development opportunities for water investments.

4.6 Perspectives offered by the development of sustainable finance

4.6.1 Global investors are increasingly seeking to invest sustainably

Initiated some fifteen years ago, "sustainable finance" is an attempt by the financial sector, and its regulators, to create financial products dedicated to investing in environmental or other sustainable activities. Although still a niche in the sense that the amounts remain modest compared to global finance, the sector is growing fast. The most well-known product is the "green bond" which is a bond issued on capital markets with a commitment to use the proceeds to invest specifically in "green" businesses. The European Investment Bank was one of the first global investors to issue such a bond in 2008. Green bonds can finance new projects, refinance existing projects, or both.[12] A specific category of green bonds is sovereign green bonds that have been used by States to finance their water investments in specific cases.[13] Other kinds of bonds include blue bonds to conserve sea and marine

[12] The Climate Bonds Initiative is a nongovernmental organisation reporting on the developments in green bond markets all over the global world, see www.climatebonds.net.

[13] For instance, the Dutch Sovereign Green Bond issued in May 2019 is used to finance the Delta Program against sea level rise.

resources for sustainable development, as well as sustainability bonds and catastrophe bonds to insure against water flooding. By signaling an environmentally targeted use of proceeds, issuers can hope to attract a specific category of investor, those who are more interested into this type of investments. Such investors are so-called "impact" investors who make having an impact on the environment a specific objective of their investment. In some countries, financial regulators have given tax or interest rate incentives that make those financial products comparatively attractive. However, because these markets are still small and relatively new, those markets generally bear illiquidity and uncertainty risks. Besides, despite financing projects with various environmental benefits, including climate change adaptation, green bonds have so far not necessarily translated into comparatively low or falling carbon emissions at the firm level (BIS Quarterly Review, 2020).

Investment funds with Environmental, Social and Governance (ESG) objectives are another category of sustainable investors. Such funds, whether investing in equity, bonds or both, purchase paper from companies with high environmental (and/or social and governance) scores. Those scores are given by the fund itself or an external ESG data/scoring provider. Such funds may push corporates to enhance their ESG scores by investing more sustainably. Water utilities, if they manage to have their clean water and sanitation investments reflected in the ESG scoring, can see the market value of the paper they issue increased because of higher demand from ESG funds.

The volumes raised on green bonds markets, and the volumes of assets under management with ESG objectives have been steadily rising over the past decade, including since the beginning of the Covid-19 pandemic as constraints on spending triggered an increase in assets under management. Investors are showing a strong appetite to invest with an ESG sustainability marker. This appetite could stem from increased awareness of climate change and environmental degradation and a desire to contribute to their mitigation, or from a desire to protect portfolios against changes in valuations due to climate change and climate change policies, or both. Investment funds can adopt active engagement strategies, i.e., gather enough voting rights in companies to promote Board resolutions. Thus, sustainable finance can be a driver to push the corporate world to adopt more sustainable practices and strategies, including in the water sector, or in sectors or projects that affect water demand and availability, or exposure and vulnerability to water-related risks.

4.6.2 The EU taxonomy and the "Do no harm" concept: raising the profile of water in nonwater investments

While sustainable investment has been a market-led initiative, financial regulators have been engaging recently in providing some clarity and transparency to these markets. The European Union has been at the forefront of this regulatory effort by issuing in June 2020 the EU Taxonomy Regulation that defines which economic activities are environmentally sustainable (European Commission, n.d.). It defines environmental sustainability as a combination of six environmental objectives: climate change mitigation, climate change adaptation, water, waste, pollution, and ecosystems. More precisely, "water" in the regulation refers to "the sustainable use and protection of water and marine resources." In order to be taxonomy-compliant, a given activity must "substantially contribute" to one of the six environmental objectives, while "doing no significant harm" to any of the other five. For each activity (for instance, farming), the regulation will define detailed Technical Screening Criteria according to which the activity contributes substantially to climate change mitigation while not harming the protection of water. It will also define which economic activities contribute substantially to the sustainable use and protection of water and marine resources. The full set of Technical Screening Criteria is to be developed by the end of 2021, for application starting in 2023. The Regulation is not an obligation to invest, but it will feed into various regulations of the European financial system that will contain obligations to report on the environmental features of economic activities that corporates and financial market participants (asset managers, asset owners) invest in. Therefore, this Regulation will highlight for financial markets where and how to invest to protect water. Assuming that the detailed criteria that will be developed for water encompass the full spectrum of water investments, including access to water, the EU taxonomy could therefore play a role in increasing water investments in the EU. It will be important also to check that carbon finance does not crowd out water finance. In any case, the taxonomy regulation cannot by itself make water investments more profitable, or resolve the issue of a high cost of capital compared to low profitability. For this, economic regulation focusing on valuing water and pricing the negative externalities of excess water consumption and water pollution is necessary. One must hope that the increasing evidence of the impacts of recent droughts on European economies will lead regulators to increase their action in this domain, and the markets to accelerate the uptake of more sustainable investment practices.

4.7 Conclusions

Clean water and sanitation are essential elements of a sustainable life on Earth, for mankind and for global ecosystems. Water scarcity and pollution are immediate threats that climate scientists are well aware of. While there is an increasing awareness of those crucial problems in the general public and in the economic world, the financial ecosystem as it is today is largely unfit to provide for the massive infrastructure investments that are required to address the problem, both in advanced and emerging countries. Drastic and fast action is needed at both the level of pricing incentives and financial market structures. One of the reasons why this is slow to happen is the prevalence in the global world of economic and financial theories that assume that optimal outcomes are provided by free markets. The need for regulatory action to price externalities linked to water use is not an agreed notion amongst economists. The same goes for the financial ecosystem. Its shortcomings in delivering sustainable development outcomes, in both advanced and emerging economies, are far from being recognized by a majority of financiers or government authorities. An important contributing factor to this situation is the inadequacy of economic theory and modeling (Grandjean and Giraud, 2017). Most macroeconomic models do not contain variables reflecting finance and natural resources. Prices are generally assumed to adjust to supply and demand, providing a "general equilibrium" in modeling that averts any significant crisis. Physical models of climate change and environmental degradation are sending very worrying and urgent signals, which are voiced by the community of climatologists. But when those models are translated into macroeconomic models of climate change, the signal gets significantly weakened because of the nature of economic modeling and the underlying economic theory. As a result, policymakers do not perceive the same threat that climate scientists do. As for the effect of the financial ecosystem on the economy and society, economic models do not inform them much either. The development of new economic models, integrating finance and natural resources, should be an exciting task for the younger generation of economists. As the effect of climate change and resource depletion becomes more apparent, and as financial imbalances impact more and more the economy and society, we must hope that the attractions of the "business as usual" approach will weaken, and that we will collectively find our way out of the very dangerous path of unbalancing the Earth ecosystem.

References

BIS Quarterly Review, 2020. Green bonds and carbon emissions: exploring the case for a rating system at the firm level. Bank for International Settlements, Basel https://www.bis.org/publ/qtrpdf/r_qt2009c.htm.

BNP Paribas Group, 2019. Universal Registration Document, p. 424, https://invest.bnpparibas.com/sites/default/files/documents/bnp2019_urd_en_20_03_13.pdf (accessed November 2020).

European Commission, n.d. https://ec.europa.eu/info/business-economy-euro/banking-and-finance/sustainable-finance/eu-taxonomy-sustainable-activities_en (accessed December 2020).

European Investment Bank, 2020. Investment Report 2020/2021. European Investment Bank, Luxemburg, pp. 54–55. http://www.doi.org/10.2867/904099.

IFC, n.d. International Finance Corp., Washington, DC. https://www.ifc.org/wps/wcm/connect/corp_ext_content/ifc_external_corporate_site/solutions/products±and±services/syndications/preferred-creditor-status (accessed December 2020).

Grandjean, A., Giraud, G., 2017. Comparaison des modèles météorologiques, climatiques et économiques: quelles capacités, quelles limites, quels usages? Working Paper, Chaire Energie et Prospérité, Paris http://www.chair-energy-prosperity.org/publications/comparaison-modeles-meteorologiques-climatiques-economiques/ (accessed November 2020).

Jorda, O., Schularick, M., Taylor, A.M., 2016. The great mortgaging: housing finance, crises and business cycles. Economic Policy 31 (85), 107–152. https://doi.org/10.1093/epolic/eiv017.

Macrotrends, n.d. https://www.macrotrends.net/2596/cac-40-index-france-historical-chart-data (accessed December 2020).

Morse, J., 1983. Do we need a new bretton woods? Fiscal Studies 4 (2), 8–18. www.jstor.org/stable/24434550. (accessed February 2021).

Monnet, E., 2018. Controlling Credit, Central Banking and the Planned Economy in Post-war France, 1948-1973. Cambridge University Press, Cambridge.

MSCI, n.d. https://www.msci.com/documents/10199/db217f4c-cc8c-4e21-9fac-60eb6a47faf0 (accessed December 2020).

Schildbach, J., 2013. Bank Performance in the US and Europe, an Ocean Apart. Deutsche Bank Research, Frankfurt am Main September 26, 2013, ISSN 1612-314X https://bit.ly/3oIjOHG .

Tradingeconomics, n.d. https://tradingeconomics.com/country-list/rating (accessed July 2020).

Turner, A., 2016. Between Debt and the Devil: Money, Credit and Fixing Global Finance. Princeton University Press, Princeton, NJ /Oxford.

Werner, R., 2003. Princes of the Yen: Japan's Central Bankers and the Transformation of the Economy. M.E. Sharpe, Armonk, NY.

Suggested additional reading

Abdelal, R., 2007. Capital Rules: The Construction of Global Finance. Harvard University Press, Cambridge, MA.

Keen, S., 2011. Debunking economics - revised and expanded edition: The naked emperor dethroned? Zed, London.

Mian, A., Sufi, A., 2015. House of Debt. University of Chicago Press, Chicago, IL.

The World Bank, 2017. Public-Private Partnerships Reference Guide. Washington, DC. https://library.pppknowledgelab.org/documents/4699/download.

CHAPTER FIVE

Critical disconnections between donor and domestic realities

Martin S. Baker[1]
Counsel, Dentons

Definitions
- Discontinuity: differences in opinion or criteria.
- Project criteria: the elements of a project.
- Project financing: In matters of general infrastructure, the financing of a standalone project which has revenue directly associated with it, although in water and sanitation, a project is usually a part of a system without directly associated revenue.
- Financing criteria: the elements required to financing a project.
- System financing: the financing of a portion of a system.

5.1 Introduction

Some cities fight fires with drinking water. Some cities wash their streets with drinking water. Some cities have no sanitation. Some cities have no clean, treated drinking water. Some cities have extensive water available but no distribution of that water. Some cities have substantial "non-revenue water" which is available in the system but is "lost" through leakage, diversion, non-payment or otherwise.

Some areas are depleting their surface or groundwater supplies. This has an influence on the selection and design of projects and on their financing. It is likely to indicate a lack of attention to the management of the system. In many places poor management leads to poor service, poor planning and poor cost recovery. Such lack of attention is an element in the management that is important for those undertaking financing to understand.

[1] Martin S. Baker is an environmental attorney. He is Counsel to the international law firm Dentons. He has assisted in the financing of water and other infrastructure projects in more than 20 developing countries. He is an advisor to the Water Finance Facility and its project in Kenya, the Kenya Pooled Water Facility described elsewhere in this book. He was for 17 years Director of the New York State Environmental Facilities Corporation, an independent governmental agency providing all water and sanitation financing for villages, towns, and cities in New York State.

Financing Investment in Water Security: Recent Developments and Perspectives.
DOI: https://doi.org/10.1016/B978-0-12-822847-0.00006-5

SDG 6 mandates water and sanitation for everyone. This is an accepted international standard, well known amongst development banks, nongovernmental organizations and international organizations. In many developing counties, a lack of priority and attention to this sector undermines efforts to advance toward this goal. Does this mean that there should be more effort on the part of international organizations to advance an understanding of the value of water and sanitation in places where they are not considered priorities? Does it mean accepting a very slow process of development and adjusting ambitions and pace of water financing? It is an area where the development finance institutions can better "coordinate" with water and sanitation NGOs promoting water and sanitation advances in developing countries.

These observations suggest that there is an opportunity for the international community including financial institutions and private sector entities to devote attention to advancing an understanding of the health, societal, economic, and political benefits of improving access to water and sanitation. Their efforts might help increase the priority given to these issues in domestic institutions where they are not current priorities.

As individuals, organizations, and the international development aid community address water and sanitation needs, it is appropriate that they consider the many differences that exist within individual countries. It is also useful to consider the differences in concepts, working processes, and language between those who wish to "help" and those in direct contact with the local issues. This chapter identifies and discusses such discontinuities. These differences have manifested themselves in the author's experience in the financing of water and sanitation and other infrastructure facilities in more than 20 countries over several decades. Some of these efforts were successful. Some were not successful. Lessons have been learned from each. This is not an amalgam of professional literature or of accounts of others.

5.2 Project definition

5.2.1 Differences in types of "projects" needing financing

The term "project" has many meanings. This is especially so in the water and sanitation arena. It may refer to the acquisition of pumps, computers, or supplies for the maintenance and operation of a system. Larger, long-life projects can be an addition to a system, such as a new well, a new reservoir, a new water supply conduit or a waste treatment facility. It does not "stand alone." It is part of a system. It normally does not have revenue directly associated with it. It relies on the whole system's revenue from tariffs or other sources.

The way a project is defined, designed, or scoped drives financing options. This is consequential for financing collateral projects. For example, the delineation of a specific city or town by a legal definition or regulation influences the delineation of nearby rural areas. A more expansive definition of an urban area might permit adjacent areas to be incorporated into a development financing program.

Water projects are different from infrastructure projects for which common "project financing" techniques are used. As that term is used broadly in infrastructure financing, such projects have direct revenue that can be ring-fenced, such as landing fees at airports and tolls on highways. In these cases, the project's revenue is usually the credit for the financing. Only rarely does a water "project" have a distinct revenue associated with it. It might be a desalination project which involves a high-tech desalination plant that typically falls outside the routine technical competence of a water utility or where the water is used by a specific user which pays directly for the water produced.

Water and sanitation projects are usually part of systems for the delivery of water and sanitation services to an area, normally an administrative division like a village, a town, or a city. As such, they must be recognized as part of the whole water system, and be financed in the context of the whole system. That is, they are not financed on the revenue generated solely by the project itself because there is none, but by the revenue of the whole system. That revenue might be from tariffs, from general taxes or from transfers from national governments to local governments or to water utilities.

Projects may address short-term needs and have short useful life. For example, they could include computer control equipment, individual meters for households, or replacement pumps. They are useful to the water department operations for the systems they manage.

Projects may also address major capital improvements. They may be additions to the water supply system, such as a new reservoir or aqueduct. Frequently, such additions to municipal water systems may be physically located outside the municipality for which the project serves. They may be local in nature such as a sewer connection to a neighborhood that is not serviced. They may be the series of connections that constitute the "last mile" that is, which connect individual users to the water or sewer system.

Projects might benefit low-income neighborhoods. They might serve neighborhoods that are not presently served. They may be of service to industrial or government users. They might include local sewers or waste treatment facilities. They might even be related to watershed protection or water supply sources. They may be large scale or small scale.

It follows that there is no generic "water project." Almost any water project can be imagined, framed, engineered, and designed in many ways. The way they are organized will influence financing options. For example, financing options will differ for small and large projects, and for ones in urban and rural areas. Joining suburban and urban areas into unified projects may generate economies of scope and scale. Such economies may make the whole project financially more feasible. In some political situations, it may permit cross-subsidies from urban to rural areas or the bundling of several financing mechanisms. Considering rural areas in isolation limits options for sustainable financing but may allow simpler engineering solutions.

Each of these issues needs to be considered in designing an appropriate financing mechanism for each specific "project." These issues largely determine whether a project can be financed with any kind of debt. The financing technique must recognize each of these issues and accommodate them. Where there are several financing sources, the conditions of each must accommodate the project definition, the sources of existing financing for the system, and the needs and conditions of those participating in the financing.

5.2.2 Objection to "privatization"

There has historically been popular objection to "privatization" in the water and sanitation sector. If water is "free," private sector involvement is thought to be anathema. Public objection to private involvement in the water sector is based on the concept that any private sector involvement—even in financing—is allowing a "profit" from something that should be "free." This popular position does not distinguish between private operation, and financing in the private sector capital markets.

Water operations involving private entities are fundamentally different from debt financing. There is a belief that water operations by private operators involve the private entities taking a profit from the public service. This is different from securing loans from private sources, which loans reflect not a profit from operations but repayment on predetermined terms for the money provided.

For domestic water supply and for sanitation purposes, the capital needs and costs are largely for system improvements related to collection, treatment, and distribution. Raising capital to debt-finance water projects is different from private equity financing. If a utility cannot or will not pay its debt service, the financing entity has certain rights to available monies, but rarely has any rights to operations or to decisions regarding the provision of service. In the event of defaults, conflicts will arise between interest groups wanting to keep "outsiders" away from system finances, resulting in complexities in dispute resolution.

In countries with capital markets, pension funds and insurance companies have substantial capital which they are willing to invest in domestic infrastructure. That money is available to vastly expand the facilities of the water and sanitation sector. The distinction between debt financing and private ownership of water resources needs to be kept in proper perspective where financing for new system projects is needed and available from domestic capital market participants. Where financial institutions can participate in part of the collective financing of many projects, such as by financing a part of a pool or other legal aggregation of projects, the financial structure mitigates the risk of individual projects.

5.2.3 Differences in understanding project solutions and appropriate technologies

There are different views held by donors, project planners and designers, financial participants and water service providers in developing countries. Simple, low-technology approaches may be more appropriate than sophisticated advanced treatment. New technologies, such as intelligent metering, information technology to identify leaks in distribution systems, and solar desalinization may be appropriate but may not be familiar to the financing entities or to the water system planners and engineers.

Attention to these differences can have a real impact on the definition of a project and on the project's cost. Water managers' and financiers' conservatism may foster unnecessarily complex solutions. Weak water system managers may seek familiar providers or solutions which impede innovation and the seeking of solutions appropriate for the specific need.

The parties involved in financing water and sanitation projects need to be aware that if projects are required to be designed to the highest engineering standards, simple approaches may be missed. In such situations, mistakes in understanding with local water service providers are likely to result in complex, sophisticated solutions that are beyond the experience or ability of utility managers. Costs may become inappropriately high.

Innovation can and does occur in low-technology situations. At a simple level, water pushed in wheel-barrow type containers is more efficient than carrying water from wells on people's heads. Drinking water can be made substantially safer by boiling or the addition of simple additives, rather than requiring extensive water treatment facilities. Such innovations should be encouraged, even where there may be high-technology solutions that are more "elegant" to the technologically sophisticated professionals.

The highest standards for planning and design of projects may be normal in countries where donors or international development entities are located. These standards are frequently not understood by small and medium-sized water service providers in developing countries. Such solutions may not be appropriate because they assume certain standards of construction, of operation and of capital maintenance. They may not be consistent with the technical standards with which the operating employees are familiar.

Where international financial institutions can do so, they might seek amalgamation of small municipal water systems to facilitate better development of technical capacity. That amalgamation may facilitate the financial capacity to capture economies of scale and recruit and retain technical expertise, e.g., on management, hydraulic engineering, construction, or laboratory functions. It is important not only for the utilities but also for the financiers.

5.2.4 Unintended or unexpected conditions or consequences

There are many unintended or unexpected events or consequences that can affect the financing or the financing decision. A clear example is the Covid-19 epidemic, which has caused extensive delays in designing projects and in securing the necessary approvals for financing. Delays are accommodated in certain contracts by the provision for permissible delays for unanticipated situations. But the understanding of unanticipated situations differs under different legal systems. They need to be understood both under the law of the project and under the law of the financing entity to evaluate foreseeable risks.

Project design may have unintended consequences to the environment. The withdrawal of water from groundwater or from rivers or lakes may have ecological consequences. These consequences might include, for example, wetlands degradation, the reduction of water available for agricultural or industrial uses, or environmental impacts on estuaries or lakes.

The disposal of untreated wastewater into existing waterbodies is environmentally detrimental. The treatment of wastewater needs to be subject to a level of treatment that is consistent with the existing and anticipated status of receiving waters. For example, the disposal of highly treated wastewater, sometimes referred to as tertiary treated water, into saline water, or of desalination brine into seawater can be considered an adverse environmental impact or considered inconsistent with religious or cultural values.

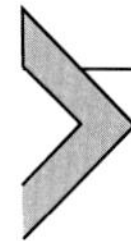

5.3 Financing considerations

5.3.1 Appropriate financing techniques

The definition of the project or the type of project should determine the appropriate financing technique. An understanding of the nature and useful life of each particular project is essential to understanding what is a reasonable, efficient, low-cost way of financing such project. An understanding of the nature and useful life of each particular project is also essential to permit a rational application of conditionalities to the financing of such projects.

There is a substantial difference in the financing of short-term items (3–5 years) and long-term physical developments (generally more than 10 years). There are substantial differences in types of projects for which different financing approaches are appropriate. For example, construction of new reservoirs or aqueducts are clearly long-life projects which are integral to the improvement of whole systems. They are appropriately financed for their useful life using long-term financing techniques. Projects addressing losses in distribution systems are important capital maintenance issues for which operating budget financing may be appropriate. The purchase of new computers or other equipment with short useful lives is appropriate for short-term financing.

Each of these items has a different useful life. They have differences in operational and maintenance needs. They have different meaning in financial statements and projections. They usually have different implications for tariff and other revenue sources.

These differences are important elements in choosing which techniques are appropriate for financing each project. The distinction between capital construction and operation and maintenance needs to be considered in all financing and assistance. Project design can also affect the allocation of costs between construction and operational needs. It is therefore an important consideration when the overall project costs and benefits are determined. Differences in understanding financing terminology where there are several techniques to be used in financing a water project or program are important. All participants need to have a similar understanding of the terms being used. The technical meaning of many programmatic terms are not well understood by most well-intentioned and interested parties to such transactions. Terms such as "pooled" financing or "revolving funds" are examples. Similar examples involve "gross pledges," "compensating balances," "first calls," "conditional guarantees," "reserves," and "reserve funds."[2]

The misapplication of such terms can cause substantial difficulties. One of the more obvious examples of such situation is where a financing is reasonably ready to occur and then it becomes apparent in the contract drafting that the parties have different understandings of such terms. The normal process for resolving such misunderstandings before the financing closes is during negotiation. The process for resolving them after financial closure can involve construction delays, defaults, and judicial determinations.

The role of transaction lawyers is important, but there is frequently a disconnect between the lawyers representing international financial sources and domestic lawyers representing recipients of such financing. They have vastly different experiences and begin to address financing of an individual project from the experience of the documentation which has been used for previous projects. Such documents from previous projects are likely to skew the expectations of the transactional lawyers, a consideration that can have substantial impact on complexity and negotiations regarding specific project financings.

5.3.2 Selection of the currency in which financing occurs

The currency in which water financing occurs is very important to the economics and financial feasibility of the project. Where projects are financed, secured or guaranteed in more than one currency, such as a local currency and an international currency, the differences in inflation and currency fluctuations make resolution of financial problems difficult. This is due to

[2] "Pooled financing" refers to the aggregation of individual project loans by the legal entity which raises the money for the individual project loans.
"Revolving funds" are pools in which the repayment of individual project loans is made available for new project loans.
"Gross pledges" are legal requirements that all of the revenue of the entity receiving the project loan be available for the repayment of the project loan.
"Compensating balances" refer to lender requirements that sufficient money for the repayment of project loans be maintained in bank accounts of the lender.
"First calls" are lender requirements that the first money received by the project borrower be used for the payment of debt service (interest and periodic repayment of principal of the loan).
"Conditional guarantees" are guarantees of loan repayment recognizing certain unknown conditions or circumstances.
"Reserves" are money or other legal commitments set aside to be used to cure defaults in obligations of the borrower.
"Reserve funds" are money set aside to be used to cure defaults in obligations of the borrower.

the changing nature of inflation, currency exchange rates and even to the barriers to the conversion of currency (see Martini, This volume).

Currency hedging mechanisms are available to mitigate such difficulties. They are expensive and are generally available only for short-term currency fluctuations. As such they are generally not used for projects with long-tenor financing.

The possibility of all financing being done in local currencies is generally not available where international entities are providing loans or credit support in international currency for long tenor, local currency financings. There may not even be sources of local currency for such financing, if there is no capital market and the national government has or is subject to currency controls.

Local currency is attractive and natural but poses its own challenges. In several countries, those supporting water and sanitation financing seek to do the financing in local currency and have developed mechanisms to accommodate hard currency techniques into their programs (van Oppenraaij et al., This volume). An important World Bank publication suggests that borrowing in hard currency tends to be more expensive than local currency financing due to the risks inherent in currency devaluations (World Bank/UNICEF, 2017).

5.3.3 Blended finance

The concept of "blended finance" is conceptually attractive because it addresses the aggregation of different sources of funding. Each specific source must be carefully considered because each brings its own specific constraints. Each of the participants in such aggregations has his own agenda and conditions that he imposes on his participation in financing a project. When a project has to meet each of the conditions of each participating entity in a blended finance situation, the financing increases in complexity and difficulty.

This generally means that the whole project must meet each of the participants' requirements. This is not necessarily so. Each participating entity could achieve its conditions for a part of the project if the whole project can be designed in parts. In such situations, each individual part may not need to meet all participants' conditions.

Large projects may be able to accommodate complexity in their structuring. Small projects are likely to be burdened by complexity. For example, a blended finance project might include participation from a donor, a soft-money investor, an institutional lender or buyer of securities backed by project loans, and an entity providing credit support for such private sector financing. In this situation, the donor may have one set of economic development interests that it will insist on being met by the project. The soft-money investor may have environmental or social interests that need to be served. The private sector institutional investor requires that the return on the investment be financially secure. Domestic private sector institutional investors such as local pension funds or insurance companies normally only seek credit ratings or financial security as defined under local law and in practice in the local capital market. The entity providing the guarantee will have other requirements, such as independent reserves for loans or participation or backing by the national government.

Projects may be disaggregated into parts, with each financing participant financing a physical part, rather than each participant financing a part of the whole. In the practice of some international financial institutions "mingled financing" and "parallel financing" are distinguished. In "mingled financing" situations, all financing elements and supervision are managed by a lead financier. The legal relationship must provide if different sources want their funds directed at certain categories of expenditures, such as social components, and must provide for disbursement practices. "Parallel financing" implies that each financing entity finances one part of the project but all parts are undertaken under the province of a joint policy objective and possibly other shared approaches, principles and work plan.

These approaches are rarely contemplated or used as they require that several "sub-projects" be defined, which is often difficult to do or not politically correct. Such approaches require the engineering design of the subprojects to be distinct. Such approaches may require separate documentation and approvals for each, may involve different govermental subdivisions or approvals, and may make oversight and management of the whole project difficult in practice.

5.3.4 Cost of financing

There are substantial differences in understanding the cost of financing between project implementers, governmental approval entities, cost accountants, investment bankers, and providers of financing. Non-financial professionals frequently speak of interest rate as the indicative cost of financing. Such understanding does not take into account the conditions that can be placed on financing, such as the gross pledge of all revenue, compensating balances, requirements that other financing services be bundled with project loans, collateral services, auditing, and information reporting, which all have an effect on the overall cost of financing. If there are private participants in a project financing, they must at least achieve an interest rate return equal to their weighted average of the cost of capital, which is the lender's cost of capital determined by market forces. It is not at the discretion of the financial source itself. It is also true that certain sources of financing provide technical assistance with project development or with financial management, valuable services that are not often quantified in consideration of financing costs.

5.3.5 Financing terms should match the useful life of the project

Banks generally do not provide long-tenor financing, although they frequently assert that short-term financing may be rolled over or refinanced with additional short-term loans at the end of the initial term. They rarely emphasize that the rate and other terms of the refinancing loan will be at the terms of the market at the time of the refinancing and may be substantially different than the terms of the original loan. This can get complicated where the initial loan does not fully amortize the principal of the loan.

For facilities that have relatively short-term useful life expectancy, such as computers or pumps or vehicles, short-term financing is usually available from banks and other commercial lenders.

Commercial short-term lending, such as bank loans, often require compensating balances, gross pledges of all revenue, other banking accounts to be maintained with the lending institutions, cash management services, and other elements of banking, which are a condition of the lending but which are not thought of as a cost of the lending.

Long-term financing is best for capital projects which are parts of water and sanitation systems. In urban water supply as well as wastewater management, there are many parts such as pipes, aqueducts, canals, buildings, tanks, and reservoirs have a 30–50 year useful life period. Within treatment facilities proper about half of the capital expenditures are for electromechanical parts that have 10–15 year useful lives. Many electric parts have a useful life shorter than that.

Long-term capital projects are usually a part of a whole water and sanitation system. As such they are in service as part of the municipal system. A long useful life suggests that repayment over the long life will spread out the payment obligation to users over the useful life, rather than burden users in the first few years to pay for the whole project.

Importantly, to the rate payer, long-term financing is cheaper annually than short-term financing. This has an important effect on tariffs or other sources of financing.

There is a mistaken assumption that the interest rate is the determinate of annual financing cost. Annual debt service payment amount rather than interest rate is determinate. Annual debt service payments are the periodic repayments of the sum of the periodic loan repayment and the applicable interest payment. Long-term financing, such as 12–20 year loans, will have a significantly lower annual debt service payments than short-term loans. This is because the portion of the debt being repaid is stretched. Irrespective of interest rate, long-term financing is cheaper to the users. For example, the annual debt service of a US$ 1,000,000 4-year loan at 10% is approximately US$ 300,000 per year while the same amount financing a project with a 20-year term at an interest rate of 15% is approximately US$ 160,000 per year.

Long-tenor financing of long-lived projects is also attractive because it is fairest to the system budgets and where applicable, to ratepayers. This means that the project sponsor retains more money each year that it can apply productively to other projects or other system needs. Over the useful life of the project, all the users pay for it rather than only the population that benefits from it during the early years. Such financing permits the maximum infrastructure that can be afforded by the system. It also encourages long-term planning and accounting for capital replacement by recognizing that depreciation and amortization are proper financial considerations in municipal budgeting.

Domestic pension funds and insurance companies are interested in long-tenor, local-currency, high-credit-standing domestic infrastructure investments. In developing countries' water sector, there has been little financing by institutional investors because the credit risks are perceived as substantial due to the political interference and uncertainty about tariffs and other revenue sources. Such projects cannot alone achieve the necessary credit quality without guarantees, reserves, or other credit enhancement. That credit enhancement involves additional parties to the financing transaction, increasing the complexity, and cost of the financing transaction. These credit enhancements are generally only suitable for very large projects such as those of the largest cities in developing countries.

For medium and smaller water service providers, such projects are also generally not large enough to attract institutional investors or to be able to absorb the transactional costs of the financings. Accordingly, though the largest cities in developing countries are able to access the capital markets or directly issue loans or securities for their large projects, small and medium-sized municipal systems have neither the scope nor the professional expertise to do so.

However, pooled financing approaches have worked where project loans are organized into pools or other aggregations appropriate under local laws. Such approaches have been shown to be a successful approach to financing in local domestic capital markets. The pool itself reduces the political risk of the individual loans. Credit support such as reserve funds or guarantees at the pool level can result in financial security sufficient for bonds or other securities to be attractive to the institutional investor.[3] The Kenya Pooled Water Fund described elsewhere in this book is intended to be structured in this manner.

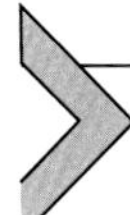

5.4 Domestic issues

5.4.1 Ownership of water

There is a general concept that "water is free." Certainly when an individual collects rainwater for personal use, it is "free," probably under most legal systems. Water is located in many places. Water is in the ground. Water is in rivers. Water is in lakes. Water is in reservoirs. Water is stored in tanks and other structures. Water is cleaned and reused. Similarly, water is used for many purposes. It is, by volume, primarily used for agriculture. It is used for industrial purposes. A relatively small amount is used for human consumption.

This concept influences all financial, political, and legal structuring in the cause of financing water and sanitation projects. Financing must resolve the distinction between the ownership of water and the infrastructure needed it deliver properly treated water to users. Political issues must be resolved especially where the location of the infrastructure spans more than one jurisdiction giving rise to competing claims to the delivery of water to users. Legal structuring must accommodate the domestic legal distinctions between those with claims to the water, those providing the infrastructure and those with responsibility for the payment of debt service for infrastructure projects.

[3] An early example is the Water and Sanitation Pooled Fund of the Tamil Nadu (India) Urban Infrastructure Financial Services Ltd. (www.tnuifsl.com) described in the World Bank's Case Studies in Blended Finance for Water and Sanitation, August 2016.

Some countries have legal systems in place that define who "owns" the water. Some do not. Water "ownership" is a very sensitive political topic in most countries and it gets easily wrapped up in the water privatization controversy. This is critical and central to the question of water security and water infrastructure financing as it directly controls the rights and obligations of a water "owner," the entity providing the infrastructure for its use, the financing entity, and the water user. For financing, there must be a predictable, secure, and sustained access to the water sufficient to support the revenue stream needed to repay the investment.

The legal systems often differ by historic location of the water, the location of transportation infrastructure for the water and the location of the use of the water; financiers should be aware of these legal issues. Water may not be "owned" by anyone, being common to all. Elsewhere, water may be "owned" by the national government. It may be owned by the municipal government. It may be owned by the first entity to have taken water from a particular resource, like a lake, a river or groundwater.

While water may not be "owned," many legal systems recognize the ownership of "water rights", the legal right to "use" the water. The right to use it may be based on the first user, or in a river by the upstream user. There are differences in rights to use surface water and groundwater and between agricultural users, industrial users, and domestic users. In many places without legal frameworks, there are likely to be cultural or historically defined water use rights. Financial entities, especially international entities, need to be very aware of such domestic legal isssues in the contemplation and design of financing transactions.

The domestic legal system in each country is the context for which all discussion of water financing, all legal decisions, and the development of water resources for human consumption and sanitation takes place. But in a large number of developing counties the domestic legal system does not have, or has only rudimentary legal provisions for the financing of municipal infrastructure generally and for water and sanitation infrastructure particularly.

Where water is "owned" or legally controlled by one entity and infrastructure improvements are sought to facilitate the use of it by another entity, there are often complex regulatory and intergovernmental processes that must be met. For example, where groundwater in a rural area is sought to be used by taking it from wellfields and transporting the water to an urban area, there are many potential conflicts. Agricultural users in the rural area may be disenfranchised as governmental jurisdictions access the resource and actively transfer the water to the urban water system for its population. The legal conflicts might arise from the agricultural users' historic rights to use the water and the taking of "their" groundwater because the taking of the groundwater may increase the depletion of the resource. That would pit rural and urban interests against each other. Related conflicts arise as the "taking" jurisdiction seeks to use the land over which the water is to be conveyed.

The urban government entity seeking the funding for the project might be required to show it "owns" or at least has an unalienable right to the water for which the project is being financed.

In many countries, these cases present novel and complex legal issues. They are further subject to intergovernmental political complexity. Each potential legal or customary rights holder's interest must be accounted for in the structuring of the financing arrangements. Financing decisions need to be made with the proper account of each of these potentially conflicting rights.

In many financing situations, there are legal voids or conflicts between those with legitimate claims to ownership of the water and to the regulation of its use. These differences must be taken into account early in the design of specific financing deals as they present distinct risks which will add complexity to design of the legal structure of the financing. If not properly dealt with during the structuring and documentation of the financing, these complexities may give rise to conflicts and potential defaults after financing.

It is most important that the international development aid participants properly understand the details of ownership in each situation for which financing or other assistance is contemplated.

5.4.2 Political and governmental interests

In any given country, many governmental agencies are likely to be involved in water or sanitation financing. Such agencies include the entity administrating the project and the system in which it occurs, the political subdivision in which the administrating agency resides, the national government agency responsible for water and infrastructure, the national Treasury which is likely to be required to recognize the financing, and the national legal department which is likely to be required to recognize the legality of the proposed transaction. Capital market regulators may be involved. These are in addition to the retained engineers, bankers, and lawyers participating in such decisions. Frequently, these entities will have divergent political or policy views about the topic in general and about their proper role.

The interests of domestic political leaders are different from the interests of international aid donors and participants. The political importance of water and sanitation projects to many local leaders often differs from that of the international community promoting SDG 6 objectives. Politicians are more willing to finance highly visible infrastructure projects but not as ready to finance projects that are inherently not "visible" to their electorate. Patronage and other personal interests of political leaders may influence the choice of infrastructure projects to finance.

The long time required for project development usually means that projects will not be completed until after an elected official's term in office. This influences the political attractiveness of projects to local officials.

In some countries, including India, Canada, several European countries and the United States, there are independent water and sanitation or general infrastructure finance authorities that enjoy a continuous existence, independent of the election cycle. They have designated responsibility for water and sanitation infrastructure planning, finance and operations. In many such situations, political influence is reduced and fiscal and physical planning can operate outside the strict political calendar. From the perspective of the financial provider, the credit stature of the independent entity is easier to ascertain and is thought to have a longer prospect than that of political subdivisions.

The existence of such institutions with continuing access to funding from the capital market and public sector sources encourages water service providers to more easily plan for long-term system needs. They can do so either with advance funding for planning and design of future projects or in the reasonable certainty that such funding will be available.[4]

5.4.3 Appropriation of projects and concepts

There are many examples where good projects are developed by interested entities only to have them appropriated by others. Appropriation occurs by local water service providers, domestic banks and other agencies, and by international entities.

Projects are often developed with the intention of having them financed by the government or by domestic or international financing entities, such as development banks or others. After investing in the preparation phase,

[4] The author was a Director of the New York State Environmental Facilities Corporation for many years. That entity, governmental but not directly political, gave municipalities in the State of New York the reasonable certainty that properly planned future projects would be financed, thus encouraging planning and management practices in the context of long term needs.

these projects can reach the stage of planning and engineering design where they are ready for financing and construction. Such projects then attract the attention of those who are willing to make a grant to finance a project. The water service provider may take the grant and reject its implicit or legal commitment to have the project financed by the entity that was subsidizing the planning and engineering. The project is appropriated from the entity that has sponsored it or developed it.

Because these projects are fully developed and ready for construction, they are attractive to grant providers who are desirous of doing projects immediately. They can be attractive alternatives to the original project sponsor because the project may be built now with grant money rather than by taking a loan for the project. This is attractive to the water service provider but disadvantages the entity paying for the project preparation which presumed it would be reimbursed for the costs of engineering and project preparation as part of the anticipated project loan. One effect of this appropriation is that it undermines the incentives to those financing project development, thus contributing to the lack of bankable projects in the water and sanitation sector.

It has also been observed that good financing programs that are developed and supported by international entities are appropriated by domestic entities. Domestic government agencies and financial entities such as banks and investment advisors have been known to take the project put forth by the international entity which has developed the concept. They have been known to appropriate the project even in the face of the original proponent having spent considerable time and money developing, designing, and legally organizing such a facility. Some appropriation is based on the perception that the governmental agency should be the entity to get the political credit for a program's success.

The effect of this appropriation is to freeze out the international entity in favor of the appropriating entity and to discourage financing entities from using their expertise and experience in defining and engineering developing countries' projects. That expertise has been shown to be a very important element in the appropriate design of water and sanitation projects and of financing entities.

5.4.4 Tribal, ethnic and religious issues

Tribal, ethnic and religious issues are not often considered by NGOs and multilateral agencies. These issues need to be better understood in the context of financing water projects. Water resources may be located in places that have different religious or tribal identification. Moving water from or between such places involves cultural or ethnic issues that are generally not considered by international entities. They also may not be considered by the international engineering and professional firms that are used to explore, identify and design projects that, irrespective of these issues, seem natural and needed. Sharing of such water resources may evoke generations of relationships between such groups.

Such cultural preferences also matter when it comes to basic questions like whether the prospective "beneficiaries" actually value the new service at all. In rural areas, the premise that "clean, non-odorous water" is an advance can be wrong where many people may actually prefer the pronounced taste of their traditional water source even if that source is deemed "unsatisfactory" by international standards.

Similarly, concepts of sanitation differ greatly between those held by international financial participants and citizens in many countries. In many countries, there are religious traditions that involve the handling and reuse of wastewater. These traditions can directly influence how treated wastewater may be disposed. Even if it is tertiary treated to "pure" water standards, it may be impossible to dispose of such treated wastewater for any use, even for agricultural use or to replenish groundwater. Such treated wastewater may not even be disposed into saline ocean waters as it can be deemed a "pollutant".

A decision of where to run new water lines, in which neighborhoods to connect to sewerage systems, and the extent of engineering that is needed for a specific project, may give rise to difficulties that are not routinely apparent to financing entities. There is a good opportunity for those advocating increased water and sanitation access to appreciate the local knowledge of such issues and the identification and history of resolution of such differences.

5.5 Conclusions

There are substantial differences between the views of those institutions advocating the SDG 6 standards and the realities in individual countries. These stem from cultural and political differences. They stem from a different understanding of technologies and systems for delivering and paying for the cost of such service. They stem from differences in understanding of needs. They stem from conditionalities that are part of the culture of the international institutions but are not part of the politics or culture of the intended "beneficiaries."

It is important to be aware of and address these differences. Such understanding is necessary to form the basis upon which suitable financing approaches can be adapted to specific contextual realities.

It is certainly the experience that successful financing solutions can be achieved in specific situations. For example, carefully designed and in accommodating places, such pooled private financing or bundled financing is possible. In India, the capital-market financing was successfully provided for the Water and Sanitation Pooled Fund of the Tamil Nadu Urban Infrastructure Financial Services Limited to fund a group of projects. In Colombia, the success of the issuances of Grupo Financiera de Infraestructura and the organization of the Kenya Pooled Water Fund show that the structure can be achieved and the projects identified but that the political and institutional acceptance takes time.

The need in developing countries for water and sanitation for all is real, current, and urgent. As such, the efforts of international institutions that seek to assist are needed and important. Whether such institutions need to focus on simpler projects or simpler financing techniques is for them to decide. Certainly easier projects and easier solutions can form the basis of successes which will lead to more complex projects, larger financing, and more successful financing.

Reference

World Bank/UNICEF (United Nations Children's' Fund), 2017. Sanitation and water for all: priority actions for sector financing. The World Bank, Washington, DC.

PART II

Investment Needs and Financing Challenges

CHAPTER SIX

Characterizing financing needs and financing capacities in different regions: a global perspective on water-related financing flows and drivers for investment needs

Richard Ashley[a] and Bruce Horton[b]
[a]Emeritus Professor University of Sheffield, Department of Civil and Structural Engineering, United Kingdom
[b]Environmental Policy Consulting and Stantec, United Kingdom

6.1 Introduction

This chapter considers the needs and patterns of investment in water supply and sanitation (WSS) across different regions of the globe. In Section 6.2, we briefly summarize the existing need for and trends in investing in WSS assets. In Section 6.3, current investment and financing flows are outlined, highlighting the extent to which these meet (or not) current needs. In Section 6.4 we consider the drivers for and scale of investment needs into the future, together with opportunities for innovation. Finally, in the concluding Section 6.5, we discuss what future investment is needed and how this might be delivered in the most sustainable way.

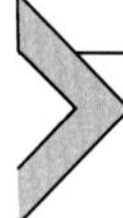

6.2 The need for water supply and sanitation assets and trends in investment

6.2.1 The economic case for water-related investment

In this millennium significant advances have been made in ensuring "universal" access to WSS[1] under the Millennium Development Goals (MDGs),

[1] Different publications refer variously to WSS (water supply and sanitation) or to WASH (water, sanitation and hygiene); often neither of these will include surface water drainage. Here the terms are used interchangeably, dependent on the source of the information.

Financing Investment in Water Security: Recent Developments and Perspectives.
DOI: https://doi.org/10.1016/B978-0-12-822847-0.00007-7

with 2.6 billion people gaining access to safe water and 2.3 billion to basic sanitation (e.g., Hutton and Varughese, 2016).

As WSS access has improved, global water use has increased by a factor of six over the last century, with demand increasing by some 1% per annum UN (2018) due mainly to population growth, economic development, and changing consumption patterns. It is expected that industrial and domestic demand for water will continue to increase much faster than in other sectors. UN (2021) highlights some of the key regional disparities, with levels of water stress increasing significantly in some regions. These changes are taking place simultaneously with the earth's rapidly changing climate, creating more extremes, resulting in increasing heat waves, droughts, and floods (e.g., Berndtsson et al., 2019). The resulting deficits in service provision, largely though not exclusively in the developing world, sit alongside the near-universal challenges of decay of existing assets and underfunding of the needed remediation (OECD, 2020a) approaching a crisis. "This underfunding of water infrastructure is putting many countries at worse risk in the COVID-19 crisis" (Our Future Water, 2020).

The overexploitation of the natural world, especially for irrigation needs, continues, with water shortages mainly in the southern hemisphere, and polluting discharges poorly controlled in most countries (e.g., Kern et al., 2020). The EU is perhaps an exemplar for improving natural waters, despite various limitations especially in regard to agriculture (EEA, 2020). Contemporary challenges, alongside climate change, include the need to manage microplastics and complex chemicals from human activity entering the natural environment (e.g., Eerkes-Medrano et al., 2015) and more recently the need to plan and manage shocks like pandemics (e.g., Cheshmehzangi, 2021).

Water crises are rated highly in global risk assessments (WEF, 2020), disproportionately impacting the less-well off, whilst wealthier countries often underinvest in WSS infrastructure (e.g., Libey et al., 2020). WSS costs are frequently seen as a burden rather than a privilege. Even in so-called developed countries like Ireland (an EU member state), the prolonged unwillingness by water users to pay for water and wastewater infrastructure has been linked to recurrent outbreaks of cryptosporidium infections amongst the population that have continued for more than two decades[2].

The economic importance of water and the impacts from shortages and excesses are evident across the world. The EU's water-dependent sectors

[2] Health Protection Surveillance Centre: https://www.hpsc.ie/az/gastroenteric/cryptosporidiosis/publications/.

generate EUR 3.4 trillion or 26% of the bloc's annual Gross Value Added (BLUE2, 2019). Sadoff et al. (2015) suggest that annual global economic losses related to water are USD 260 bn from inadequate WSS; USD 120 bn from urban property flood damages; USD 94 bn from water insecurity to existing irrigators; and losses in agriculture, health, income, and property could result in a decline by as much as 6% of GDP by 2050 and lead to sustained negative growth in some regions of the world.

Predicting trajectories of WSS development at the global scale is notoriously difficult, largely because of the number and diversity of drivers. Numerous estimates of future needs have been summarized by Winpenny (2015). Nonetheless, the estimated shortfall in investments needed to address the gaps in provision and the challenges faced, i.e. additionally, up to 2030, is some USD 1.7tn, around three times current investment levels (Hutton and Varughese, 2016; Hutton, this volume). Projections of global financing needs for water infrastructure in 2015 range from USD 6.7 tn by 2030 up to USD 22.6 tn by 2050 (e.g., Winpenny, 2015).

6.2.2 Types of water supply and sanitation systems

WSS investments depend on the context of how urban areas are laid out and managed, as land use, planning, living conditions, and lifestyles lie at the heart of what type, and how best to provide WSS services. Traditional centralized WSS (water collection, treatment, supply and wastewater collection, treatment and return to rivers) with "corporate" operators have developed to serve the form and layout of industrialized "western" urban areas and have protected human health for more than a century (Ashley et al., 2020), being "translated" to many developing countries as the "gold-standard".

WSS service provision and resilience in urban areas have been further challenged by the global Covid-19 pandemic (Ashley et al., 2020; Ashley, 2020). New forms of urbanism now aim to create more "human" walkable and accessible spaces with human-scaled design as part of the enhancement of human health (Iravani and Rao, 2020) having profound implications for WSS. This brings new opportunities for alternative WSS provision, managing rainfall better where it lands and opening up opportunities for local, "decentralized" WSS. These societal changes alter the potential for how WSS can be provided in both developed and developing cities (Kang et al., 2020; Cook and Makin, 2020).

Alternative provisions for WSS include a variety of different configurations that require more engagement with and acceptance by communities than the traditional demand-led and centralized approach. Table 6.1 provides some examples of types of WSS in use around the world. There is no one type of system that is "ideal." Each has specific challenges dependent on context, especially population demands and cultures, land accessibility, types of institution. Because of this, costs and appropriateness of systems vary and each situation needs to be assessed individually. However, by utilizing integrated systems not only for water but across a range of utilities and services and matching the type of WSS to the community needs, WSS can be both

Table 6.1 Types of water supply and sanitation systems, traditional, and examples of alternatives.

Type of system	Characteristics	Examples of where used	Potential for wider applicability and cost savings
Traditional			
Water resource abstraction, storage, treatment, distribution. Wastewater collection, conveyance, treatment, and return to rivers or alternatives	Considered the "normal" type of system and commonest in developed countries (for economies of scale). Centralized, large-scale, usually operated by single service provider. Water supply and wastewater systems separated. Increasingly energy is recovered from residual wastewater solids	Across the EU, e.g., London, Paris; also elsewhere, including New York, Sydney. These systems have been utilized for decades or even millennia in many cities. The cost of constructing them anew would now be prohibitive for most[a]	This system is that aspired to by many of those lacking basic services. Costly to construct, operate, and maintain, and hard to adjust to shifting circumstances
			Smart and digitalized operation could bring savings on current operational costs (Poch et al., 2020)

(continued on next page)

Table 6.1 Types of water supply and sanitation systems, traditional, and examples of alternatives—cont'd

Type of system	Characteristics	Examples of where used	Potential for wider applicability and cost savings
Fully decentralized			
Local rainfall use, recovery, recycling, reusing, and minimizing freshwater use	Maximized decentralized system. Point of use—rainfall-runoff, local treatment and waterless toilets. May have numerous service operators.	Typically only individual properties, or isolated communities; e.g., Swedish ecovillages (Magnusson, 2018). Systems may be part of cross-utility operations, comprising energy recovery and fertilizer production[b]	Where there is no existing water supply and sanitation, this is likely to be more cost-effective, but will require much more responsibility by users (Sapkota et al., 2018).
Centralized dual-reticulation			
As (1) above, but wastewater treated and returned into "used" supply	Part-decentralized.		
	Wastewater treated centrally and returned to supply system as alternative supply. Usually single service operator	Melbourne (different classes of water[c]), Singapore (NEWater[d]). Water reuse in this way is growing[e]	Similar to (1) above, although users need to ensure the correct supply type is used for purpose intended

[a] For example, in United Kingdom, the current replacement cost valuation of the companies' assets is estimated at about £224 billion. (https://www.ofwat.gov.uk/publications/rd-0410-regulatory-capital-values-2010-15/).
[b] https://run4life-project.eu/.
[c] https://www.melbournewater.com.au/water-data-and-education/water-facts-and-history/where-your-sewage-goes/producing-recycled-water.
[d] https://www.pub.gov.sg/watersupply/fournationaltaps/newater.
[e] https://www.waterworld.com/international/wastewater/article/16201136/water-reuse-its-time-to-wake-europes-sleeping-giant.

affordable and sustainable. For example, Wang (2020) shows how alternative configurations of WSS can be matched to demand–supply needs and energy use by catering for scale (e.g., catchment, precinct, neighborhood).

Increasingly, different approaches to WSS are being considered in terms of how best to integrate across the entire service and utility provision, and to increase the resilience of vital systems (e.g., Heinzlef and Serre, 2020). This is in response to the numerous shocks imposed over recent decades in terms of, e.g., flooding, droughts, earthquakes, and will also need to consider the implications of the Covid-19 pandemic and how living patterns could alter, with implications for WSS (e.g., Dobson et al., 2021). This new understanding of the need to integrate is helping to break down barriers between services and providers, encouraging greater cooperation and sharing of resources. For example, the human health benefits of access to green and blue spaces are well-known (e.g., Carmona, 2019). Such spaces are increasingly being used to manage surface water, utilizing, and coping with rainfall where it lands and bringing a myriad of added benefits (e.g., Ashley et al., 2018). As a result, what were once budgets only for parks or for managing stormwater, can now be pooled with health budgets to create better places, providing opportunities for WSS to manage flooding, promote and sustain human health and wellbeing (e.g., Lovell et al., 2020).

6.3 Current investment and financing flows for WASH

A summary of WASH budgets by region is shown in Table 6.2. In total, around USD 85 billion is budgeted for WASH in 57 countries (largely, though not exclusively, developing countries) representing a population of 4.4 billion and reporting through the UN GLAAS (Global Analysis and Assessment of Sanitation and Drinking-Water) initiative. This includes both capital and O&M (operation and maintenance) expenditures. On average, annual WASH expenditure is around USD 19 per capita and 0.42% of GDP, although there are large regional differences (see also Chapter 2.2 in this volume).

Breakdowns show a much lower average per capita WASH expenditure for countries in low- and lower-middle-income categories as compared with upper middle- and high-income countries, but a higher proportion of WASH expenditure as compared with GDP for low- and lower-middle-income countries. Annual government WASH budgets are increasing at an annual average rate of 4.9% as countries take on board the SDGs (WHO, 2017).

Table 6.2 Summary of WASH budgets, by SDG region.

SDG region	No. of countries	Aggregate WASH budget (USD mln)	WASH budget per capita (USD)	WASH budget as % of GDP
World	57	85,809	18.97	0.42
Sub-Saharan Africa (exc. South Africa)	16	2,078	3.88	0.27
Latin America & Caribbean	15	18,007	33.23	0.42
Eastern & South Eastern Asia (exc. China)	7	1,967	8.68	0.19
Central & Southern Asia	8	5,274	3.10	0.20
Europe & Western Asia	5	231	9.76	0.19
Oceania	4	185	100.67	3.36

Source: UN Water and WHO (2019).

However, some regions are not well represented in this regard. Further, even within regions, there are large disparities in investment (see also Chapter 2.6 in this volume). For example, OECD (2020a) documents current levels of expenditure for WSS in the European Union. Overall, baseline estimates point to an annual average expenditure of EUR 100 billion across the 28 EU member states, with the greatest share attributable to the longest-standing members of the EU-15 (Germany, France, United Kingdom, and Italy in particular). Around two-thirds of this investment goes toward water supply. This split is mirrored in developing countries. The UN notes that "despite lagging levels of sanitation coverage, non-household expenditures (government and external support) for sanitation have typically been half the level of expenditures for drinking-water services" (WHO, 2017).

Globally, there are three major sources of funds for WSS, known as the three Ts. They include:

- taxes from individuals and businesses (public funding),
- tariffs paid by households, businesses, and governments, and
- transfers, encompassing overseas aid, remittances, or loans.

Water finance and investment needs are heterogeneous. For this reason, OECD (2018) goes beyond this categorization, considering a typology of water security investments and a typology of financiers to help match specific investment needs with the most appropriate financing available. Potential classifiers for water investments include scale (from watershed to household); function (water supply, wastewater management, flood protection, etc.); and operating environment (ownership, governance, and regulation).

The GLAAS country survey (UN Water and WHO, 2019) included responses from 35 countries. This found that 66% of WASH funding originates from household sources, 22% Government, 9% repayable finance, and 3% external sources; representing USD 52 billion in annual expenditure for WASH. Government funding via taxes comprised the majority of non-household financed WASH expenditures.

Even though domestic and nondomestic customer charges are used to pay for much of the investment in most parts of the world, in over 50% of countries, household tariffs are insufficient to cover operation and maintenance costs, leading to an increase in disrepair and service failure (UN Water & WHO, 2017).

Private (repayable) funding (e.g., debt, equity) plays a smaller role globally, although this is more important where WSS services have been privatized, although in much of the world funding is financed through bond issuance and loans via (private) capital markets, repaid from the revenue of utilities through customer charges. The private sector is also increasingly involved in directly supporting water programs. For example, through the Sustainable Water Fund, companies based in the Netherlands and partner countries are involved in 22 water projects in 17 countries. These public-private partnerships have generated more than EUR 60 million of additional investment in WASH in developing countries.

Like investment, sources of funding can vary widely by country, even within regions (WHO, 2017). Some countries report major contributions from households (e.g., Brazil, Costa Rica, Serbia, Uruguay), others report more reliance on external aid (e.g., Kenya, Lesotho, Tajikistan), and a few countries report that national finance supports the majority of WASH expenditures (e.g., Bhutan, Fiji, Pakistan, Peru). Repayable sources of finance were considered separately to highlight the relative level of borrowing compared to other revenue sources. Repayable sources of finance are significant in a number of countries including Nepal, Paraguay, South Africa, and Uruguay.

International flows of finance are small in comparison with other sources of finance. However, they can be important in supporting the extension or enhancement of WSS service provision in certain regions and countries and may even be the largest source of WASH financing (WHO, 2017).

Globally, over USD 11 billion in ODA (official development assistance) grants and loans (USD 7.4 billion), non-concessional loans/credits (USD 3.4 billion), and other funds (over USD 300 million) from high-income countries (bilateral aid, multilateral development banks, NGOs, and private foundations) were disbursed per annum (i.e. spent) on water and sanitation in 2015 (WHO, 2017). Such financing allows governments and utility borrowers to distribute payments for capital infrastructure investment over time and finance repayment through future taxes, fees and tariff revenue.

International flows of finance can also play an important role in supporting more innovative approaches to service provision, as they are increasingly tied to delivery of more sustainable solutions, and to projects with a greater proportion of operation and maintenance (O&M) costs. For example, nature-based solutions (NbS) are cited by OECD (2020a) as a means of stimulating innovation and minimizing future financing needs. However, globally, only 5% of global investments in water resources are directed at NbS[3].

The amount of future funding committed through aid is declining. While international aid spending on WASH increased from USD 6.3 billion to USD 7.4 billion between 2012 and 2015, future commitments declined from USD 10.4 billion to USD 8.2 billion in the same period (WHO, 2017). This may stem from a gradual shift from sector-specific funding toward a preference for general budget support.

6.4 Drivers for and scale of future investment needs

There are several drivers for investment needs in different parts of the world, including economic development. These are summarized with examples in Table 6.3.

The drivers in Table 6.3 cover a wide range of topics and vary in terms of relevance to future needs and funding options for WSS. There are clear regional groupings: (1) where there are already significant WSS, such as for much of the EU, and (2) regions where WSS provision is partial, poor, or lacking. Within these groupings there are areas where significant investments are needed to maintain or provide essential WSS for the next few decades. The other major distinction is between rural and urban areas. The migration of rural populations into urban areas has been a feature of

[3] https://www.watersciencepolicy.com/2020/05/22/is-green-the-new-grey-if-not-why-not/.

Table 6.3 Main drivers for water supply and sanitation investment needs in various parts of the world (not ranked, and with overlaps in categories).

Driver	Relevance/pressure
	A. ENVIRONMENTAL FACTORS
1. Climate change (impacts relevant to WSS)	Water and climate change are linked (e.g., UN, 2020). Although the pressures vary globally, temperature increase will lead to greater variability and extremes of weather-related phenomena in all parts of the world. Ashley and Cashman, (2006) estimated that climate change would potentially add some 33% to the cost of water services provision in OECD countries by 2025. Water supply and sanitation will need to be adapted to cope with the changes in climate as these are both impactors on the drivers of climate change and recipients of the impacts. An overview of the potential impacts on water systems in developed western nations is given by U.S. Department of the Interior Bureau of Reclamation (2013) and in developing countries by UN (2020)
a. Heat stress	Many nations will require more irrigation, air conditioning and firefighting resources. Water supply and sanitation will be required for air conditioning waste, where the need for AC is increasing due to rising mortality pressures. See also 1(h)
b. Flooding	Global pressures will require added investment to manage all forms of flooding. Some countries like Bangladesh, are especially vulnerable to increased flows and sea level rise, as below (1d). As the rural poor rely more on agriculture, the impacts tend to be more profound (UN, 2020)
c. Droughts	Many regions of the world are beginning to require ever more resources and inter-regional water transfer. Increased floods and droughts lead to soil erosion loss, desertification and ecological impacts. By 2025, some 1.8bn people are expected to be living in water scarce regions and two-thirds of the world population could be under water stress (Our Future Water, 2020). Irrigation investments are estimated at some USD 960bn to expand and improve irrigation up to 2050 in 93 developing countries, compared with 2005–2007 investments (UN, 2020)
d. Sea level rise	Implicated in increasing flood risk as in (1b) but also affects land-inward systems. Globally changing at different rates, compromising some river outfalls and necessitating more pumping assets. Also contaminating groundwater that is a major resource. Malta for example, experiences flash flooding[a] but has a water supply compromised by saltwater intrusion and as a consequence has had to resort to wastewater reuse (60%)[b]

(continued on next page)

Table 6.3 Main drivers for water supply and sanitation investment needs in various parts of the world (not ranked, and with overlaps in categories)—cont'd

Driver	Relevance/pressure
e. Water quality, ecology and agriculture	Water quality will be affected by higher water temperatures, reduced dissolved oxygen and self-purifying capacity of freshwater bodies (i.e. reduced resilience). Water pollution and pathogenic contamination caused by flooding or by higher pollutant concentrations during drought are also possible (UN, 2020). In urban areas this is manifest through stormwater runoff as increased diffuse pollution due to more intense rainfall. Changes in species, damage to agricultural land and desertification (1c), may require responsive changes in water supply and sanitation. The most severe threat is to agriculture and food production especially for the poorest communities
f. Population displacement	Each of the above will require additional water supply and sanitation in resettlement areas. These may be in already urbanized areas or in temporary displacement camps. Migrations may occur within a country or transnationally. Numbers and rates of migration are difficult to predict (UN, 2020), but these will be of the orders of up to 1bn worldwide (Cattaneo and Peri, 2015)
g. Economic development	Globally, climate change will affect this in different ways depending on the impact type above. For developed countries climate change is unlikely to adversely affect economic growth, but lead to new business opportunities from green growth initiatives, which ideally should be shared with the poorer countries. Changes in water supply and sanitation types of provision may bring significant opportunities as expected by the Netherlands government in funding their wastewater resource recovery program, ERMF[c], with estimated overall annual revenues from the ERMF program as some EUR 14 per cap/year by 2030.
h. Responses	Climate action failure is seen to be the greatest global risk today in terms of both likelihood and impact (WEF, 2020). Changes to the way in which water supply and sanitation are provided are aiming to reduce carbon use (mitigation) (e.g., Liu et al., 2020) and implement circularity in provision and service operation, mainly impacting how services are provided in the developed world (van Leeuwen et al., 2018). There is also a drive to ensure resilience of service provision, with cross-utility resilience being an aspiration, at the risk of greater costs, depending on the sectors (e.g., Winpenny, 2015). A multiple functioning perspective is being adopted across water supply and sanitation, with more nature-based systems being used and promoted as almost a "magic bullet" by some[d]. Specific responses include greening urban areas to address the growing heat island effects (e.g., Santamouris, 2020). An overview of impacts and potential responses specifically regarding water supply and sanitation in developed countries is given by Hulsmann et al. (2015).

(continued on next page)

Table 6.3 Main drivers for water supply and sanitation investment needs in various parts of the world (not ranked, and with overlaps in categories)—cont'd

Driver	Relevance/pressure
2. Environmental drivers (other)	Increasingly there are conflicting needs of water for society against those of environmental systems. Water supply and sanitation has to balance environmental flow needs against abstraction to meet demand in order to support natural systems. Recycling may provide some of these needs. In the Murray–Darling basin in Australia, water is one of the most valuable assets owned by licensed irrigators. The disbursement between human users and abstractors, and in places, commodification of the water rights jeopardizes entitlements for environmental water "rights" (e.g., Seidl et al. 2020).
	B. HUMAN FACTORS
3. Economic capacity and growth	Growth can lead to added demand to water supply and sanitation, putting stress on existing systems and or requiring new assets. Conversely lack of water can result in economic recession, and is highlighted as a very significant risk in future (UN, 2020). May also drive innovation and alternative water services (Table 6.1, and (13) in Table 6.3) especially in developed countries. In developing countries the main pressure is on the economic capacity to fund and provide WASH and then enhanced water supply and sanitation. Everywhere there is a need to find a way to balance growth with environmental protection, particularly as the world bounces back from the Covid-19 pandemic, when there could be a tendency to promote growth at all costs (e.g., OECD, 2020b).
4. Commercial and industrial activity	Related to (3, 5 and 7 especially), this influences the demand for water and also the potential for impacts from industrial emissions. It will also be the means of fulfilling the demand for water supply and sanitation. There will be a need to ensure appropriate environmental protection and also control of local flood risk. Increasingly aiming at carbon reduction. Worldwide the greatest demands are from agricultural practices. Industrial decline and shifts from one dominant industry to another can stress water supply and sanitation, as is the continuing creation and use of new materials, including micropollutants and plastics. The dangers from pharmaceuticals in the environment have been long known (e.g., Chitescu et al., 2016), but the realization that the entire planet is rife with plastic particles of varying sizes is only recent (e.g., EC, 2018). Dealing with this is adding considerable costs to every aspect of water supply and sanitation. For example, Added costs in Switzerland are estimated at some EUR 2.8bn (Bieber et al., 2018).

(continued on next page)

Table 6.3 Main drivers for water supply and sanitation investment needs in various parts of the world (not ranked, and with overlaps in categories)—cont'd

Driver	Relevance/pressure
5. Demographics	Globally, there are regions with population growth, others with decline. Linked with urbanization (6) this may expand, densify or denude population centers and or the land occupied. For example, the 'shrinking city' phenomenon has been known for some decades, especially in Japan (e.g., Jiang et al., 2020), but occurs for various reasons, even in Germany (Flyn, 2021). Existing and required water supply and sanitation will be impacted by either increased demand or by abandonment and redundancy. This will also be influenced by urbanism, i.e. how people live (7). In general, more people usually means more demand for water supply and sanitation and fewer people lower demand. However, lifestyles also influence the demand by using more or less water.
6. Urbanisation and creep	Often occurs as a consequence of growth in (5), demographic movement into urban areas, densification, sealing of surfaces and suburban spread. These all exert new demands on water supply and sanitation either overloading existing assets or requiring new assets.
7. Lifestyles and security	How communities choose or are forced to live is one of the primary drivers of water supply and sanitation needs. In developed cities, for example, trends for fewer persons per household impact on water use, with higher per capita usage. Elsewhere multi-person households require lower supplies. What types of water supply and sanitation are acceptable are also a key driver. Post Covid-19 developed cities may have fewer people in centers, fewer commuters and more space to unseal soil surfaces to provide new green spaces, some of which will require irrigation, but also flood storage. For many, lifestyle is fixed by poverty or as refugees and there are many informal settlements (e.g., in Lagos >50% live this way, Oberg et al. 2020) and worldwide some 1bn live in urban slums, expected to rise to 2bn by 2030 (Ross et al., 2020). These variations in living require appropriate types of water supply and sanitation. Significant in this driver category is affordability of water supply and sanitation. Although dependent on relative wealth, lifestyle and attitudes to payments for services and tolerance of risk are among the most significant.

(*continued on next page*)

Table 6.3 Main drivers for water supply and sanitation investment needs in various parts of the world (not ranked, and with overlaps in categories)—cont'd

Driver	Relevance/pressure
	C. SERVICE PROVISION
8. Lack of services (impact on water supply and sanitation)	Primarily in the developing parts of the world. Hutton and Varughese (2016) provide estimates of need. Although aspirations are in many instances to emulate developed world water supply and sanitation, this is not going to deliver fast enough nor at an affordable cost (Oberg et al., 2020). Public-private initiatives can be useful in funding water supply and sanitation investments. Winpenny (2015) report that the Moroccan Government signed a 30-year concession contract with LYDEC in 1997 to improve the water supply and wastewater services, and provide infrastructure for stormwater and flood defence in the Greater Casablanca region, which has been successful; and in the last 11 years the contributions from developers to the total investment spending has gradually increased to more than 50% of the total.
a. Equity	The SDGs set environmental and social targets, targeting equity for the world's citizens. Promoting universal access to fundamental services, including water supply and sanitation. WASH form a central part of the SDG6 goal and assist in reducing health risks for SDG 3. Questions of equity in regard to all aspects of human life and water supply and sanitation related inequities are challenging some of the ongoing assumptions about not only access to services but the "secondary" effects on, e.g., mental health (e.g., Tomaz et al., 2020). With the increasing emphasis on greening urban spaces, virtually everywhere, in recognition of the health and other benefits, it is apparent that there are inequities in for example, the increase in property values that occurs and also the gentrification of former less attractive areas of cities.
Safe and secure water supply	May require new services, even in some developed countries, like Romania, but primarily in less-developed countries, or countries where this is not yet prioritized. Also in temporary displacement areas (1f). By 2030, some 2,278mln and 4,531mln respectively will need basic and safely managed water (Hutton and Varughese, 2016).
Sanitation	More than half the world's population currently live without safe sanitation[e]. Hutton and Varughese, (2016) estimate that by 2030, some 3448mln and 5309mln, respectively, will need basic and safe sanitation.

(continued on next page)

Table 6.3 Main drivers for water supply and sanitation investment needs in various parts of the world (not ranked, and with overlaps in categories)—cont'd

Driver	Relevance/pressure
Asset management	For many developing regions, the concepts of formalized asset management as developed by, e.g., ISO 55000, are difficult to utilize. Yet, for optimality, the risk-based approaches enshrined in these methodologies are likely to provide the most efficient and effective functioning of services. Due to the very limited applications it is not possible to define as yet, the potential value of formalized AM (9). At the least, formalized performance indicator systems or frameworks are needed in order the verify and identify where interventions are required (e.g., Alegre et al., 2006).
9. Maintenance of existing water supply and sanitation	In many developing countries even where there are water supply and sanitation assets, maintenance and management may be poor. For example, in Ghana, the "fix-on-failure" approach means that service provision is frequently interrupted for long periods (e.g., Kumasi et al., 2019). In the first industrializing countries many water supply and sanitation are more than a century old. Whilst services have broadly been maintained, there are many assets in a poor condition requiring immediate action. Lack of proper investment in maintaining existing assets is a problem worldwide. Even in the first industrial revolution countries, investment in capital maintenance is woefully below what will be needed. For example, UK investment in sewer maintenance needs to be increased four-fold by 2030 and six-fold by 2070 (UKWIR, 2017).
10. Regulatory processes	In countries where legal and institutional systems define standards of performance, regulations are used to set these and help in assessing outcomes. Monitoring of performance outcomes (8d) helps ensure security of water supply and sanitation. Confirmation of security of service requires well established and robust data collection and handling. Most countries with extensive assets have standards set in regulations. The EU has numerous water supply and sanitation related standards enshrined in Directives, that are among the most important drivers for member states' water supply and sanitation and a diverse variety of regulatory processes for ensuring delivery (e.g., EurEau, 2020). The delivery of services via institutions is also typically monitored (e.g., Akimov and Simshauser, 2020). Many countries lacking, or poorly served by water supply and sanitation also have weak regulatory and institutional systems. For some, there are strong regulations, but poor-quality monitoring of performance may mean that there is no confirmation of effective outcomes (e.g., Hukka and Katko, 2015).

(continued on next page)

Table 6.3 Main drivers for water supply and sanitation investment needs in various parts of the world (not ranked, and with overlaps in categories)—cont'd

Driver	Relevance/pressure
11. Professional and technical capacity	Increasingly there are demands for competent, trained and flexible professionals. The traditional segregation of knowledge and skills into "engineers" and others (silos) will not deliver the required water supply and sanitation of the future. The cross-functioning of utilities, services and infrastructure and the various nexus' extant in this, require greater breadth of knowledge and skills and cooperative or partner working. In this, meaningful engagement with communities as well as decision makers is increasingly important. Professional attitudes are often what prevent innovation by lack of willingness to do what is needed. Willets et al. (2020) describe how communities in Indonesia are part of the co-management of sanitation systems. Also the problems of setting and collecting user fees in poor neighborhoods.
	D. OTHER FUTURE CONSIDERATIONS
12. Shocks	Traditional impact preparation is influenced by adoption of the Sendai Framework for Disaster Risk Reduction 2015–2030 (UN, 2020) and most developed and many developing countries have disaster plans. Floods, droughts, wildfires hurricanes, earthquakes etc. may not be anthropogenic in origin, but are increasing worldwide, with the exception of the latter, which is the least predictable. The statistical stationarity of phenomena can no longer be presumed (e.g., Milly et al., 2008). The Covid-19 pandemic illustrated how poorly prepared most parts of the world have been to cope with even predictable shocks, with all services affected. The poorest have been the most affected (e.g., Cook and Makin, 2020; Skorka et al., 2020). The characteristics of the population, high mobility, but not population density, aided the global spread of the virus, plus, geography (via climate). In developed countries, due to more home working, associated behavior changes and lower mobility, domestic water use has increased, partly offset by reductions in commercial water use. There is a need to distinguish between the needs for water supply and sanitation during the pandemic and what the societal and other demands will be once this crisis is over (e.g., Cotella and Bravarone, 2020).

(continued on next page)

Table 6.3 Main drivers for water supply and sanitation investment needs in various parts of the world (not ranked, and with overlaps in categories)—cont'd

Driver	Relevance/pressure
13. Innovation and digitization[f]	Advances in technologies are continuous and many applicable to water supply and sanitation. Current examples include digitization with intelligent operations, developing circular water and wastewater systems via novel treatments, low cost sanitation systems, integrated and multifunctional water supply and sanitation, operated in conjunction with other services and land uses. Understandings about good governance and influencing behaviors and therefore demands are changing the way in which water supply and sanitation are being delivered. Many cost issues and uncertainties about managing water supply and sanitation can be addressed by data acquisition predictive analytics, to detect leaks, forecast usage, reduce costs. In a US survey (Black and Veatch, 2020) some 15% of respondents reported having a robust, fully integrated approach to data capture and usage. There has been an emergence of so-called "nature-based" systems, expected to be utilizable in every sphere of water supply and sanitation (UN, 2018).

[a] http://floodlist.com/europe/malta-storm-floods-february-2018.
[b] https://www.waterworld.com/international/wastewater/article/16201136/water-reuse-its-time-to-wake-europes-sleeping-giant.
[c] https://www.efgf.nl/english (van Leeuwen et al., 2018).
[d] https://naturalcapitalcoalition.org/nature-based-solutions-for-people-planet-prosperity/.
[e] https://www.un.org/en/observances/toilet-day.
[f] See also the discussion in Section 6.4.3 on how innovation may help to offset the rising costs of water supply and sanitation).

human settlements for decades (Karaman et al., 2020) and the expanding urban living exerts growing pressures on existing WSS. However, the Covid-19 pandemic has reduced the rate of migration (Kuriakose, 2020) and in some cases reversed population movements away from dense urban areas, especially in the wealthier countries. How the earlier trends for depopulation and city abandonment in many parts of the world will overlap with this is uncertain (Flyn, 2021).

Some studies consider only specific aspects of WSS, for example, the main water asset SDGs focus (SDG 6) on WASH (drinking water, sanitation, and hygiene), ignoring urban drainage other than as a co-benefit. Hence pluvial flooding is not considered explicitly. Other studies (e.g., OECD, 2020a), include urban drainage aspects of WSS. Conversely, the hygiene costs included in estimates for the SDG are not required and therefore not included in many higher and middle-income country figures where health and hygiene practices are normal for populations.

Across the global water sector, funding remains largely out of step with the requirements to invest adequately. In some instances, capital investment is lacking, but virtually everywhere, revenue and investment in maintaining assets once in place are significantly lower than required, even in the apparently best-run countries, and for many there is no consideration of end-of-life replacement (Libey et al., 2020). Hukka and Katko (2015) for example show how there is a "vicious circle" of unviable water services pricing due to a lack of understanding of the need or a lack of political will. One or other, or both of these apply in virtually every country worldwide.

6.4.1 Drivers for water supply and sanitation investment needs with existing extensive service provision

These regions include many OECD countries, especially in Europe, North America, and Australasia. Effective and near-universal WSS has been extant in many of these areas since the first industrial revolution, with more recent implementation in, e.g., Australasia, following the "western traditional model" of WSS (Brown et al., 2008). Extensive and valuable assets are in place for water supply, wastewater management, and urban runoff control. They are operated variously by private (e.g., United Kingdom), municipal (e.g., Germany) or through hybrid private-public (e.g., France, United States) arrangements (Hukka and Katko, 2015; EurEau, 2020). The governance and institutional or service arrangements, which deliver and manage the WSS assets vary considerably across these regions. However, they have in common a tendency to utilize traditional WSS (Section 6.2.2) and to be controlled by regulatory frameworks that aim firstly at maintaining public health and welfare and secondly at ensuring WSS services are affordable to the customers. They may also address polluting discharges into receiving water bodies and/or flood risk management.

In these regions, the main drivers are outlined in Table 6.3. Although the SDGs are applicable, for many countries that are already serviced with WSS, the SDG6 which specifically addresses WSS is typically considered to be already dealt with and therefore largely overlooked in plans[4], which limits the vision and initiatives to integrate WSS together with the other SDG components. The overarching drivers are managing risk to population health, and increasingly, resilience of services. There is clear evidence of the beneficial value of investments in clean WSS, traceable back to historical

[4] For example, Bristol One City Plan only briefly mentions reducing leakage and water use: https://www.bristolonecity.com/wp-content/uploads/2020/01/One-City-Plan_2020.pdf.

typhoid transmission. For example, typhoid mortality in US cities between 1889 and1931 was reduced by 5% per USD 1 per capita invested in water supplies and 6% per USD 1 per capita investment in sanitation services at the time (Phillips et al., 2020; or USD 16.13 each in 2017 prices).

In the 20th and even 21st century, sustainability of WSS is supposedly the aim, but this is proving too difficult a target and has been largely downgraded to simpler targets of robustness and resilience (Ashley et al., 2020) avoiding the complexity of "sustainability," notwithstanding the avowed SDGs. This brings changing pressures and potential impacts that could potentially compromise effective WSS and associated services.

Asset deterioration is widespread across the well-serviced regions of the world. For example, 80% of professionals surveyed in the United States believe asset deterioration is the greatest current challenge. Some 50% of Philadelphia's water mains are at least 90 years old, with some mains predating the Civil War, while the average water main's age in Baltimore is 75 years (Black and Veatch, 2020). In the next 25 years, it will cost an estimated USD1 tn to replace the necessary underground piped infrastructure (ASCE, 2017). Only a minority of countries have a realistic replacement target. As a rule, COWI (2010) recommends a 2% replacement rate at least for sewers, which could also apply to water supply network assets. Despite the challenges related to asset management, with uncertainties around maintenance, rehabilitation, and replacement economics. This topic is still in its infancy and ascertaining when, where and how it is best economically to intervene in an asset to maintain service is the subject of debate (e.g., Cabral et al., 2019).

In Europe, only Slovakia and Germany are anywhere near the 2% replacement rate. Proactive asset management strategies in these countries are mainly based on the utility's employees' experience: their intuition and tacit knowledge of the system play an essential role in decision-making (Tscheikner-Gratl et al., 2019).

Water losses and NRW (nonrevenue water, Fluence, 2019) are mainly due to leakages rather than illegal withdrawals (e.g., Press and Roberson, 2016), caused by poor maintenance, lack of renewals, or inefficient operation. As an illustration, US Water Alliance (2021) states that between 2012 and 2018 water main bursts rose by 27% in the United States; a break occurring every 2 min. Estimates in 2019 for the value of this lost treated water were USD 7.6 bn. By 2039, estimates of the annual costs of asset deterioration to US households, and water and wastewater failures will be seven times greater than they are today, rising from USD 2 bn in 2019 to USD14 bn.

Some countries have successfully tackled losses, Denmark for example, claims to have eliminated NRW, with overall losses around some 5% and per capita consumption falling from 171 to 104 liters per day (Tscheikner-Gratl et al., 2019). Alternatives to the traditional WSS arrangements that decentralize provision (Table 6.1) are usually much less prone to problems of NRW.

6.4.2 Drivers for investment in water supply and sanitation where there is limited service provision

In the developing world, some 4.5bn people defecate in the open, or have unsafe sanitation which is responsible for more than 1200 under 5-year-old deaths per day. This is the case for even middle-income countries such as India. The main drivers potentially influencing the needs for WSS in these regions are outlined in Table 6.3. Although the other drivers relevant to developed economies are also of relevance, these drivers are not as significant where there is currently a need for basic WSS provision. For example, innovation and digitization (Driver 13) are having a marginal effect on universal WSS provision, albeit numerous novel WSS are emerging, including alternative toilets, with development supported by donor organizations such as the Gates Foundation[5].

Recent analyses of the apparent changes in access to WASH in 2000–2017 (Reiner, 2020) show the wide variation in sub-Saharan Africa, for example, although there have been transitions from the most basic to improved facilities and fewer people are defecating in the open. The conclusion is that the SDGs will only be achievable through targeted sub-national initiatives.

Globally, the decline in capital investment is compensated by the steady increase in O&M funding (see also Chapter 2.2 in this volume). Overall, current financing levels were claimed to be able to cover the capital costs of providing a universal basic service based on 2015 data. It was estimated to require some USD 28.4bn per annum to provide "universal" access to WSS over the period from 2015 to 2030. The investment underway was claimed to be in line with this in 2016 (Hutton and Varughese, 2016), with the gap between ODA funds committed and those disbursed for the water sector being a relatively small USD 100mln. But by 2019, this gap had widened to USD 2.6 billion (UN, 2021). The current trajectory to deliver SDG6 is inadequate with a fourfold need to increase the rate of progress in WSS (UN, 2021).

[5] https://www.gatesfoundation.org/what-we-do/global-growth-and-opportunity/water-sanitation-and-hygiene.

It is not only the scale of investments needed, resources need to be effectively targeted to achieve the WSS SDG targets. Effecting the required provision, and in some cases, providing the required education, accessing donors, ensuring that institutions and governance are effective, and that O&M can be sustained, is challenging. Many of the populations of the countries to be provisioned are relatively poor and finding funding for both capital investment and the recurrent expenditures is difficult. This is even in the face of reducing the costs to individuals of clean water; which for many is currently purchased expensively from mobile service providers or in bottles.

Even if the SDG6 goals are met initially, many countries lack regulatory systems or the ability to monitor and verify continued functioning (e.g., UN Water and WHO, 2019). There are also revenue-raising problems, not least due to the likely high levels of income loss due to NRW, a problem in many countries with modest levels of water service provision[6]. This should not be confused with leakage losses, although these losses may also be included in the NRW budget where data are lacking.

6.4.3 Innovation and how this might offset rising costs

Even with efficiency improvements alone, it is estimated that possible global savings of 60% (or USD 1 tn/year) are possible in the major infrastructure sectors of transport, power, telecoms, and water. WSS investments could account for several hundred billion of this (Winpenny, 2015). Integrated operation may offer even more efficiency savings. Sewers and water mains are only one of many urban infrastructures. Gas distribution networks, district heating, electricity, and data communication cables are extant in urban areas. Roads, parking spaces and urban green areas are also infrastructures occupying surfaces, interacting with stormwater management systems (Tscheikner-Gratl et al., 2019).

Van Riel et al. (2017) show that the various utility players usually operate independently, using traditional decision making, although when working with others across systems, are often willing to change practice. There is an urgent need for effective transdisciplinary and trans-utility working (e.g., Leach and Rogers, 2020). As an example of co-working, Osman (2016) presents a case study for a major infrastructure corridor in Cairo, Egypt, where upgrades to road, sewer, and water mains were all

[6] Globally, on average 34% of pumped water ends up as NRW, according to the International Energy Agency. https://www.wwdmag.com/how/four-steps-reducing-non-revenue-water.

included in an optimization model for planning collectively. Cost savings of 30% were found possible, and at the same time, the associated risks of synchronizing the utility enhancements were defined to help decision-makers. The methodology was extendable to include other utilities such as electricity and telecommunications.

There are several innovation initiatives related to WSS, each of which may have the potential for cost savings at least in the medium to longer term. The awareness of the substantial potential market for WSS-related technologies (see earlier) has led to the acceleration of patents and relevant innovations, with the USA and China leading this with water demand and supply technologies at the forefront (Leflaive et al., 2020). Examples include:

- Different types of WSS, including low-cost sanitation and water reuse (e.g., Black and Veatch, 2020) and valorization of residual matter from conventional treatment processes (e.g., for recovery of nitrogen, phosphorus, and biogas) that have the potential to reduce capital or O&M costs for new installations.
- Digitization in association with a wide range of computational models, can deliver smarter systems, which have the potential to optimize performance of existing assets (e.g., Black and Veatch, 2020).
- Multifunctional and multiutility systems, that is, across services, such as running communication cables through water pipes—however, multifunctional systems bring added risks, where they are codependent and may add costs (e.g., Winpenny, 2015).
- Specific technologies are evolving continually and as well as refinements of the traditional water and wastewater treatment processes in use, aimed at efficiencies and carbon reductions, more attention is being given to local technologies and reuse processes (e.g., Bui et al., 2018).
- Although not necessarily an innovation, rather a change in practice, such as bringing WSS infrastructure on to the surface utilizing what is now increasingly termed NbS (Nature based) systems, rather than being buried, provides more opportunity for added functionality, potentially supporting environmental, ecological and ecosystem services as well as human WSS needs (e.g., UN, 2018).

6.5 Conclusions

Due to the crucial role that WSS plays in supporting people, economies, and the environment, significant investment is required. There is clear evidence that current investment is insufficient to meet existing needs, primarily in poorer and developing regions, but also, to a lesser extent and in different ways, in wealthier regions. In all regions, investment is unevenly distributed, with a focus on capital investment and on core WSS services, with operational measures and other aspects of the water cycle often overlooked.

Given the range of drivers impacting on WSS, investment needs are likely to increase in scale and scope. However, traditional WSS systems are proving increasingly unaffordable even in wealthier developed nations, with large costs associated with maintaining existing systems and high carbon impacts for their expansion. In general, current WSS investments are insufficient to maintain services into the future and there are signs of increasing unaffordability burdens that citizens are unable or unwilling to shoulder, even in affluent countries (see Chapters 2.2 and 2.6 in this volume for more detailed discussions). In most developing countries there has been a steady expansion of WASH and in some, transitions to safer WSS, despite the high costs involved. It remains to be seen whether or not these can be maintained in the longer term as donor organizations no longer provide the support and citizens are expected to pay for these services. Europe is a case in point: over the last two decades, extension of WSS networks in countries like Estonia and Lithuania was financed as being essential with EU support. Hence new national business models and financing schemes are required to operate, maintain and renew these new assets.

Innovative solutions and new technologies will be required to address these investment needs without impacting adversely on affordability. There are signs that this is happening, including the increased use of multipurpose infrastructure and nature-based solutions that can deliver more sustainable outcomes with multiple benefits, generally at lower financial and environmental cost. Such solutions need to be scaled up and accelerated, requiring adequate policy frameworks and enabling environments.

New funding structures and sources are emerging and have the potential to support the innovation and investment required. Domestic funding remains critical, but transnational flows of finance are important in some areas, particularly where service provision needs to be extended or improved quickly. Unlocking the finance needed to deliver investment will be key.

References

Akimov A., Simshauser P., 2020. 2018–02: Performance measurement in Australian water utilities: Current state and future directions (Working paper). Griffith University, Queensland. https://research-repository.griffith.edu.au/handle/10072/390492.

Alegre, H., Baptista, J.M., Cabrera, E., Cubillo, F., Duarte, P., Hirner, W., Merkel, W., Parena, R., 2006. Performance Indicators for Water Supply Services—Second Edition. IWA Publishing, London ISBN:.
ASCE (2017). Infrastructure report card. https://www.infrastructurereportcard.org/.
Ashley, R.M., 2020. Changes in the way we live and value urban spaces. J. Delta Urbanism 1 (1), 2666–7851. (2020) [Premises]. ISSN: https://doi.org/10.7480/jdu.1.2020.5456.
Ashley, R.M., Digman, J.C., Horton, B., Gersonius, B., Smith, B., Shaffer, P., Baylis, A., 2018. Evaluating the longer term benefits of sustainable drainage. Proc. Inst. Civil Engineers Water Manag. (WM2) 57–66.
Ashley, R., Gersonius, B., Horton, B., 2020. Managing flooding: from a problem to an opportunity. Phil. Trans. R. Soc. A 378, 20190214. http://dx.doi.org/10.1098/rsta.2019.0214.
Ashley, R., Cashman, A., 2006. Infrastructure to 2030: Telecom, Land Transport, Water and Electricity. Organization for Economic Cooperation and Development, Paris 2006 https://doi.org/10.1680/jwama.16.00118.
Berndtsson, R., Becker, P., Perssone, A., Aspegren, H., Haghighatafshar, S., Jönsson, K., Larssona, R., Mobinia, S., Mottaghifgh, M., Nilsson, J., Nordström, J., Pilesjöbe, P., Scholz, M., Sternuddh, C., Sörensena, J., Tussupova, K., 2019. Drivers of changing urban flood risk: a framework for action. J. Environ. Manag. 240, 47–56. https://doi.org/10.1016/j.jenvman.2019.03.094.
Bieber, S., Snyder, S.A., Dagnino, S., Rauch-Williams, T., Drewes, J.E., 2018. Management strategies for trace organic chemicals in water—a review of international approaches. Chemosphere 195 (2018), 410–426.
Black & Veatch (2020) Strategic directions: water report. https://www.bv.com/perspectives/2020-strategic-directions-water-report.
BLUE2 (2019). Summary Report. Deliverable of the BLUE2 project Study on EU integrated policy assessment for the freshwater and marine environment, on the economic benefits of EU water policy and on the costs of its non-implementation". Report to DG ENV.
Brown, R.R., Keath, N., Wong, T., 2008. Transitioning to water sensitive cities: historical, current and future transition states. In: Proc. International Conference on Urban Drainage 1-10. Edinburgh.
Bui, X-T., Chiemchaisi, C., Fujioka, T., Varjani, S., 2018. Water and Wastewater Treatment Technologies. Springer, Singapore, ISBN 978-981-13-3258-6.
Cabral, M., Loureiro, D., Covas, D., 2019. Using economic asset valuation to meet rehabilitation priority needs in the water sector. Urban Water J. 16 (3), 205–214. doi:10.1080/1573062X.2019.1648528.
Carmona, M., 2019. Place value: place quality and its impact on health, social, economic and environmental outcomes. J. Urban Des. 24 (1), 1–48. doi:10.1080/13574809.2018.1472523.
Cattaneo, C., Peri, G., 2015. The migration response to increasing temperatures. In: NBER Working Paper 21622. Sustainability.
Cheshmehzangi, A., 2021. Revisiting the built environment: 10 potential development changes and paradigm shifts due to COVID-19. J. Urban Manag. 10 (2), 166–175. https://doi.org/10.1016/j.jum.2021.01.002.
Chitescu, C.L., Lupoae, M., Alina, M.E., 2016. Pharmaceutical residues in the environment—new European integrated programmes required. Rev. Chim.(Bucharest) 67 (5), 1008–1013.
Cook, J., Makin, L., 2020. Covid Water Use and the Impact of Poverty in the UK. NEA & Waterwise, Newcastle & London.
Cotella G. & Bravarone E.V. (2020). Questioning urbanisation models in the face of Covid-19. TeMA Special Issue | Covid-19 vs City-20, 105-118. ISSN 1970-9889, e-ISSN 1970-9870. DOI: 10.6092/1970-9870/6869
COWI (2010) Compliance costs of the urban wastewater treatment directive. Final Report. Document no. 70610-D-DFR, Lyngby.

Dobson, B., Jovanovic, T., Chen, Y., Paschalis, A., Butler, A., Mijic, A., 2021. Integrated modelling to support analysis of COVID-19 impacts on London's water system and in-river water quality. Front. Urban Water doi:10.3389/frwa.2021.641462.

EC (2018). A European strategy for plastics in a circular economy. {SWD (2018) 16 final}. https://ec.europa.eu/environment/circular-economy/pdf/plastics-strategy.pdf

EEA, 2020. The European environment — state and outlook 2020: knowledge for transition to a sustainable Europe. European Environment Agency https://www.eea.europa.eu/soer/2020.

Eerkes-Medrano, D., Thompson, R.C., Aldridge, D.C., 2015. Microplastics in freshwater systems: a review of the emerging threats, identification of knowledge gaps and prioritisation of research needs. Water Res. 75, 63–82. doi:10.1016/j.watres.2015.02.012.

EurEau (2020) The governance of water services in Europe 2020 edition. https://www.eureau.org/resources/publications/5268-the-governance-of-water-services-in-europe-2020-edition-2/file

Fluence (2019) What is non-revenue water?. https://www.fluencecorp.com/what-is-non-revenue-water/.

Flyn, C., 2021. Islands of Abandonment. William Collins, London ISBN 978-0-00-832976-1.

Heinzlef, C., Serre, D., 2020. Urban resilience: From a limited urban engineering vision to a more global comprehensive and long-term implementation. Water Security 11, 100075.

Hukka, J.J., Katko, T.S., 2015. Appropriate pricing policy needed worldwide for improving water services infrastructure. Journal—AWWA, 107 (1), E37–E46. https://doi.org/10.5942/jawwa.2015.107.0007.

Hulsmann, A., Grutzmacher, G., van den Berg, G., Rauch, W., Jenssen, A.L., Popovych, V., Rosario, M., Lyrondia, S.V., Savic, D.A., 2015. Climate Change, Water Supply and Sanitation: Risk Assessment, Management, Mitigation and Reduction, 14. IWA Publishing https://doi.org/10.2166/9781780405001.

Hutton, G., Varughese, M., 2016. The Costs of Meeting the 2030 Sustainable Development Goal Targets on Drinking Water, Sanitation, and Hygiene. The World Bank doi:10.1596/K8543.

Iravani, H., Rao, V., 2020. The effects of new urbanism on public health. J. Urban Des. 25 (2), 218–235. doi:10.1080/13574809.2018.1554997.

ISO 55000 Asset management—management systems—requirements (ISO 55001:2014, IDT)

Jiang, Z., Zhai, W., Meng, X., Long, Y., 2020. Identifying Shrinking Cities with NPP-VIIRS Nightlight Data in China. ASCE J. Urban Plann. Dev. 146 (4), 04020034. doi:10.1061/(ASCE)UP.1943-5444.0000598.

Kang, M., Choi, Y., Kim, J., Lee, K.O., Lee, S., Park, I.K., Park, J., Seo, L., 2020. COVID-19 impact on city and region: what's next after lockdown? Int. J. Urban Sci. 24 (3), 297–315. NO. https://doi.org/10.1080/12265934.2020.1803107.

Karaman, O., Sawyer, L., Schmid, C., Wong, K.P., 2020. Plot by plot: plotting urbanism as an ordinary process of urbanisation. Antipode 52 (4), 1122–1151. doi:10.1111/anti.12626, 2020 ISSN 0066-4812.

Kern, J.D., Su, Y., Hill, J., 2020. A retrospective study of the 2012–2016 California drought and its impacts on the power sector. Environ. Res. Lett. 15, 094008.

Kumasi, T.C., et al., 2019. Rural water asset management practices in Ghana: the gaps and needs. Water Environ. J. 33 (2019), 252–264 © 2018 CIWEM.

Kuriakose, B., 2020. Deurbanisation: 'New India' to Live in Small Towns and Villages. Delhi Post https://delhipostnews.com/deurbanisation-new-india-to-live-in-small-towns-and-villages/.

Leach, J.M., Rogers, C., 2020. Embedding transdisciplinarity in Engineering approaches to infrastructure and cities. In: Proc. Institution of Civil Engineers, London. J. Smart infrastructure and construction In press.

Leflaive, X., Krieble, B., Smythe, H., 2020. Trends in water-related technological innovation: Insights from patent data. In: OECD Environment Working Papers, No. 161. Paris. OECD Publishing.

Libey, A., Adanka, M., Thomas, E., 2020. Who pays for water? Comparing life cycle costs of water services among several low, medium and high-income utilities. World Dev. 136, 105155. https://doi.org/10.1016/j.worlddev.2020.105155.

Liu, F., Tait, S., Schellart, A., Mayfield, M., Boxall, J., 2020. Reducing carbon emissions by integrating urban water systems and renewable energy sources at a community scale. Renew. Sustain. Energy Rev. 123 (2020), 109767. https://doi.org/10.1016/j.rser.2020.109767.

Lovell, R., White, M.P., Wheeler, B., Taylor, T., Elliott, L., 2020. A rapid scoping review of health and wellbeing evidence for the Green Infrastructure Standards. Natural England Other Parties. ISBN 978-1-78354-648-0 http://publications.naturalengland.org.uk/publication/4799558023643136.

Magnusson, D., 2018. Going back to the roots: the fourth generation of Swedish eco-villages. Scottish Geographical J. 134 (3-4), 122–140. doi:10.1080/14702541.2018.1465199.

Milly, P.C.D., Betancourt, J., Falkenmark, M., Hirsch, R.M., Kundzewicz, Z.W., Lettenmaier, D.P., Stouffer, R.J., Feb 2008. Stationarity is dead: whither water management? Science 319 (5863), 573–574. doi:10.1126/science.1151915.

Oberg, G., Metson, G.S., Kuwayama, Y., Conrad, S.A., 2020. Conventional sewer systems are too time-consuming, costly and inflexible to meet the challenges of the 21st Century. Sustainability 12, 6518. doi:10.3390/su12166518. www.mdpi.com/journal/sustainability.

OECD (2018). Financing water—investing in sustainable growth. Policy Perspectives. OECD Environment Policy Paper No. 11. ISSN 2309–7841. https://www.oecd.org/water/Policy-Paper-Financing-Water-Investing-in-Sustainable-Growth.pdf

OECD (2020a) Financing water supply, sanitation and flood protection: challenges in EU Member states and policy options. OECD Studies on Water, OECD Publishing, Paris, https://doi.org/10.1787/6893cdac-en.

OECD (2020b). Building back better: a sustainable, resilient recovery after COVID-19. https://read.oecd-ilibrary.org/view/?ref=133_133639-s08q2ridhf&title=Building-back-better-_A-sustainable-resilient-recovery-after-Covid-19.

Osman H. (2016). Coordination of urban infrastructure reconstruction projects, Struct. Infrastruct. Eng., 12 (1), 108–121. doi:10.1080/15732479.2014.995677.

Our Future Water (2020). Investing in a water secure future. Climate Markets Invest. Assoc. https://www.ourfuturewater.com/investing-in-a-water-secure-future/

Phillips, M.T., Katharine, A.O., Grenfell, B.T., Pitzer, V.T., 2020. Changes in historical typhoid transmission across 16 U.S. cities, 1889-1931: Quantifying the impact of investments in water and sewer infrastructures. PLoS Negl. Trop. Dis. 14 (3), e0008048. https://doi.org/10.1371/journal.pntd.0008048.

Poch, M., Garrido-Baser, M., Coromina, L., Perelló-Moragues, A., Monclúsa, H., Cermerón-Romero, M., Melitas, N., Jiang, S.C., Rosso, D., 2020. When the fourth water and digital revolution encountered COVID-19. Sci. Total Environ. 744, 140980. https://doi.org/10.1016/j.scitotenv.2020.140980.

Press, E., Roberson, J.A., 2016. The financial and policy implications of water loss. American Water Works Assoc. 108, 2.

Reiner R.C. (2020). Mapping geographical inequalities in access to drinking water and sanitation facilities in low-income and middle-income countries, 2000–17. www.thelancet.com/lancetgh Vol 8

Ross, A.G., Rahman, M., Alam, M., Zaman, K., Qadri, F., 2020. Can we 'WaSH' infectious diseases out of slums? Int. J. Infectious diseases. Perspective 92, P130–P132. https://doi.org/10.1016/j.ijid.2020.01.014.

Sadoff, C.W., Hall, J.W., Grey, D., Aerts, J.C.J.H., Ait-Kadi, M., Brown, C., Cox, A., Dadson, S., Garrick, D., Kelman, J., McCornick, P., Ringler, C., Rosegrant, M.W., Whittington, D.,

Wiberg, D., 2015. Securing Water, Sustaining Growth: Report of the GWP/OECD Task Force on Water Security and Sustainable Growth. Oxford University, UK, 180 pp. http://www.gwp.org/Global/About%20GWP/Publications/The%20Global%20Dialogue/SECURING%20WATER%20SUSTAINING%20GROWTH.PDF.

Santamouris, M., 2020. Recent progress on urban overheating and heat island research. Integrated assessment of the energy, environmental, vulnerability and health impact. Synergies with the global climate change. Energy Build. 207, 109482. https://doi.org/10.1016/j.enbuild.2019.109482.

Sapkota, M., Meenakshi, A., Malano, H., Moglia, M., Sharma, A., Pamminger, F., 2018. Understanding the impact of hybrid water supply systems on wastewater and stormwater flows. Resources, Conservation Recycling 130 (2018), 82–94.

Seidl, C., Wheeler, S.A., Zuo, A., 2020. High turbidity: Water valuation and accounting in the Murray-Darling Basin. Agricultural Water Management. Volume 230, 105929. 1 March 2020 https://doi.org/10.1016/j.agwat.2019.105929

Skorka, P., Grzywacz, B., Moroń, D., Lenda, M., 2020. The macroecology of the COVID-19 pandemic in the Anthropocene. PLoS One 15 (7), e0236856. https://doi.org/10.1371/journal.pone.0236856.

Tomaz, P., Jepson, W., Santos J de, O., 2020. Urban household water insecurity from the margins: perspectives from Northeast Brazil. Professional Geographer 72 (4), 481–498. doi:10.1080/00330124.2020.1750439.

Tscheikner-Gratl, F., Caradot, N., Cherquic, F., Leitão, J.P., Ahmadie, M., Langeveld, J.G., Le Gath, Y., Scholten, L., Roghani, B., Rodríguez, J.P., Lepot, M., Stegeman, B., Heinrichsen, A., Kropp, I., Kerres, K., do Céu Almeida, M., Bach, P.M., de Vitry M, M., Marque, A.S., Simõe, N.E., Rouaul, P., Hernandez, N., Torres, A., Werey, C., Rulleauh, B., Clemens, F., 2019. Sewer asset management – state of the art and research needs. Urban Water Journal 16 (9), 662–675. doi:10.1080/1573062X.2020.1713382.

UKWIR, 2017. Long term investment in water and sewerage networks. UK Water Industry Research Ltd., London 7/RG/05/47.

UN (2018). The United Nations world water development report 2018: nature-based solutions for water. https://unesdoc.unesco.org/ark:/48223/pf0000261424

UN (2020). The United Nations world water development report 2020: water and climate change. https://unesdoc.unesco.org/ark:/48223/pf0000372985.locale=en

UN (2021) UN-Water, 2020: Summary Progress Update 2021 – SDG 6 – water and sanitation for all. Version: 1 March 2021. Geneva, Switzerland.

UN Water & WHO (2019) National systems to support drinking-water, sanitation and hygiene: Global Status Report 2019.

U.S. Department of the Interior Bureau of Reclamation (2013). Literature synthesis on climate change implications for water and environmental resources. 3rd Ed. Technical Memorandum 86-68210-2013-06. https://www.usbr.gov/climate/docs/ClimateChangeLiteratureSynthesis3.pdf

U.S. Water Alliance (2021). Recovering stronger: a federal policy blueprint. http://uswateralliance.org/sites/uswateralliance.org/files/publications/Recovering%20Stronger%20Federal%20Policy%20Blueprint_0.pdf

van Leeuwen, K., de Vries, E., Koop, S., Roest, K., 2018. The Energy and raw materials factory: role and potential contribution to the circular economy of the Netherlands. Environ. Manag. doi:10.1007/s00267-018-0995-8.

Van Riel, W.J., 2017. A gaming approach to networked infrastructure management. Struct. Infrastructure Eng. 13, 855–868. doi:10.1016/j. watres.2011.07.008.

Wang, J., 2020. The Water-energy Nexus Framework in Urban Water Management at Precinct Scale. PhD thesis Monash University, Melbourne.

WEF (2020) Global risk report. world economic forum. 15th ed. http://www3.weforum.org/docs/WEF_Global_Risk_Report_2020.pdf

Willetts, J., Mills, F., Al'Afghani, M., 2020. Sustaining community-scale sanitation services: co-management by local government and low-income communities in Indonesia. Front. Environ. Sci. 8, 98. doi:10.3389/fenvs.2020.00098.
Winpenny, J., 2015. Water: Fit to Finance?, World Water Council. OECD https://www.worldwatercouncil.org/sites/default/files/Thematics/WWC_OECD_Water_fit_to_finance_Report.pdf.
WHO (2017): UN-Water Global Analysis and Assessment of Sanitation and Drinking-Water (GLAAS): Financing universal water, sanitation and hygiene under the Sustainable Development Goals

CHAPTER SEVEN

SDG 6 global financing needs and capacities to ensure access to water and sanitation for all

Guy Hutton
Senior Adviser for Water, Sanitation and Hygiene, UNICEF New York, NY, United States

7.1 Introduction

To support the achievement of any development goal or target, it is important to know what the spending needs are and how these compare to the current financing available. There are several reasons why these numbers are useful. They help communicate to policymakers and financiers what are the funding amounts that need to be raise, and thereby inform financing strategies. They are also used to inform policies, norms, and standards. If target population coverage rates or service standard levels cannot be afforded with the financing envelope available, then the targets and standards themselves may need to be adjusted to make them achievable. Finally, projections of spending needs help to inform the planning process where budgets are requested and allocated.

In the past 20 years, since the Millennium Development Goals (MDGs) were inaugurated, there have been a series of global costing studies of water, sanitation and hygiene (WASH), being the focus of MDG Target 7c. The World Water Council (2006) published a review of six global cost studies of MDG target 7c[1,2]. Later, under the auspices of the high-level panel on Financing Infrastructure for a Water-Secure World, World Water Council and OECD (2015) compared eight cost studies, with some overlap with their earlier review. While at first glance the studies give very different global cost estimates, the studies found that if the results are analyzed on comparable bases, they appear quite similar. As stated by the study, the capital costs of

[1] MDG Target 7c: "By 2015, halve the proportion of people without sustainable access to safe drinking water and basic sanitation." Baseline year was the year 1990.

[2] Later studies were published re-estimating the needs as the MDG endpoint drew nearer; however, these are not the focus of this chapter.

Financing Investment in Water Security: Recent Developments and Perspectives.
DOI: https://doi.org/10.1016/B978-0-128-22847-0.00001-6

achieving MDG target 7c focused on low- and middle-income countries, and were as follows:

- First, approximately US$ 10 billion was required per year from the outset of the MDG period to extend low-cost water and sanitation services to people who were not supplied in the year 2000 (Water Supply and Sanitation Collaborative Council, 2000; World Health Organization, 2004), and US$ 27 billion was needed per year to achieve universal coverage of improved water supply and sanitation (World Health Organization, 2012);
- Second, a further US$ 15 to 20 billion a year was required to provide a higher level of service and to maintain current levels of service to people who are already supplied (Water Academy, 2004);
- Third, an additional US$ 100 billion per year was required for collecting and treating household wastewater (Global Water Partnership, 2000) and US$ 15 billion for integrated water resources management and ecological sanitation solutions (Stockholm Environment Institute, 2004).
- A later study that included high-income economies estimated that US$ 205 billion per year was needed until 2050 for global, universal coverage of safe water supply, sanitation, and sewerage (Lloyd Owen, 2011).

Hence, even in the early 2000s, costing studies were going beyond the requirements for MDG target 7c alone, exploring higher service levels and environmental protection. Several studies provided a breakdown by region and distinguished the costs of water supply, sanitation, and broader water resource. However, these estimates are now almost 2 decades old and hence out-of-date; also, at the time there were fewer costing studies conducted on the broader aspects of water resources such as water efficiency and water for nature. Also, the majority of studies have a relatively short planning horizon of 10–20 years and did not go beyond the year 2030, except Lloyd Owen (2011) and World Bank (2010) which adopted a 2050 timeframe.

The water- and sanitation-related targets within the sustainable development goal (SDG) framework are much more extensive and even more ambitious than the MDG target, with targets contained not only in SDG 6, but there are also explicit water-related targets within SDG 1 (target 1.4 on basic services), SDG 3 (target 3.8 on quality health care services) and SDG 4 (target 4a on education infrastructure). From being an "improved facility" in the MDG framework, the standard has been augmented to a "safely managed service" within targets 6.1 and 6.2. Hygiene has been explicitly added, with an indicator in target 6.2. The phrasing of "Water for All" implies beyond the household setting hence institutions, workplaces, and public spaces. And most importantly, targets 6.3 to 6.6 extend beyond narrow WASH to broader water resource quality and water use for both human and non-human beneficiaries.

World Bank (2010) assessed the costs of adapting specified types of water infrastructure to climate change (coastal zone protection, water supply, flood protection) resulting in an estimate of up to US$ 100 billion additional costs per year. This significant cost of adapting to climate change suggests that an important new agenda for climate change resilience will need to be urgently confronted. Among others, it will mean managing the increased risk of water scarcity, coastal storm surges, flooding and heat stress, and greater consequences—thus demand much greater financial resources, as indicated by World Bank (2010).

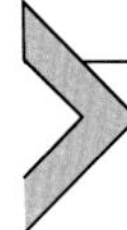

7.2 Studies assessing financing needs to cover the capital costs of achieving SDG6

This chapter focuses on studies that have estimated the global costs of achieving some or all targets of SDG6. Most studies included only low- and middle-income countries while some included all countries. Several studies present regional breakdowns of costs. Five global studies are described below, and summarized in .

1. World Bank (2016), authored by Hutton and Varughese, published a report that estimated the costs of providing universal basic and safely managed WASH[3] in households of 140 low- and middle-income countries to meet the SDG targets 6.1 and 6.2. The estimates covered the 15-year period from 2016–2030 and provided a regional breakdown.
2. Sustainable Development Solutions Network (2019), authored by Sachs et al., published a comprehensive SDG costing study with analysis of financing options for 59 low and lower-middle-income countries. For SDG 6 they included household WASH and irrigation. The household WASH estimates used World Bank (2016) numbers.
3. World Bank (2019), authored by Rozenberg and Fay, estimated infrastructure costs related to selected SDGs. They estimated costs of 3 scenarios—a minimum spending scenario (less ambitious goals, high efficiency) as well as higher ambition and lower efficiency scenarios. The study provided regional breakdowns. The minimum scenario is presented here for SDG 6. The household WASH estimates used World Bank (2016) numbers.

[3] Annex 7.1 provides definitions of "basic" and "safely managed" water and sanitation services.

4. International Monetary Fund (2019), authored by Gaspar et al., included development areas for which public intervention is critical: education, health, roads, electricity, water, and sanitation. Their estimates cover 121 countries that include 49 low-income countries (LICs) and 72 emerging market economies. For SDG 6, only targets 6.1 and 6.2 are included. The household WASH estimates used World Bank (2016) numbers.
5. World Resources Institute (2020), authored by Strong et al., presents the costs of meeting all SDG 6 targets, covering WASH, water pollution (industrial and agricultural), and water scarcity; they also included water management associated with these prior water challenges. It includes all countries and covers the period 2015 to 2030. The household WASH estimates used World Bank (2016) numbers, and costs were added for high-income economies.

As noted, all studies draw on World Bank (2016) report numbers to produce estimates for household WASH in low- and middle-income countries. No studies to-date have estimated the regional or global costs of institutional WASH, such as WASH in healthcare facilities, WASH in schools or WASH in workplace settings.

Given the different units targeted to achieve the six targets, different methodologies have been employed for each target, covered below.

7.3 Methodologies used by costing studies

To estimate the costs of meeting the WASH targets 6.1 and 6.2 of the SDGs, World Bank (2016) builds on the earlier methodology employed in the MDG era by World Health Organization (2004) and later updated by World Health Organization (2012). While the former studies published by WHO included only "improved" water and sanitation as per MDG indicator under Target 7c, the World Bank study captured the costs of achieving both safely managed water and safely managed sanitation. While treatment of excreta is included, full costs of wastewater treatment are not captured in this study. The algorithm employed is a simple sum of the estimated households without coverage multiplied by the unit costs of different levels and types of WASH services. A mix of target interventions is used. The time dimension is handled by assuming a linear increase in service coverage from the baseline year to the target year when universal coverage is achieved. The population to be served until the year 2030 includes population growth, with all additional populations assumed to have no current coverage (hence it assumes household size remains constant over time). World Bank (2016) estimates both capital and O&M costs over this period, with capital costs including program management and behavior change costs. Capital costs were estimated as cost per capita for each service type, and the total population to be served from 2016 to 2030 was divided into 15 equal tranches and applied to each year.

To estimate costs of achieving target 6.3, wastewater treatment, World Resources Institute (2020) estimate the additional costs of industrial and agricultural wastewater treatment (in addition to the septage treatment and sewage management in target 6.2), using their own methodology, as follows:

- Industrial wastewater: based on available evidence, the study estimates industrial water withdrawals and return flows to identify the amount of industrial wastewater entering water bodies in 2010 and 2030, and the proportion already treated. They apply existing studies on cost per m^3 to secondary wastewater treatment standards. To meet the implied Target 6.3, they applied an additional cost per m^3 to raise the treatment from secondary to tertiary wastewater treatment standards.
- Agricultural wastewater: the full reduction of agricultural sources of nitrogen and phosphorus to acceptable concentrations in all water bodies is estimated by applying agricultural best management practices for nutrient reduction. Total global nitrogen and phosphorus loading in water bodies in 2030 is combined with projected 2030 available blue water in water bodies, natural and maximum nutrient concentrations in water bodies, and weighted average unit costs to remove nitrogen and phosphorus.

To estimate costs of achieving target 6.4, water efficiency measures, World Resources Institute (2020) drew on data sets with catchment-level gaps in water supply and demand, efficiency targets per sector, and cost per m^3 of efficiency savings. Cost curves were used to estimate the cost per sector. The projected gap (percentage of total sector withdrawal to be saved) was used to integrate the cost curve per sector. The final estimated cost is the sum of the cost to achieve agricultural, domestic, and industrial water savings as well as to add additional supply infrastructure when needed.

To assess the governance and management capacity gap, a target expenditure rate (as a percentage of total required investments) was applied to the aggregate of other estimated costs. Based on historical data from high-income economies, a standard percentage modifier of 20% was taken for achieving the management conditions necessary to deliver sustainable water management (target 6.5).

No estimates have been made to date for the costs of achieving target 6.6, protecting and restore water-related ecosystems.

On the other hand, water-related targets in other SDGs have been addressed as follows:

- Irrigation is a key intervention to achieve SDG 2 (End hunger, achieve food security and improved nutrition and promote sustainable agriculture). Sustainable Development Solutions Network (2019) uses the GLOBIOM partial equilibrium model with irrigation module from Palazzo et al (2019). They analyzed moderate and high public support for irrigation through different subsidies. World Bank (2019) also estimates the infrastructure investment costs of irrigation in low- and middle-income countries.
- Flood protection is one key measure to achieve SDG 13 ("Take urgent action to combat climate change and its impacts"). Sustainable Development Solutions Network (2019) estimates the costs of achieving coastal and river flood protection under SDG 13 in 59 low- and middle-income countries, covering capital, operating and maintenance costs. The study uses the DIVA model to minimize total costs and risk associated with coastal flooding. For river flood protection, the authors used the GLOFRIS global flood risk model to also identify needs to minimize total cost and risk associated with river flooding.

7.4 Global spending needs

The main results of these five studies are presented in Table 7.1. The majority of estimates provided include only capital costs. For targets 6.1 and 6.2, the annual capital costs amount to US$ 114 billion for low- and middle-income countries. An additional US$ 139 billion is estimated for high-income countries, using the difference between World Bank (2016) and WRI (2020). This amounts to 0.3% of global product.

Given uncertainty in many of the inputs to the cost models, World Bank (2016) estimates a lower and upper bound by varying three important variables: life span of infrastructure, type of technology chosen per service type, and discount rate used for future costs[4]. The results are shown in Fig. 7.1—with a variation in total costs in 140 low- and middle-income countries of between 0.26% and 0.54% of the gross product. Regional variations are presented in Section 7.5.

The annual costs of meeting other SDG 6 targets are shown in Table 7.2. World Resources Institute (2020) is the most comprehensive study, with

[4] Refer to Table 2.2 of World Bank (2016) for a full list of variables and their levels of uncertainty.

Table 7.1 Overview of main global cost results from available studies.

Detail	World Bank (2016)	Sustainable Development Solutions Network (2019)	World Bank (2019)	IMF (2019)	WRI (2020)
Study details					
Countries included	140 LMICs	59 LMICs	137 LMICs	49 LICs and 72 EMEs	All countries
Years covered	2016–2030	2019–2030	2015–2030	2016–2030	2015–2030
Year of prices	2015	2019	2015	2016	2015
Targets 6.1 and 6.2: Water, sanitation and hygiene (WASH)					
Total capital cost	$ 114 bn per year	$ 37 bn per year	$ 116 bn per year	–	$ 263 bn per year
Value as % of GDP	0.39% GDP of all 140 countries	1.2% GDP of LMICs; 2.7% GDP of LICs	0.32% GDP of all 137 countries	0.4% GDP LICs; 0.1% GDP EMEs	0.30% of Global Product
O&M cost	$128.8 bn per year in 2029 for all added infrastructure	–	–	–	–
Target 6.3: Water pollution					
Total value	–	–	–	–	$ 153 bn per year[a]
Target 6.4: Water use efficiency					
Total value	–	–	–	–	$ 445 bn per year[b]

(continued on next page)

Table 7.1 Overview of main global cost results from available studies—cont'd

Detail	World Bank (2016)	Sustainable Development Solutions Network (2019)	World Bank (2019)	IMF (2019)	WRI (2020)
Target 6.5: Integrated water resources management					
Total value	–	–	–	–	$ 172 bn per year[c]
Other targets: Irrigation					
Total value	–	$7.1 bn per year	$43 bn ($2.9 bn per year)	–	–
Value as % of GDP	–	0.2% GDP LMICs; 0.5% GDP LICs	0.12% GDP	–	–
Other targets: Flood protection					
Total value	–	$16.5 bn per year	–	–	–
Value as % of GDP	–	0.5% GDP LMICs; 1.2% GDP LICs	–	–	–
Total					
Total value	–	–	–	–	US$ 1.04 trillion per year
Value as % of GDP	–	–	–	–	1.21% of global product

LMICs, lower-middle income countries; *EMEs*, emerging market economies; *LICs*, low-income countries.
[a] Includes industrial and agricultural.
[b] Water scarcity.
[c] Water management.

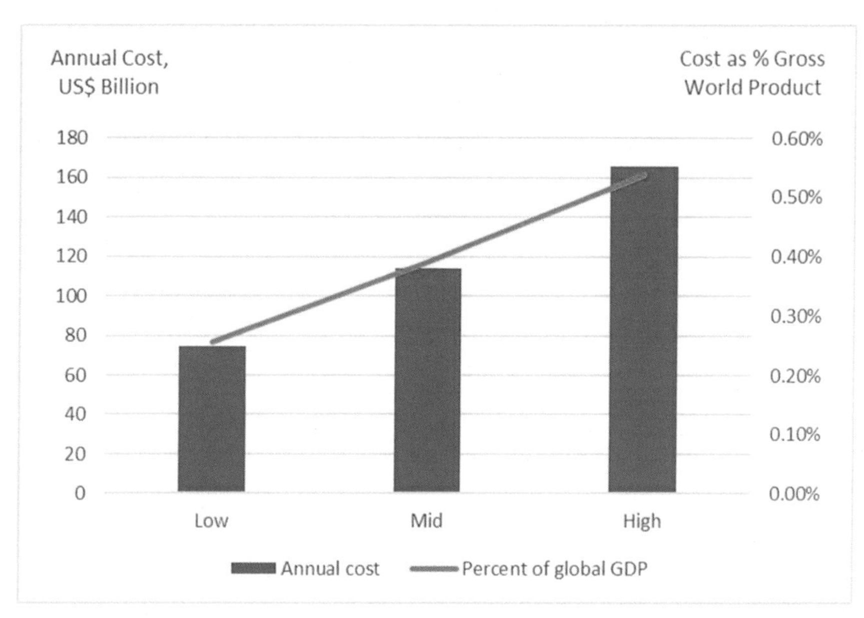

Figure 7.1 Global annual costs of achieving WASH targets 6.1 and 6.2 (left axis) and costs as a percent of global world product (right axis) for 140 low- and lower-middle income countries (source: World Bank, 2016).

total costs at just over 1 trillion dollars per year, or 1.21% of global gross product (Fig. 7.2). When irrigation and flood protection costs are added from Sustainable Development Solutions Network (2019), the global cost estimate increased to US$ 1.054 trillion per year. However, Sustainable Development Solutions Network (2019) costs only included 59 low- and middle-income countries, the global costs of irrigation and flood protection will be significantly greater, and therefore a higher proportion of total cost than these estimates suggest. On the other hand, some costs are omitted from Targets 6.4 and 6.5, as the interventions costed in the targets are not complete to meet the target.

In addition to progressive investment in capital over the 15-year period to 2030, significant funds are needed to operate and maintain water and sanitation services. Indeed, the spending required for operations and maintenance (O&M) increases over time as the capital stock is extended, hence becoming a very significant financial requirement. Fig. 7.3 shows indicative spending on capital investment compared with that on the O&M required to continue providing safely managed services from 2015 to 2030 (World Bank, 2016). The estimates are based on linear growth in coverage, with 15

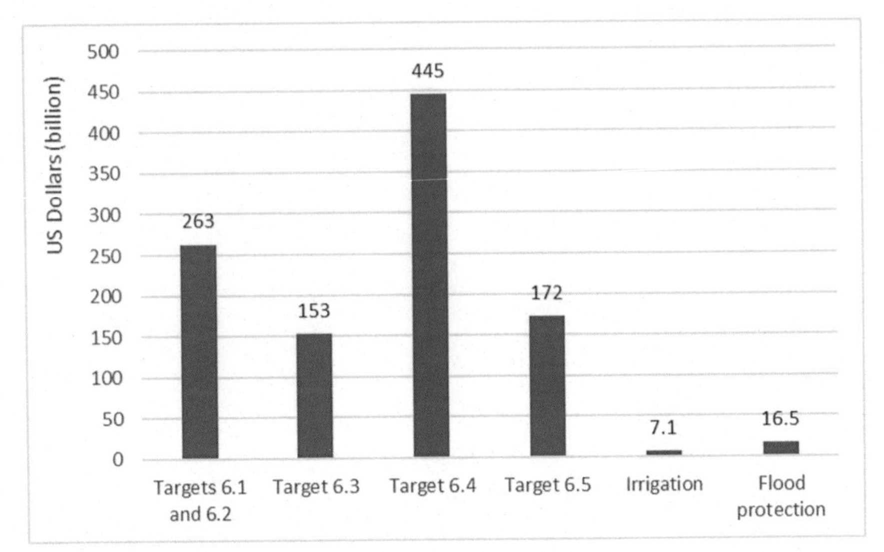

Figure 7.2 Annual global costs of meeting SDG 6 and related water targets (source: WRI, 2020, for targets 6.1–6.5, and Sustainable Development Solutions Network (2019) for irrigation and flood protection.

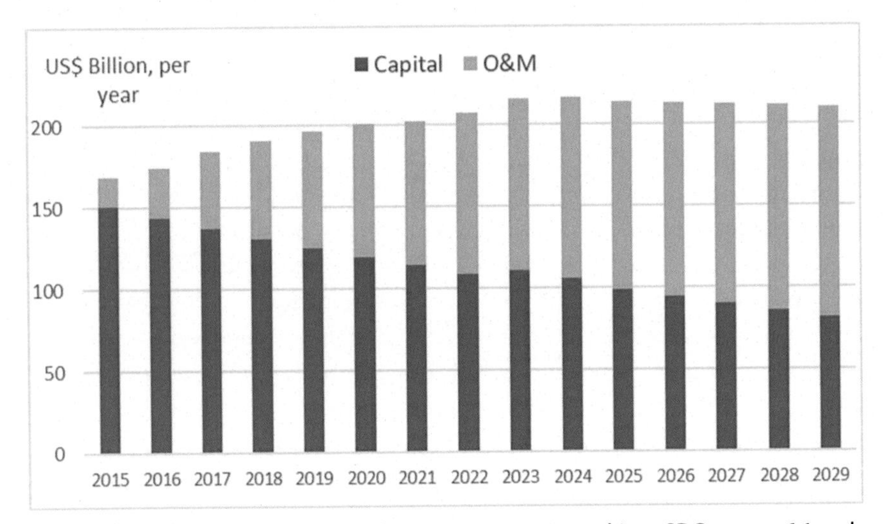

Figure 7.3 Time series of total costs from 2015 to 2029 to achieve SDG targets 6.1 and 6.2, comparing capital and O&M costs (source: World Bank, 2016).

equal tranches of unserved population gaining access each year and future costs discounted by 5% per annum (hence the apparent decline in capital costs over time).

A major observation is that, despite the discounting of all future costs at 5% per year, the spending requirements are increasing over time because of the growing needs for O&M as infrastructure is added, and more services are provided. To achieve SDG targets 6.1 and 6.2, the global O&M costs must increase gradually from $18.0 billion in 2015 to $128.8 billion in 2030. By 2029, spending on O&M for the newly served from 2015 to 2029 will outweigh capital costs by 1.6 times. Note that replacement of infrastructure prior to 2015 is omitted, which would require further public budgets and user charges.

Given the world is now five years into the SDG period, updates of previous cost studies are needed, given the spending needs will vary due to several factors. First, progress toward the 2030 targets may not be happening at the required pace, hence the annual requirements will increase as 2030 draws closer. Second, some assumptions in the cost studies may not hold, such as unchanging unit prices. For example, technological advances or market developments might lead to reductions in unit costs. On the other hand, when fully considering the speed of climate change, solutions will be more costly than anticipated.

For the State of Sanitation Report 2020 (WHO, UNICEF, 2020), World Bank (2016) report was updated for Target 6.2 for 13 years from 2017 to 2030. The model was updated to account for more recent global coverage estimates and the fewer years the world now has until 2030 to meet the sanitation target 6.2. The same 140 countries were included in this update as the original study. The following four variables were updated from 2015 to 2017: sanitation coverage estimates, unit costs (using the GDP deflator for each country), the Gross Domestic Product of each country, and the population estimates and projections. Fig. 7.4 shows that the annual costs for achieving target 6.2 have not changed since the original cost study (UNICEF, 2020), remaining at about US$ 69 billion per year. The O&M costs add an average of US$ 36 billion per year, giving a total of US$ 105 billion per year for the sanitation component of target 6.2. Regional breakdowns are shown in the following section.

7.5 Regional spending needs

Several of the global costing studies presented above provide a regional breakdown of total costs. Drawing on the World Bank (2016) study for the costs of achieving targets 6.1 and 6.2, the breakdown between the five main regions of the US$ 114 billion global annual costs are as follows:

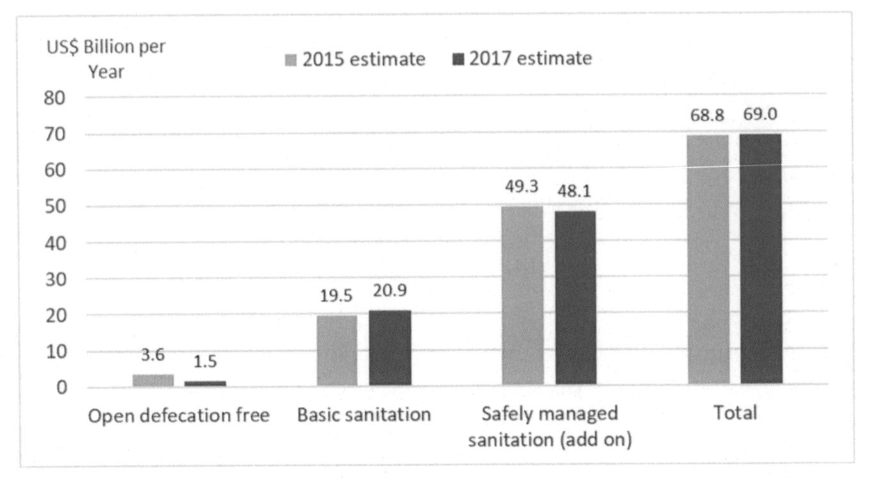

Figure 7.4 Comparison of global annualized costs between previous (2015) and current (2017) estimates (source: figures assembled from World Bank (2016) for 2015 estimates and from UNICEF (2020) for 2017 estimates. See Annex 7.1 for what interventions are included under each service level).

- Sub-Saharan Africa: 31% of global cost, or $35.5 billion per year.
- Southern Asia: 22% of global cost, or $24.5 billion per year.
- Eastern Asia[5]: 14% of global cost, or $15.9 billion per year.
- South-Eastern Asia[6]: 9% of global cost, or ($10.4 billion per year).
- Latin America and the Caribbean: 12% of global cost, or $14.0 billion per year.

Fig. 7.5 provides a breakdown by urban and rural area. Sub-Saharan Africa accounts for an even bigger proportion of rural costs, at 38% of the global cost, followed by Southern Asia at 33%. In terms of urban costs, Latin American and the Caribbean and Eastern Asia become relatively more important, although Sub-Saharan Africa still dominates.

When compared with the gross product of each region, the results vary significantly around the global averages. The regional and global costs of basic and safely managed services as a proportion of the gross product are shown in Fig. 7.6, with an indication of the uncertainty levels based on a sensitivity analysis as described in World Bank (2016).

The region with the highest capital costs to achieve universal basic WASH as a proportion of gross regional product (GRP) is Sub-Saharan

[5] China, Democratic People's Republic of Korea, Mongolia.

[6] Cambodia, Myanmar Indonesia, Lao People's Democratic Republic, Malaysia, Philippines, Thailand, Timor-Leste, Vietnam.

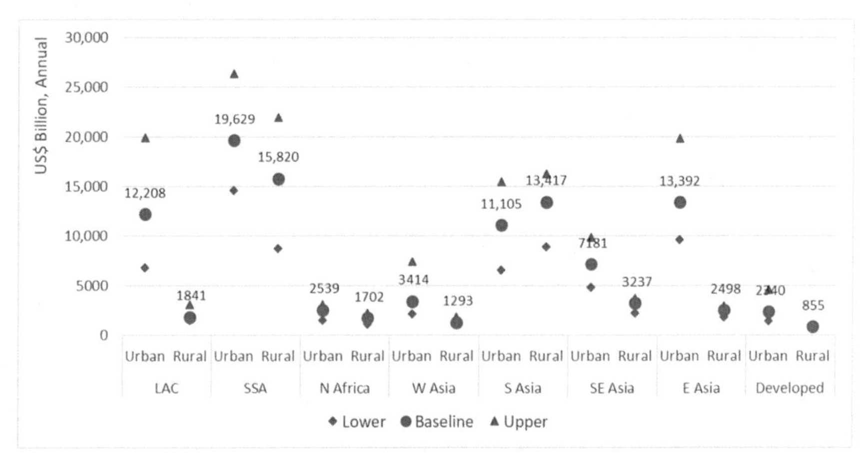

Figure 7.5 Annual capital costs of meeting SDG targets 6.1 and 6.2 by world region and urban-rural area. Note: See appendix 1 of World Bank (2016) for regional abbreviations and groupings. Some regions with small costs are omitted. *SDG*, sustainable development goal; *LAC*, Latin America and the Caribbean; *SSA*, Sub-Saharan Africa. Source: World Bank (2016).

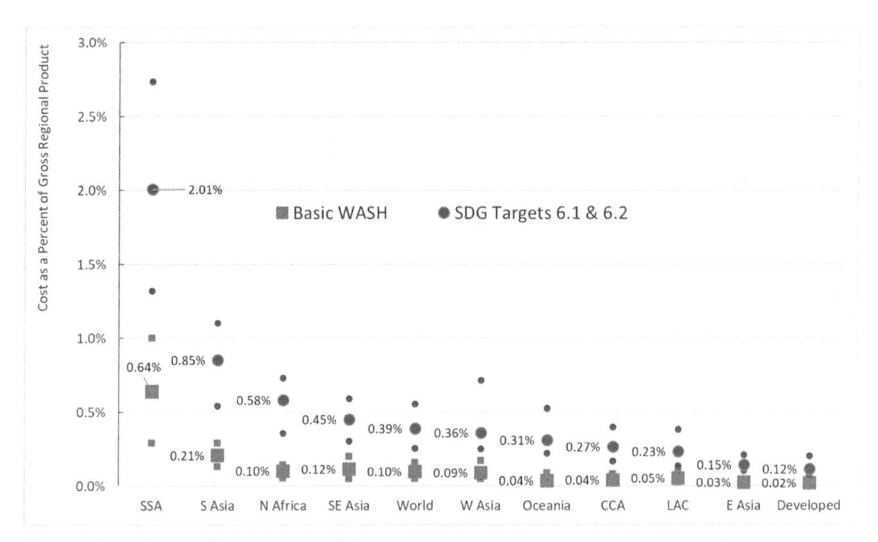

Figure 7.6 Costs of basic and safely managed services as a percentage of gross regional product (*GRP*) by world region, with uncertainty range. World regions: *MDG*, millennium development goal; *WASH*, water, sanitation, and hygiene; *SDG*, sustainable development goal; *SSA*, Sub-Saharan Africa; *LAC*, Latin America and the Caribbean; *CCA*, Caucasus and Central Asia. Gross regional product is based on the aggregated GDP of countries in each region. An economic growth rate of 5% is assumed across all regions. Source: World Bank (2016).

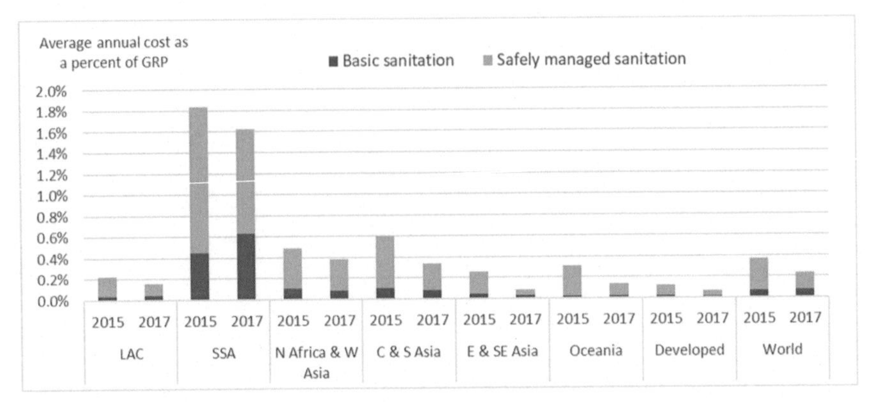

Figure 7.7 Comparison of universal sanitation costs as percentage of GRP, by service level. Source: UNICEF (2020).

Africa, with basic WASH costing 0.64% (range: 0.29%–1.0%) of GRP. The other region well above the world average is Southern Asia, with capital costs of 0.21% (range: 0.13%–0.29%) of GRP for basic WASH.

From the update on sanitation costs (UNICEF, 2020), Fig. 7.7 shows the implications of the update from 15 years to 13 years. For all regions, the overall percentage has reduced between the 2015 and 2017 estimates. For example, for sub-Saharan Africa the requirement has reduced from 1.8% to 1.6% of GRP, while for Central and Southern Asia it has reduced from 0.6% to 0.34% of GRP. Given the slow progress in many world regions, this might come as a surprise. One of the main reasons is that the study only covers a 2 year update period from 2015 to 2017, and second, that the baseline values for safely managed sanitation coverage used by World Bank (2016) were pessimistic and much lower than the baselines once they were published by WHO/UNICEF Joint Monitoring Programme in 2017.

Fig. 7.8 shows the regional breakdown for the total cost results of achieving SDG 6, by SDG 6 target. As can be seen, two world regions not covered by other studies are North America and Europe, which add significantly to the global costs. East Asia and Pacific have the highest regional costs at over US$ 260 billion per year, accounted for mainly by water scarcity (SDG target 6.4).

7.6 Estimated financing gaps

Meeting the WASH-related SDG targets will require considerably more capital resources in all regions than is currently being spent, even in

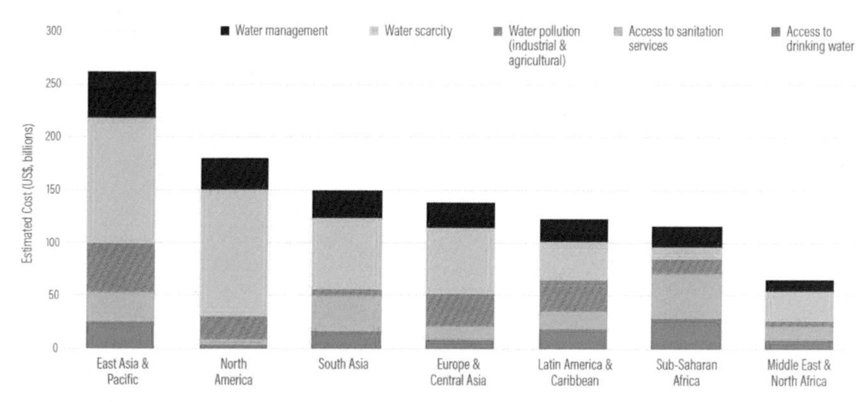

Figure 7.8 Annual costs by world region and SDG 6 target. Source: WRI (2020). This publication is licensed under the Creative Commons Attribution 4.0 International License: http://creativecommons.org/licenses/by/4.0/ and this figure is available to copy and redistribute in any medium or format.

Table 7.2 Sufficiency of funds to meet national WASH targets, by sub-sector.

WASH area	Percentage of countries reporting sufficient finance[a]
Urban/rural drinking water ($n=78$)	21%/15%
Urban/rural sanitation ($n=74$)	14%/8%
Hygiene ($n=67$)	4%
WASH in health care facilities ($n=69$)	12%
WASH in schools ($n=71$)	8%

Source: GLAAS 2018/2019 country survey.
[a] In the GLAAS 2018/2019 country survey, sufficient finance was defined as more than 75% of what is needed to meet national targets

high-income economies. World Bank (2016) estimates that spending needs to increase by an additional three times to meet the SDG targets 6.1 and 6.2 in households, rising from approximately 0.12% to 0.39% of the gross product of 140 low- and middle-income countries. Summary results from the UN-Water and WHO (2019) GLAAS report confirm this, indicating that over 80% of countries reported insufficient spending to meet their national WASH targets, as well as targets for WASH in health care facilities and schools (Table 7.2).

Fig. 7.9 shows the funding gap for 20 reporting countries from UN-Water GLAAS report (2019). For these countries, a WASH funding gap of 61% exists between identified needs and available financing for WASH. A growing number of countries are drawing on comprehensive costing exercises and financial tracking exercises such as public expenditure reviews

Annex 7.1 Service level definitions.

Service	"Basic" service	"Safely managed" service
Water	Improved drinking water sources are those which by nature of their design and construction have the potential to deliver safe water. If the improved source does not meet any one of the "safely managed" criteria, but a round trip to collect water takes 30 min or less, it will be classified as a basic drinking water service (SDG 1.4). If water collection from an improved source exceeds 30 minutes, it will be categorized as a limited service	In order to meet the criteria for a safely managed drinking water service (SDG 6.1), people must use an improved source meeting three criteria: • it should be accessible on premises, • water should be available when needed, and • the water supplied should be free from contamination
Sanitation	Improved sanitation facilities are those designed to hygienically. If the excreta from improved sanitation facilities are not safely managed, then people using those facilities will be classed as having a basic sanitation service (SDG 1.4). People using improved facilities that are shared with other households will be classified as having a limited service	There are three main ways to meet the criteria for having a safely managed sanitation service (SDG 6.2). People should use improved sanitation facilities that are not shared with other households, and the excreta produced should either be: • treated and disposed of in situ, • stored temporarily and then emptied, transported and treated off-site, or • transported through a sewer with wastewater and then treated off-site
Hygiene	Households that have a handwashing facility with soap and water available on premises will meet the criteria for a basic hygiene facility (SDG 1.4 and 6.2). Households that have a facility but lack water or soap will be classified as having a limited facility, and distinguished from households that have no facility at all	

Source: WHO and UNICEF (2017), pages 8-9.

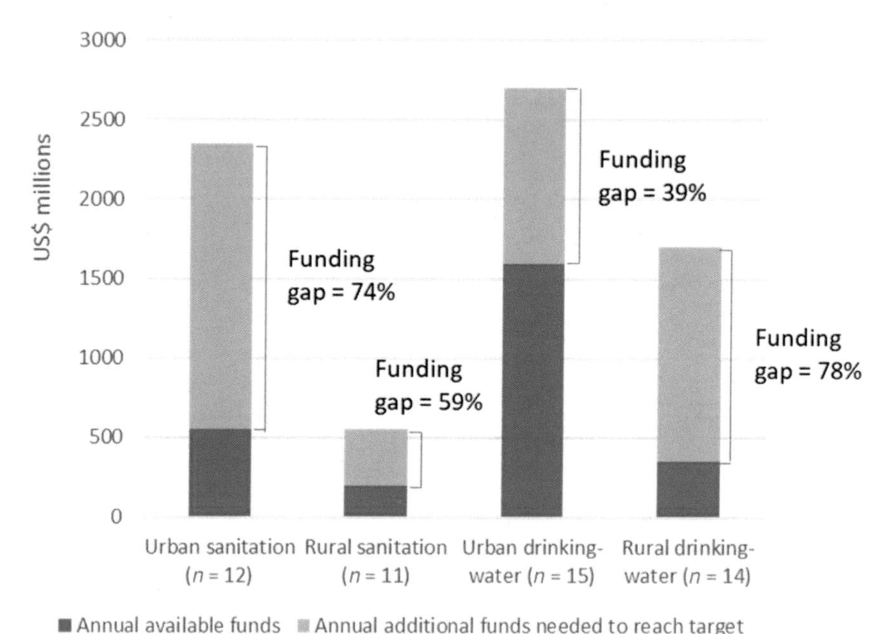

Figure 7.9 Available funds versus funds needed to reach national targets (per year). Source: GLAAS 2018/2019 country survey.

or UN-Water's "TrackFin" initiative, hence giving better quality estimates. However, only 20 countries were able to report on this question out of 103 responding to the GLAAS survey, which suggests that many countries either do not know total sector funding needs, or total sector funding available, or both. Hence, considerable further efforts are needed to conduct costing studies as well as to collect information on current volumes of spending on WASH, as a precursor to developing national financing strategies.

In some regions, the capital cost of meeting targets 6.1 and 6.2 appears feasible when compared with overall economic output (see section 7.5), varying from 0.12% of gross domestic product in countries classified as high income, 0.15% in Eastern Asia, to 0.23% in Latin America and the Caribbean (World Bank, 2016).

However, in some regions considerably more funds as a proportion of gross product are required. For example, 0.45% of gross product is needed in South-eastern Asia and 0.85% in Southern Asia. In Africa, 0.58% of gross product is required in Northern Africa and 2.0% is required in Sub-Saharan

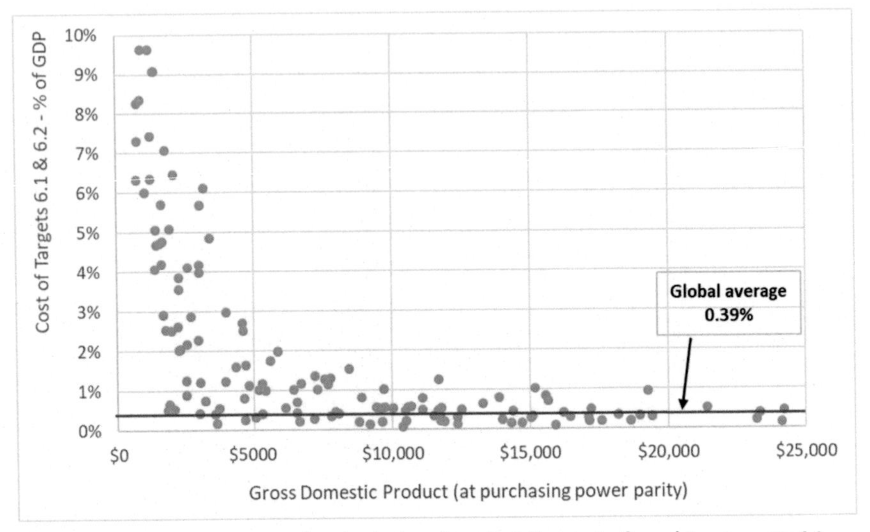

Figure 7.10 Country income level tabulated against the cost of reaching targets 6.1 and 6.2, as a proportion of GDP. Source: World Bank (2016).

Africa. Even these regional averages hide considerably greater variation at the country level.

Fig. 7.10 shows that the costs of achieving SDG targets 6.1 and 6.2, with each country represented by a point. All countries with costs in excess of 3% of GDP are in Africa. For countries contributing the largest share of global costs, the capital costs of meeting SDG targets 6.1 and 6.2 in the first year as a proportion of current GDP is estimated to vary: 0.20%, China; 0.27%, Brazil; 0.29%, Mexico; 1.0%, India; 1.7%, Nigeria.

For countries with higher cost burdens when compared with GDP levels, it is especially critical to choose capital investments that can be operated and sustained, when considering the O&M costs and the ability of the users to cover the user charges. However, not all populations will be able to cover the tariff levels, even if they appear reasonable. Hence an understanding of willingness to pay and affordability is needed (see next section).

7.7 Affordability

Global cost estimates give the major financiers insights into where the financing needs and priorities are, whereas the cost per person served indicates the likely affordability to the population of different service levels. Countries often have very different policies on financing which dictate

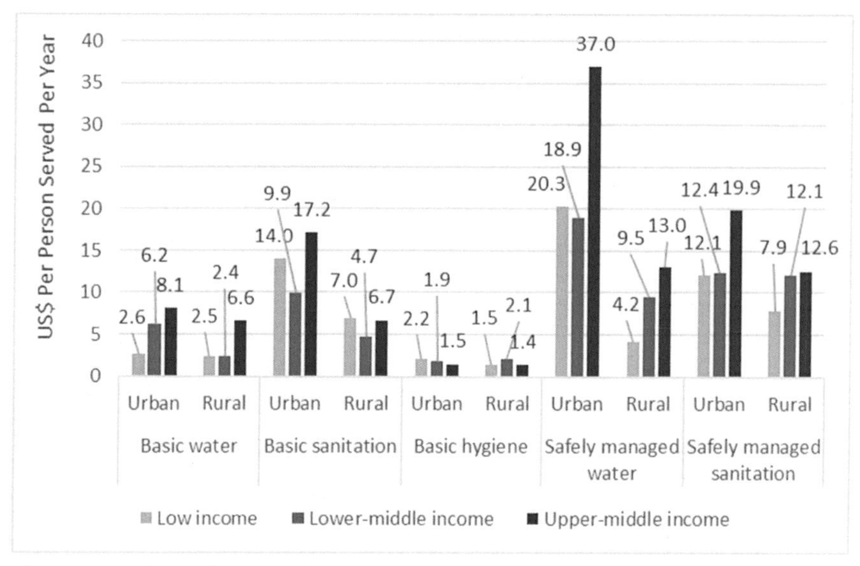

Figure 7.11 Costs of water and sanitation per person served per year (capital and O&M) by service and country income grouping. Note: O&M = operations and maintenance. Source: World Bank (2016).

what the user is expected to contribute toward both capital and recurrent costs of water supply and sanitation services. Larger capital expenditures do tend to be financed from public or donor funds, but households are more typically likely to pay part or the full recurrent (O&M) costs. There are few examples of fully autonomous utilities in low- but also middle-income countries, meaning that the tariff covers not only O&M costs but also capital replacement and environmental charges. For example, in 2005 Costa Rica expanded the use of water payments by revising its water tariff and introducing a conservation fee earmarked for watershed conservation (Fallas, 2006). In one example of its application, the town of Heredia established an "environmentally adjusted water tariff," the proceeds of which are used to pay landholders to maintain and reforest watershed areas.

As shown in Fig. 7.11, annual costs per person are strongly related to a country's income level. Urban areas also have a higher cost per person than rural areas. For sanitation, this is partly explained by the assumed higher technology requirements in urban areas. Fig. 7.10 also shows that the costs of basic sanitation exceed those of basic water, especially in urban areas. Hand washing is the lowest-cost service. In rural areas, the annual cost per

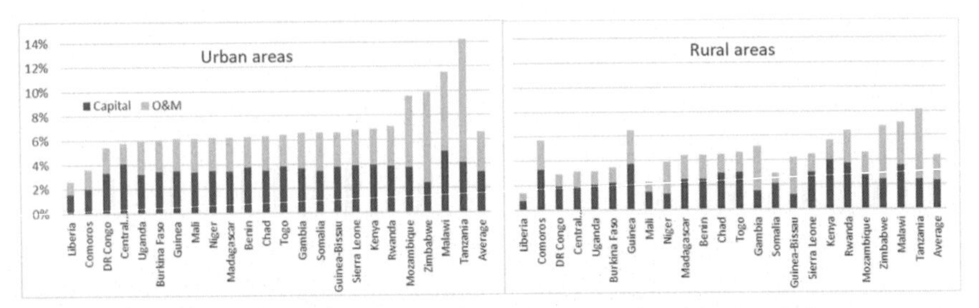

Figure 7.12 Proportion of poverty threshold income spent on basic WASH in low-income countries of Sub-Saharan Africa, separating capital and O&M Costs and with urban-rural breakdown. Notes: *WASH*, water, sanitation, and hygiene; *O&M*, operations and maintenance. Poverty income of $1.90 at purchasing power parity (PPP, 2011) adjusted to 2015 and compared with cost per person expressed in PPP. Source: World Bank (2016).

person of basic WASH is approximately $11 in LICs and $9 in lower-middle-income countries (LMICs), whereas in urban areas the costs are just under $20 in all low- and middle-income countries. A cost of $11 for basic WASH corresponds to less than 2% of the average income in LICs. However, because of the highly unequal distribution in incomes, the affordability of WASH services needs to be assessed, specifically for the poor.

An assessment of affordability requires a comparison of the costs paid by households with the income of those households. Commonly total household expenditure is used to proxy income due to the difficulties of ascertaining the latter. Fig. 7.12 shows the annual costs of basic WASH services as a proportion of the World Bank's lower poverty income threshold ($1.90 at purchasing power parity—PPP—in the year 2011[7]), also breaking down annual spending requirements on capital versus O&M costs. The calculations are based on the low-technology cost option for basic services, and include water supply, sanitation, and handwashing. When estimated in annual equivalent values, the capital costs are slightly higher than the O&M costs in the urban areas of most countries. In rural areas, the capital costs constitute an even larger share. In most low-income African countries, the capital and O&M costs combined exceed 5% of a poor person's annual income in urban areas. In rural areas, capital costs account for at least 2% of poverty income. If higher-technology options are chosen for basic WASH, or safely managed water and sanitation are targeted, the percentages are a multiple of those shown in Fig. 7.12.

[7] These values have been updated to 2015 using the average growth of the poverty threshold from 2005 (when it was $1.25 per capita per day) to 2011.

There are weaknesses associated with this simple comparison of WASH expenditure and total expenditure of households (UNICEF and World Health Organization, 2021). First, many households do not have WASH that meets the national standard, so the costs they pay do not correspond to what would be needed to access at least the national standard. Second, expenditure surveys include very few questions to capture WASH costs and hence omit several cost categories, especially when there may be more than one paying water source and other costs such as soap, water purifiers, and capital items including maintenance are excluded. Third, non-financial costs are not included, such as the opportunity cost of time to haul water or travel to place of defecation outside the home.

Fourth, these simple comparisons of WASH expenditure with total expenditure that generate an expenditure ratio ignore issues associated with household poverty and agrarian economies, such as irregularity or seasonality of income and the non-monetized economy (e.g., subsistence production) that pervade rural societies in many less developed economies. Fifth, the costs paid by other agencies are omitted from these expenditure ratios, such as government or donor subsidies, hence the long-term sustainability is not fully considered (e.g., where the withdrawal of one funding source might put at risk the financial viability of delivering a service). Finally, WASH affordability needs to be understood in the context of the affordability of a range of other essential goods and services—and the shares of household income that should be preserved for those. There are ongoing efforts to understand all these issues and their implications for global and national monitoring of WASH affordability to enable more accurate assessment of whether human rights are being upheld and understanding the very real constraints households have to achieve the national standards (UNICEF and WHO, 2021).

7.8 Conclusions

This chapter has reviewed the global cost studies on SDG 6, compared their findings and analyzed some of the policy implications. Most studies included targets 6.1 and 6.2, while other targets of SDG 6 have been assessed by fewer studies. Several gaps remain in the studies, in terms of addressing the full breadth of interventions implied in the phrasing of the targets and indicators. Furthermore, unit costs used by these studies for capital and O&M costs are not necessarily fully reflective of the range of costs in setting up and running a service, such as administration, tariff collection, and water quality testing, to name a few. If not considered, the omission of these costs would have a major consequence for the running of these services.

The very significant cost of achieving SDG 6—over US$ 1 trillion globally per year—is a multiple of the current financing dedicated to closing the gap in each of the targets. However, this gap will vary significantly between countries and regions, as indicated by the total WASH costs as a percentage of gross product (Fig. 7.6). The technologies, service levels chosen, and financing options to fill the gaps will also vary by country and by subnational context such as income levels or population density. Hence, there is no blueprint for closing the financing gap for SDG 6, and to be implementable, financing strategies will need to be formulated for each country and fine-tuned for each sub-national jurisdiction.

An important conclusion of the costing studies of targets 6.1 and 6.2 is that a greater share of future costs will be in urban areas, and sanitation costs will outweigh water costs (World Bank, 2016). Urban sanitation alone accounts for 44% of the capital costs of basic WASH globally. Another important finding from the costing of targets 6.1 and 6.2, and one that is equally relevant for all the SDG 6 targets, is the growing importance of operational and maintenance costs over time. Hence, when raising funds to pay for the capital costs, realistic plans also need to be made for cost recovery of O&M costs as well as replacement, to ensure continuity of service.

Meanwhile, at least half of the resources need to be spent on the bottom 40% of the population. Thus, the allocation of scarce public and donor finances should be decided based on finding the balance between attempting to recover costs by charging a full cost tariff to households that can afford it, and ensuring subsidies are targeted to those living in or close to poverty and those facing unreasonably high costs for the initial capital investment. When affordability is considered from the household perspective, even meeting O&M costs alone can place a significant burden on a poor household's income. Because of affordability concerns and the lower coverage of basic WASH services among the lower-income groups, a significant share of public funds should target poor and marginalized population groups. Donors also need to reconsider which countries they support. All this rethinking will require tough choices between achieving basic WASH for the unserved versus bringing better services to those already with basic services.

Although understanding costs is an important part of planning and implementing WASH services to reach universal coverage, financing is only part of broader systems strengthening that includes technology development, incentivization of private suppliers and providers, policy reform, institutional strengthening, regulatory measures, and improved monitoring and evaluation. Overall systems strengthening will increase the efficiency of services, provide cost savings, raise the demand for services, provide access to new (e.g., commercial) sources of funding, and stimulate service providers. These aspects are largely covered under what has been termed "means of implementation" in SDG goal 17, but they will require further definition of what components are prioritized.

This chapter has focused on exploring cost studies available on achieving SDG 6, and revealed many evidence gaps. Future work on global financing needs should focus on a fuller assessment of the intervention options that will meet the global targets covering water and sanitation under SDG 6 and other relevant SDGs. A wider definition of cost categories is needed for an impactful implementation of the interventions and ensuring both operational and environmental sustainability. Also, the actual costs of delivering on these targets in challenging contexts—such as slum populations or remote communities—need more exploration to ensure the vision of leaving no-one behind is concretely included in the policy dialogue and practical actions resulting. In particular, a deeper examination is needed of different proposals for appropriate technological solutions that meet the needs and expectations of these disadvantaged populations.

References

Fallas, J., 2006. Identificación de zonas de importancia hídrica y estimación de ingresos por canon deaguas para cada zona. FONAFIFO, San José.

Global Water Partnership, 2000. Towards Water Security: A Framework for Action. GWP, Stockholm.

IMF, 2019. Fiscal policy and development: human, social, and physical investment for the SDGs. In: Gaspar, V., Amaglobeli, D., GarciaEscribano, M., Prady, D., Soto, M. (Eds.), IMF Staff Discussion Note Eds. International Monetary Fund, Washington, DC.

Lloyd Owen, D., 2011. Infrastructure needs for the water sector. Unpublished paper commissioned by. In: the OECD in preparation of the OECD report Water Security for Better Lives. OECD Publishing, Paris, https://books.google.com/books?id=z0sWCAAAQBAJ&pg=PA29&lpg=PA29&dq=lloyd+owen+infrastructure+needs+for+the+water+sector&source=bl&ots=r8BqDUHvs5&sig=ACfU3U0RVntpqk9IlIfJYexZ_2yGJ5ObJw&hl=en&sa=X&ved=2ahUKEwjJteqvqo72AhXSl4kEHYhvAkQQ6AF6BAgrEAM#v=onepage&q=lloyd%20owen%20infrastructure%20needs%20for%20the%20water%20sector&f=false.

Sustainable Development Solutions Network, 2019. In: Sachs, J, McCord, G, Maennling, N, Smith, T, Fajans-Turner, V, SamLoni, S (Eds.), SDG Costing and Financing for Low-Income Developing Countries. Sustainable Development Solutions Network (SDSN), New York, Paris, and Kuala Lumpur, https://sdgfinancing.unsdsn.org/static/files/sdg-costing-and-finance-for-LIDCS.pdf.

UNICEF, 2020. Global and Regional Costs of Achieving Universal Access to Sanitation to Meet SDG Target 6.2. UNICEF, New York, NY Authors: Hutton G. and Varughese M.

UNICEF and World Health Organization, 2021. The measurement and monitoring of water supply, sanitation and hygiene (WASH) affordability: a missing element of monitoring of sustainable development goal (SDG) targets 6.1 and 6.2. WHO/UNICEF Joint Monitoring Programme and UN-Water Global Assessment and Analysis of Sanitation and Drinking-Water. UNICEF, New York and World Health Organization, Geneva.

Water Academy, 2004. A Review of Various Estimates and a Discussion of the Feasibility of Burden Sharing. Académie de l'Eau, Paris.

Water Supply, Sanitation Collaborative Council, 2000. Vision 21: a shared vision for hygiene, sanitation and water supply and a framework for action. WSSCC, Geneva.

World Resources Institute, 2020. Achieving Abundance: Understanding the Cost of a Sustainable Water Future. Authors: Strong, C., Kuzma, S., Vionnet, S., Reig, P. World Resources Institute, Washington D.C.

World Water Council, 2006. Costing MDG Target 10 on Water Supply and Sanitation: Comparative Analysis, Obstacles and Recommendations. Author: Toubkiss, J. World Water Council, Marseille, France.

World Bank, 2010. The cost for developing countries of adapting to climate change. New Methods and Estimates. The Global Report of the Economics of Adaptation to Climate Change Study. The World Bank, Washington, DC.

Authors World Bank, 2016. The costs of meeting the 2030 sustainable development goal targets on drinking water, sanitation, and hygiene. In: Hutton, G., Varughese, M.C. (Eds.), Water and Sanitation Program World Bank, Washington, DC.

Authors World Bank, 2019. Beyond the gap: how countries can afford the infrastructure they need while protecting the planet. In: Rozenberg, J., Fay, M. (Eds.), Sustainable Infrastructure Series World Bank, Washington, DC.

World Health Organization, 2004. Evaluation of the Costs and Benefits of Water and Sanitation Improvements at the Global Level. WHO, Geneva Authors: Hutton, G and Haller, L.

World Health Organization, 2012. Global costs and benefits of drinking-water supply and sanitation interventions to reach the MDG target and universal coverage. WHO, Geneva Author: Hutton, G.

World Health Organization and UNICEF, 2017. Progress on Drinking Water, Sanitation and Hygiene. 2017 Update and SDG Baselines. Geneva: WHO; New York: UNICEF.

World Water Council and OECD, 2015. Water: Fit to Finance? Catalysing national growth through investment in water security. Report of the High-Level Panel on Financing Infrastructure for a Water-Secure World. World Water Council, Marseille and OECD, Paris.

CHAPTER EIGHT

Financing water for growth and development in Africa

Alex Simalabwi
Global Water Partnership

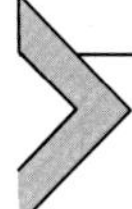

8.1 Africa's water finance challenge

8.1.1 Water resources and uses in Africa, now and in the future

Many African countries already are suffering from water scarcity and water insecurity although the continent is endowed with abundant water resources; the total annual freshwater withdrawal to support socioeconomic activities is estimated at 236 billion cubic meters, about 4.2% only of the long-term average annual total renewable water resources. However, regional averages mask wide disparities. Africa's water resources are unevenly distributed in space and time. African countries share 63 transboundary river basins that are home to 77% of the continent's population, as well as 38 shared aquifers. The Nile, Niger, Congo, and Zambezi, and large natural lakes including Victoria, Tanganyika, and Malawi, are among the world's largest water bodies (UNEP, 2010). The continent has 83 groundwater aquifers (IGRAC, 2014) with more than 50% of the continent's water resources concentrated in Central Africa, and less than 3% in North Africa.

Agriculture accounts for 78.7% of water withdrawals in the region, followed by municipal services (14.6%) and industry (6.7%) (FAO, 2021a). Over 400 million people do not have access to clean drinking water and lack basic water supply services, while 779 million people live without access to basic sanitation, but figures differ starkly between countries (Fig. 8.1) (WHO/UNICEF, 2021). Lack of access to clean water and sanitation has a major negative impact on poverty reduction and economic development. Water, sanitation, and hygiene also hold significant potential for job creation, economic growth, and health. Sub-Saharan Africa reportedly loses 5% of its GDP annually because of a lack of water and poor sanitation. In addition to this, each year 40 billion hours of otherwise productive time is spent just collecting water (AfDB/ICA 2019).

Financing Investment in Water Security: Recent Developments and Perspectives.
DOI: https://doi.org/10.1016/B978-0-12-822847-0.00017-X

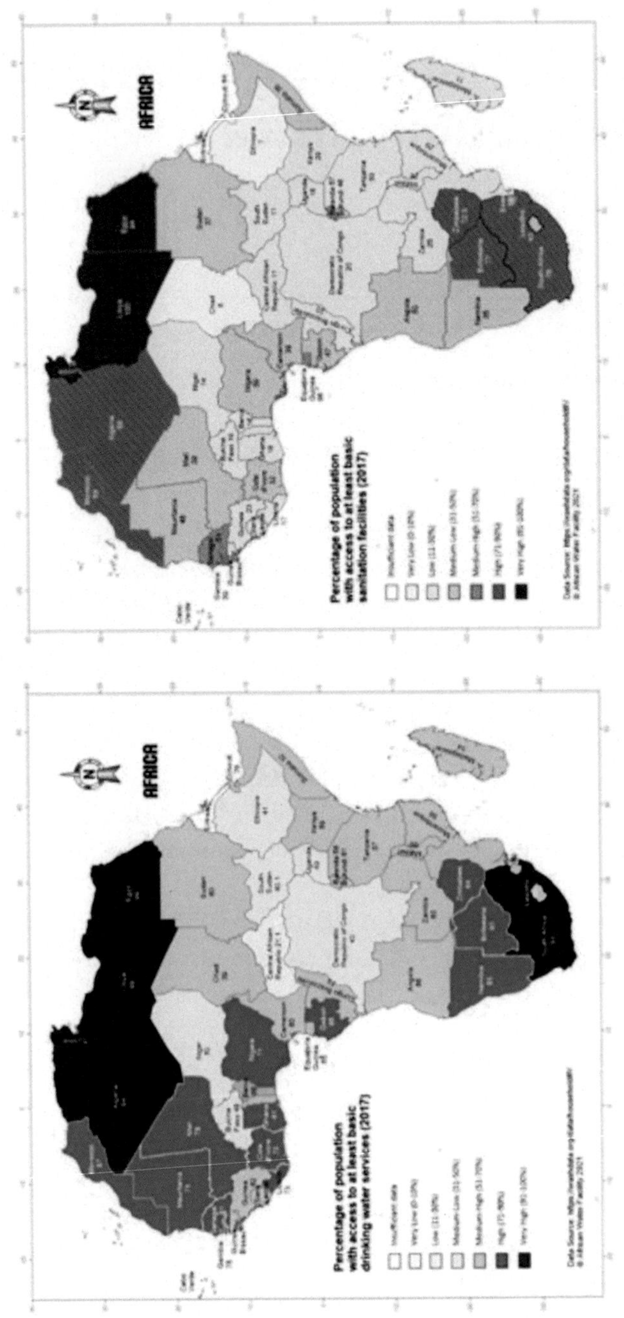

Figure 8.1 Access to at least basic water supply, and at least basic sanitation facilities in Africa. *Source:* JMP (2021).

The considerable disparity also exists between rural and urban areas, with rural areas typically receiving half the service levels of urban areas. Compared to Sub-Saharan Africa, North Africa has much higher coverage of water supply and sanitation provision with 90% of the population having access to water supply services, 83% to sanitation facilities, and about 70% to hygiene services (WHO/UNICEFJMP, 2021).

Moreover, Africa's water and environmental resources that are critical to sustaining growth and development, are faced with severe pollution due to inadequate sanitation. Ninety percent of wastewater is discharged directly into rivers and lakes without treatment. Africa's energy deficit is stifling its economic growth and impacting job creation, agricultural transformation, and improvements in health, education, and living standards. Hydropower is considered an increasingly important energy source thanks to its renewable and pollution-free nature, long economic life, and, depending on the case, low energy unit costs. The installed hydropower capacity in Africa in 2020 stood at 38 GW, or 2.8% of the world's installed capacity (IHA, 2021). It accounts for over 70% of the renewable electricity share and about 16% of the total electricity share. Hydropower's share of total electricity is predicted to increase to more than 23% by 2040 following efforts to achieve universal access and a low carbon energy transition (IHA, 2021)

Water use in the agricultural sector accounts for almost 80% of the total water withdrawals in Africa. According to estimates by FAO, 153 million people, about 26% of the population above 15 years of age in Sub-Saharan Africa, suffered from severe food insecurity in 2014. With over 95% of African food production being rain-fed, Africa's poor subsistence farmers are in addition among the most vulnerable to climate variability. Food production is hampered by this high dependence, inadequate irrigation schemes, and widespread watershed degradation. In three-quarters of African countries, less than 200,000 ha is equipped for irrigation. To address the food security challenge, many countries are now planning to significantly expand agricultural irrigation, which is expected to further sharply increase water demand from agriculture (FAO, 2021a).

8.1.2 Two compounding factors: COVID-19 and climate change

Water insecurity is compounded by climate change, the uneven distribution of water resources across the continent, and the high variability in hydrology. A great challenge to water security, however, is economic water scarcity

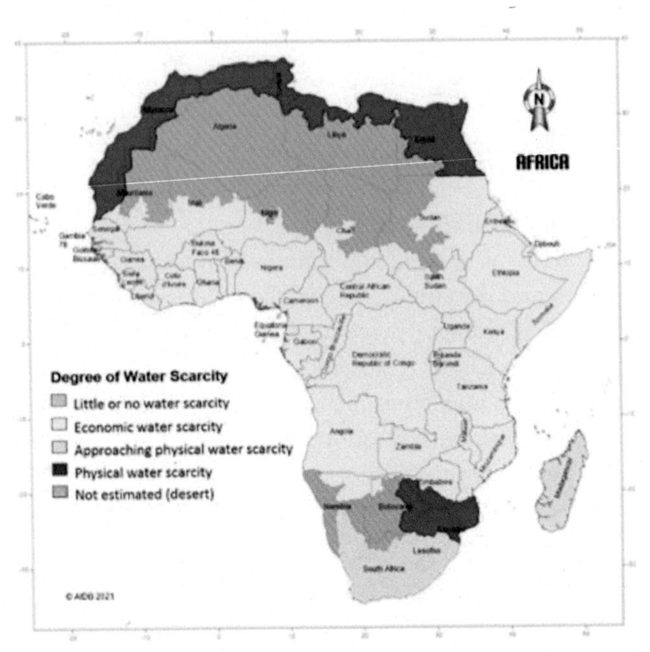

Figure 8.2 Areas of physical and economic water scarcity. *Source:* Molden (2007).

due to inadequate investment in water infrastructure (Fig. 8.2) (Molden, 2007). A key indicator of economic water scarcity is the low total capacity available for water storage to ease scarcity when natural supply is limited. On the continent, the 320 dams and reservoirs provide an average African per capita water storage of 796 m^3/year, compared to the 6150 m^3/year in North America.

The COVID-19 pandemic has exposed weaknesses in the systems of water governance and service delivery. COVID-19 has exposed glaring inequalities in access to water and health services. Water, sanitation, and hand hygiene, together with physical distancing, are central to preventing the spread of COVID-19, and the first line of defense against the threat to lives, economies, and health systems. Both the COVID pandemic and climate change demand radical and coordinated action and increased financing of investments in water security to address systemic challenges.

The impact of climate change on water resources is urgent already and continues to increase. Climate change risks and hazards such as droughts and floods compound the challenge by increasing water insecurity across the continent. Accelerating investments in climate-resilient water infrastructure (including nature-based solutions), information systems, and institutions, is the only sustainable route to secure water resources for the long term. With climate change, COVID-19, an increasing population, and an ambitious industrialization agenda, Africa's water finance needs are mounting. Climate change impacts such as floods and droughts, and water pollution, are persistent threats to water resources in Africa. According to the Global Centre on Adaptation (GCA, 2021) assessment for adaptation finance in Africa, there is a pressing need to accelerate finance for climate adaptation in Africa over the coming decade as the National Determined Contributions (NDCs) of 40 African countries cumulatively show a need for an estimated $331 billion in investment for adaptation through 2030, with about 20% of this sum coming from their annual budgets. The GCA adaptation finance assessment projects that this would create an adaptation investment shortfall of approximately $265 billion through 2030, which needs to be met by international donors and domestic and international financiers. A strong economic case for financing water is required.

According to the 2014 Africa Water and Sanitation Sector Report submitted to the African Union (AMCOW, 2014) only less than 26% of the economic, social, and environmental water demands were satisfied. As another indicator of the economic development value of water, Sub-Saharan Africa loses 5% of its GDP annually because of a lack of water, contaminated water, or poor sanitation. And, each year, 40 billion hours of otherwise productive time is spent just collecting water (ICA, 2019). By 2030 Africa's population will reach 1.6 billion, up from 1.4 billion in 2022. Thus, Africa will need to produce at least 50% more food, and a tenfold increase in water to produce the energy to support growth and development.

Further, consistent investment in water infrastructure, its operation and maintenance, the efficient management of the water resources, and in the strengthening of policy and regulatory frameworks remains necessary to help achieve the SDGs for water and sanitation by 2030. Finally, to sustain growth under climate change conditions, African countries must urgently put in place water infrastructure and institutional measures to enhance resilience to such shocks. While innovative low-cost infrastructure would be an important instrument to provide water livelihoods for the poor at a small scale, the awareness and adoption of such infrastructure are still limited. For example, rainwater harvesting to augment supply for domestic and agricultural uses and manage stormwater is yet to be fully capitalized on. In 2014, countries reported that the contribution of rainwater to the total municipal water consumption accounted for only 1.49% in 2013, compared to the target of 10% by the year 2015, (AUC-AMCOW, 2016).

The United Nations Sustainable Development Goals (UN SDG) Global Accelerator Framework recognizes financing as critical to achieving the SDGs. Typically, every dollar invested has a US$5.50 return from improved sanitation and US$2.00 from improved drinking water. Yet at the global level, the investment gap for water and sanitation infrastructure remains huge, with US$114 billion in capital investment (excluding operation and maintenance) needed annually for the next decade to close the gap for SDG 6.1.1 and 6.2.1 alone. The investment gap is reportedly largest in Africa, with Sub-Saharan Africa requiring the most urgent action as delivery of water investments is lagging economic and social needs. The African Development Bank estimates that US$64 billion is required annually to meet the 2025 Africa Water Vision of Water Security for All; the actual amount invested stands between US$10 and US$19 billion per year. The average level of commitment in 2016–2018 period was US$13.3 billion, leaving an annual gap of US$43 to US$53 billion (ICA, 2018). Current water investments are insufficient to cover the basic social-economic needs of the continent. At current levels of progress, Sub-Saharan Africa will only reach 37% coverage for safely managed drinking water leaving behind the majority of the 1.68 billion population of Africa by 2030. To achieve universal access to safely managed services by 2030 requires an average fourfold increase in current rates of investment, but a tenfold increase in the least developed countries and even a 23 times increase in the fragile economies of the continent.

Also, financing for integrated water resources management is lagging: nearly all African countries report insufficient funding for the planned programs or projects for integrated water resource management at subnational or basin watershed and aquifer level (Fig. 8.3) (AMCOW, 2018).

8.1.3 Climate finance as an opportunity

Climate-related crises caused massive disruptions to Africa's human and economic development in 2021 and will continue to do so until the significant investment is made in Africa's adaptation capacity. According to the Global Center on Adaptation (GCA, 2021), current adaptation finance flows to Africa are insufficient to meet growing adaptation needs on the continent. The GCA assessment showed that in 2017 and 2018, annual global average of USD 30 billion in adaptation finance was tracked. Of this amount, just over USD 6 billion was tracked in adaptation finance to Africa in that period. If this trend continued through 2030, total finance from 2020 to 2030 would only amount to USD 66 billion, far short of the requirement

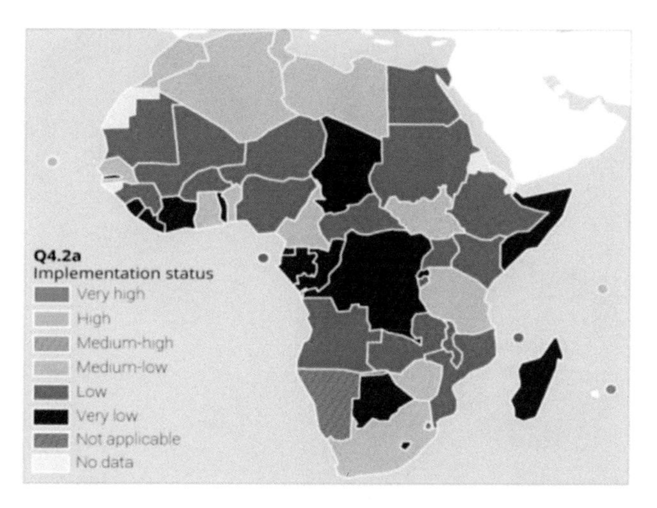

Figure 8.3 Implementation status of IWRM. *Source:* AMCOW (2018).

of USD 33 billion annually in estimated needs per stated cost estimates in NDCs. Research by the World Economic Forum, (Georgia et al. 2021), based on OECD data, found that adaptation funding to Africa amounted to less than US$5.5 billion per year between 2014 and 2018, which works out to roughly US$5 per person in Africa (Fig. 8.4).

The Glasgow Climate Pact, signed at the 2021 United Nations Climate Change Conference, offers some hope that the situation could improve for Africa, which contributes only 4% of global emissions yet is exposed to the greatest climate hazards. The Pact commits industrialized nations to at least double their collective provision of climate finance for adaptation in developing countries from its 2019 level (US$80 billion) by 2025. Africa needs a coordinated and collaborative approach to ensure that industrialized nations honor their financial commitments and that climate funding reaches vulnerable societies—specifically addressing water security, as it is central to collective social, economic, and ecological objectives.

8.1.4 The predominant role—yet at a low level—of domestic finance for water

African governments are making efforts to prioritize water investments, but resources are limited. For 2013, the total domestic expenditure in the water and sanitation sector was reported at US$18.48 billion, (ICA, 2014), that is, falling considerably short of the annual requirement of US$64 billion determined by the AfDB as the minimum required to assure the realization of the Africa Water Vision 2025. As mentioned, in 2018, the

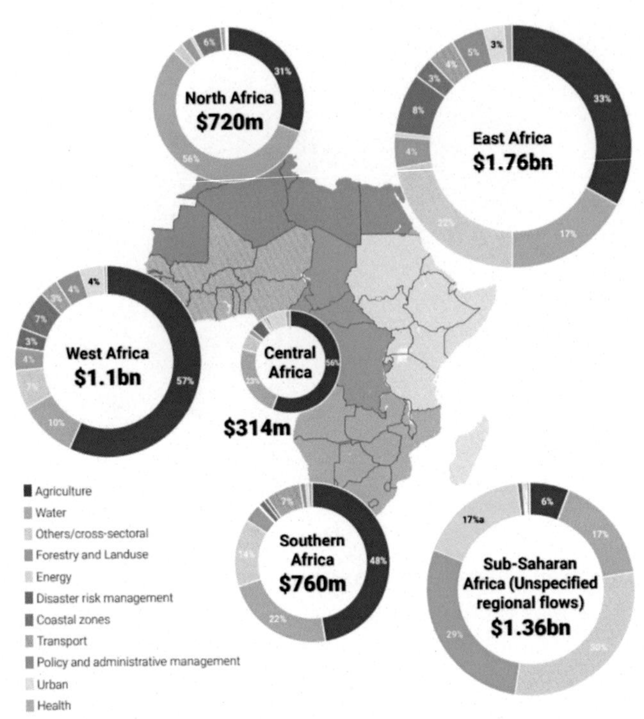

Figure 8.4 Tracked adaptation finance by region and sector (USD, 2017–18 average). *Source:* GCA (2021).

water and sanitation sector commitments were estimated at US$13.3 billion, of which US$5.6 billion were national budget allocations and US$5.1 billion contributions from development agencies and G8 members, less than those of the national governments. The remaining US$2.6 billion came primarily from bilateral and multilateral agencies and the private sector (ICA, 2018). Fig. 8.5 shows the regional distribution of water sector financing in 2018.

The large financing gap, thus, is the result from a combination of low water tariffs driven at least partially by social policies and affordability challenges, limited government fiscal capacity to assume debt, inadequate private sector financing, and limited contributions from national government funds, Multilateral Development Banks and other Official Development Assistance.

Typically, the central government budget has been the main driver of infrastructure investment. Despite some progress in domestic resource mobilisation, averaging 15% before the pandemic, improvements have been slow and uneven across countries in the face of persistent structural issues, high levels of informality, and weak reform efforts (IMF, 2018). However, this is still to be translated into a greater financial commitment to the water sector.

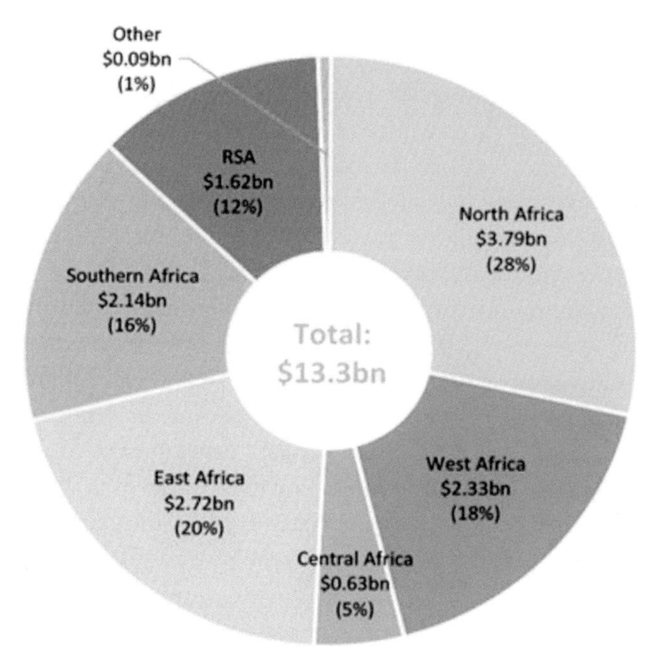

Figure 8.5 Africa water sector financing by region. *Source:* ICA (2018).

The pandemic has only added to existing financing pressures, triggering large health spending needs for prevention and treatment, and subsequently for securing vaccines. Looking forward, several new approaches are necessary to bridge the infrastructure financing gap beside increased public funding, namely structural encouragement of private finance in partnerships where the public sector retains responsibility for ownership and asset operations, and giving equal importance to the maintenance of existing infrastructural assets and the development of new infrastructure to help reduce capital replacement costs (ICA, 2018). The private sector is gradually more engaged in infrastructure; its financing of general infrastructure in Africa in 2018 stood at US$11.8 billion (ICA, 2018), about 94% of which was directed at the energy and ICT sectors, but only 2% (US$ 256 million) was committed to the water sector.

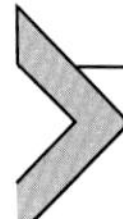

8.2 Transforming Africa's water finance and investment outlook

Solutions are direly needed to transform Africa's investment outlook and close the gap in financing water investments, as required, for example, to make progress toward the targets of the Africa Water Vision and the 2030 SDG 6 water-related targets, to adapt to climate change, and to make progress in socialeconomic development. Measures are urgently required to address the impacts of climate change on water availability, to scale up water investments, to protect and better manage Africa's freshwater resources and avert the growing risks and uncertainties to economic recovery from the regional disruptions, low labor productivity, and political instability caused by the loss of economic opportunity and unemployment.

Across Africa, the function and value of water in the economy remains insufficiently recognized. Water is still mainly seen as essential only for drinking and hygiene, and in irrigation. This narrow view and inadequate appreciation of water's role in the broader economy are reflected in policy decisions on water management. National development and economic growth strategies seldom specifically include water as a key input for economic growth and job creation. Water is often relegated to the social sector for the basic drinking and basic hygiene needs. It is rarely treated as a key pillar for social stability and for sustaining productivity and economic growth in energy, health, agriculture, industry, mining, tourism, etc. Examples abound of instances where investments in energy, agriculture, and industry have progressed, only to realize that water would not remain available in sufficient quantity or quality to sustain the investment project. Economic development pathways across the region often fail to include a basic analysis of how water shortage—because of annual or seasonal variability or because many other users will start drawing from the same source—will constrain the projected development. Water availability is often taken for granted until a "water crisis hits" in the form of a drought, flood, or larger-scale disruption such as insect infestations or pandemics. Too often when the crisis hits it is too late to provide an immediate remedy as water investments take years to plan and build.

A new narrative on water is required that recognizes the full significance of water in the economy. This new narrative should raise the appreciation of water in economic growth, industrialization, and job creation, and strengthen the business case of water in national and regional development. Aggressive efforts are required to position water better in the economic framework, and accelerate the pace of water infrastructure investment.

As argued in the World Development Report (World Bank, 2012), investing in water is investing in jobs. Jobs are the number one priority consistently expressed by the citizenry and by the policymakers of countries, independent of region or income levels. Jobs are at the center of development and lay the foundation for societal well-being. Labor is the most important asset of the poor and jobs are the most important pathway out of poverty. Development happens through innovation and labor productivity. How and whether job opportunities expand is a key determinant on how widely the benefits of innovation and growth are shared. Thus, water and jobs are fundamentally connected, yet policy and investment decisions rarely recognize their interconnectedness. Three out of four jobs in Africa are water dependent, directly or indirectly, and thus water scarcity and supply disruptions are among key causes which limit economic growth and jobs. Therefore, water scarcity—exacerbated by climate change—can contribute significantly to migration and social instability.

The investments to ensure adequate, secure water to the economy have to be financially sustainable, and must be covered either through the pricing of water to consumers, users, and beneficiaries (as in the case, e.g., of flood protection) or through public expenditure. Large-scale water infrastructure, which countries need to ensure water security, is generally expensive, sometimes requiring complex or more innovative financing arrangements. The investments need to be justified in terms of their impact on the national economy and on the water-dependent growth sectors. This requires a more in-depth analysis of the role of water in the economy than has been customary.

To appreciate the role of water in the economy, country-based integrated social-economic analysis should become standard procedure. Such analysis should generate the evidence base and data to inform governments regarding the prioritization and financing of water-related investments. The integrated social-economic analysis related to water infrastructure should address among others following questions:

- What are the roles of water in the national economy, now and in the foreseeable future?
- What impact do water pricing and availability have on different sectors in the economy and how does this affect the economy as a whole?
- Where does water insecurity (shortage of water, floods, polluted water, etc.) currently constrain the economy and where is it likely to be a constraint in the future?
- What synergies and trade-offs are to be considered when prioritizing water investments across sectors?
- What are the likely costs of the "do nothing" scenario to the water-dependent growth sectors and the economy as a whole?

- What is the next critical least-cost augmentation of the existing water supply?
- Given scarce resources, on which water infrastructure is the next investment dollar best spent?
- What regional transboundary water projects should be supported to lower the "cost of doing business" and assure water availability to the productive sectors of the economy?
- Which national and transboundary regional investments in other infrastructure sectors (energy, transport, ICT, and others) can be "bundled" with water, through a "cluster approach" to diversify and de-risk investments in water projects?

8.3 Opportunities to narrow the finance gap

Despite the described challenges, the recent developments to strengthen continent-wide water sector policy coordination, improve water sector monitoring, and mobilize more investment are encouraging. The African Development Bank has issued a new water policy and strategy with the overarching objectives to enhance Africa's water security, and to foster more sustainable, green, and inclusive social-economic growth and development. In February 2021, the Assembly of the African Union Heads of State and Government adopted the Continental Africa Water Investment Programme (AIP) as part of the second phase of the Programme for Infrastructure Development in Africa (PIDA-PAP 2). PIDA is coordinated by the African Union Development Agency—New Partnership for Africa's Development (AUDA-NEPAD). The Program wants to transform the investment outlook for water security and sustainable sanitation for a prosperous, peaceful, and equitable Africa. It aims to narrow the water investment gap by leveraging US$30 billion annually in climate-resilient water investments by 2030 and in the process create 5 million jobs.

To help target interventions and actions toward mobilizing water investments, the AIP-PIDA Water Investment Scorecard (AIP Scorecard) was launched in April 2021. The Scorecard will accompany the implementation of transboundary water projects which were prioritized under PIDA-PAP 2 and become part of the modus operandi of the AUDA-NEPAD delivery model. On February 6, 2022, the Assembly of the African Union (AU) Heads of State and Government approved the AIP-PIDA Water Investment

Scorecard, as a framework for tracking progress in the mobilization of PIDA water investments.

The scorecard is aligned with the African Ministers' Council on Water's (AMCOW) Pan-African Water and Sanitation Sector Monitoring and Reporting System (WASSMO) under the Financing Theme. The scorecard will support African countries to track progress in mobilizing investments, identifying bottlenecks, and taking action to narrow the water investment gap. It will also help make the case to mobilize political leadership and commitment to accelerate financing. Furthermore, it will promote mutual accountability through the tracking progress and help sustain political commitment to act and serve as a tool to engage with public and private investors.

While mobilizing additional concessional funds is essential and could help, this will not be sufficient. The UN and World Bank High Level Panel on Water recommends a new sector financing paradigm based on five broad themes:

- Improve sector governance and efficiency (improving creditworthiness).
- Crowd in or blend private finance (leveraging capital).
- Allocate sector resources more effectively to deliver the maximum benefit for every dollar invested (targeting capital).
- Improve sector capital planning (by pooling and clustering) to reduce unit capital costs (minimizing capital requirements).
- Readjust the strategies of financing institutions to reflect the role of sustainable water management in development (banking system being part of the transformation).

Achieving the new financing paradigm requires stronger political commitment, leadership, enabling environment, investment climate, as well as business cases for water in the economy. This calls for a more collaborative approach in which the key stakeholders play an active role, and which should include the assessment of the social and economic costs of *not* investing in water infrastructure.

Such new paradigm also needs to identify permanent revenue sources for operation and maintenance of infrastructure, preferably through the 3 Ts (tariffs, taxes, and transfers) and from those who benefit from the service; design investment pathways that maximize water-related benefits over the long term; ensure synergies and complementarities with investments in other sectors; and attract more financing by improving the risk-return ratio of water investments. Similarly, addressing the bottlenecks of project preparation and of capacity for transaction management is essential to expand the project preparation and improve the bankability of water infrastructure projects so they can attract financing. Over the past two decades, a number of project preparation facilities (PPFs) have emerged in Africa. Some focus on upstream activities related to the enabling environment. PPFs need more support, however, to help develop concepts to the prefeasibility stage and to prioritize them. Equally important, financing needs to go beyond the projects financed by large international financiers, to include smaller projects that domestic financing sources can finance. Public financial sources are clearly insufficient to bridge the huge financing gap in Africa. Dedicated actions are therefore needed to help increase financing from the private sector for infrastructure investments as well as for service delivery operations, rehabilitation, and maintenance.

Africa needs to translate high-level political commitments to action. The significance of the 2025 Africa Water Vision to the continent's aspirations for economic growth and transformation is underlined by its recognition as one of the strategic continental policy frameworks for the realization of Aspiration 1—a prosperous Africa, based on inclusive growth and sustainable development—of the African Union Agenda 2063 "The Africa We Want". In pursuit of the Africa Water Vision and 2030 SDG water-related targets, a number of decisions and declarations taken by the African Union need to be actualized in order to foster the implementation of the African Water Agenda. These include the 2008 African Union Heads of State Sharm El-Sheikh Declaration, the Commitments for Accelerating the Achievement of Water and Sanitation Goals in Africa, the 2008 Sirte Declaration on Water for Agriculture and Energy in Africa, the 2020 decision on the African Continental Free Trade Area (AfCFTA), the 2020 decision on fast-tracking commitments for accelerating agriculture transformation in Africa through a biennial review mechanism and the Africa Agricultural Transformation Scorecard, and others. In addition, programs and delivery mechanisms to mobilize action and funding toward in-country implementation of the above declarations offer opportunities such as the AfDB's Rural Water and Sanitation Initiative to redress the inequity in rural access to water supply and sanitation services; the African Water Facility that was established by AMCOW and is managed by the AfDB to improve the enabling environment and strengthen water resources management; the Regional Strategic Action Plans for integrated water resources development and management in the SADC and ECOWAS regions; the Nile Basin Initiative; the World Hydrological Cycle Observing System (WHYCOS) project; the AfDB and GCA's Africa Adaptation Acceleration Program (AAAP); the Program for

Infrastructure Development in Africa (PIDA) with its transboundary water infrastructure, energy, ICT and transport projects; and others.

Collectively, these continental initiatives combined with the delivery mechanisms that are currently being implemented, are comprehensive and hold promise for achieving the goals of the African Union Agenda 2063. But as financing water investments are still lagging behind, these declarations need to be translated into more concerted efforts and action. The financing bottlenecks that have constrained progress need to be addressed, notably water governance, planning, preparation of bankable projects, weak institutional capacity for project preparation and implementation, inadequate appetite of the private sector in water investments, and lack of a compelling business cases for water.

As multiple chapters in this volume argue, it will be essential to put water on the radar of other policy communities which are decisive as they strongly affect water demand and availability, and exposure and vulnerability to water-related risks. While there are ample opportunities to enhance policy and institutional frameworks for water resources management and water services, we now better recognize that the enabling conditions to attract finance for water security in Africa encompass other policy domains such as sustainable economic development and financial sustainability. In Africa, as in other regions of the world, this calls for ambitious reforms at the national level and for deeper regional cooperation.

References

AMCOW, 2018. Status Report on the Implementation of Integrated Water Resources Management in Africa: A regional report for SDG indicator 6.5.1 on IWRM implementation.

AUC-AMCOW, 2014. Africa water and sanitation sector report: securing sanitation in Africa, Addis Ababa-Ethiopia.

AUC-AMCOW, 2016. African water resources management priority action programme 2016 –2025 (WRM – PAP). Addis Ababa, Ethiopia.

GCA, 2021. State and Trends in Adaptation'. Global Centre for Adaptation. How Adaptation Can Make Africa Safer, Greener and More Prosperous in a Warming World.

ICA, 2014. Africa infrastructure financing trends, Infrastructure Consortium for Africa.

ICA, 2018. Africa infrastructure financing trends in Africa, 2018, The Infrastructure Consortium for Africa Secretariat c/o African Development Bank.

ICA, 2019. Africa infrastructure financing trends in africa, 2018. The Infrastructure Consortium for Africa Secretariat c/o African Development Bank.

IGRAC, 2014. Transboundary aquifers of the world map. International ...International Ground water Resources Assessment Centre (IGRAC).

IHA, 2021. Hydropower Status Report Sector trends and insights. International Hydropower Association, London.

IMF, 2018. Regional Economic Outlook: Sub-Sahara Africa, April. International Monetary Fund.

Molden, D. (Ed.), (2007). Water for Food, Water for Life: a Comprehensive Assessment of Water Management in Agriculture. London; Earthscan; Colombo: International Water Management Institute (IWMI). 645 pp.

Savvidou, G., Atteridge, A., Omari-Motsumi, K., Trisos, C.H., 2021. Quantifying international public finance for climate change adaptation in Africa. Climate Policy 21 (8), 1020–1036. http://doi.org/10.1080/14693062.2021.1978053.

UNEP, 2010. Africa water atlas'. Division of Early Warning and Assessment (DEWA). United Nations Environment Programme (UNEP), Nairobi.

WHO/UNICEFJMP, Progress on household drinking water, sanitation and hygiene 2000-2020: Five years into the SDGs. Geneva: World Health Organization (WHO) and the United Nations Children's Fund (UNICEF), 2021. Licence: CC BY-NC-SA 3.0 IGO.

World Bank, 2012. World Development Report 2013 : Jobs. Washington, DC. © World Bank.

CHAPTER NINE

Financing water security in Asia

Silvia Cardascia[a], Coral Fernandez-Illescas[a] and Xavier Leflaive[b]
[a]Asian Development Bank
[b]OECD Environment Directorate

9.1 Background and rationale

Water security features are high on the agenda of national and subnational governments in the Asia Pacific region. The combination of rising population and urbanization and climate change require particular attention and action, to mitigate exposure and vulnerability to water-related risks and to enhance resilience of economies, societies, and ecosystems.

Building on the analyses reported in the fourth edition of Asia Water Development Outlook[1], this chapter sheds light on water security, financing needs, capacities, and gaps in Asia Pacific vis-à-vis inclusive and sustainable growth. Moreover, it explores investment opportunities and innovative financing approaches toward water security in the region by illustrating case studies on green-climate finance initiatives and emerging market-based instruments for building resilient water infrastructure and enhancing holistic landscape and watershed management.

In contrast to European member states (see Chapter 2.6 in this volume), countries in the region do not share a common ambition for water security, which would be captured by common goals and converging regulations and policies. They do not share a common statistical apparatus to monitor key trends that drive financing needs and capacities for water-related investments. In that context, the global ambitions set by the 2030 Agenda for Sustainable Development—and the related sustainable development goals—and the Paris Agreement on climate change provide for a shared vision.

Disclaimer: The information and views set out in this chapter are those of the authors and do not necessarily reflect the official opinion of the European Commission, the OECD or their member countries.

[1] Since its first edition in 2007 (ADB, 2007), the AWDO has positioned itself as a flagship publication, contributing to a policy-relevant definition of water security, developing a metric to measure progress toward water security, and inspiring local and national reforms. The fourth edition of the AWDO, in 2020, includes for the first time an assessment of financing needs and capacities required to ensure water security at national level across the region.

Financing Investment in Water Security: Recent Developments and Perspectives.
DOI: https://doi.org/10.1016/B978-0-12-822847-0.00002-8

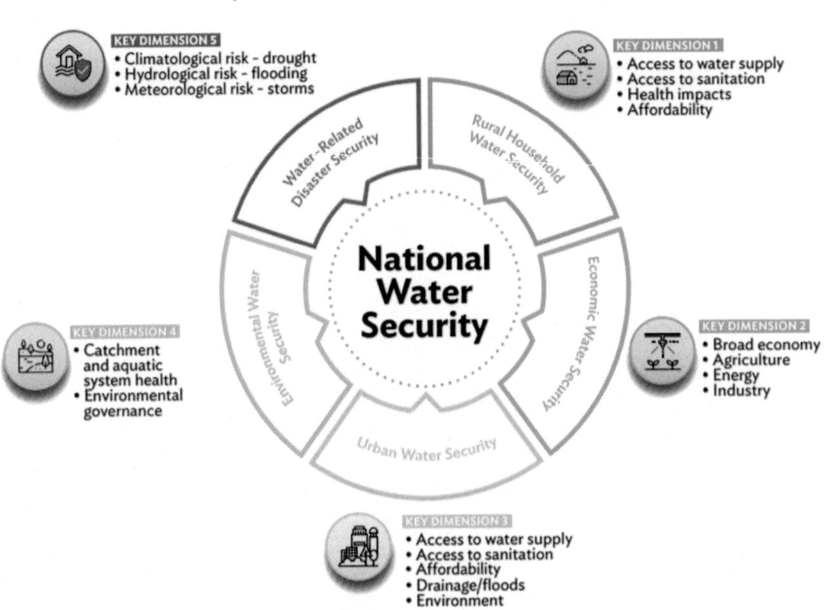

Figure 9.1 Water security framework of five interdependent key dimensions. *(Source: Asian Development Bank, 2020,* The Asia Water Development Outlook, *4th edition, ADB, Manila, The Philippines.).*

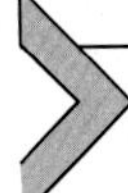

9.2 Financing water security in the Asia Pacific region. Needs, capacities, and gaps[2]

The economic and social case for investing in water security in the Asia-Pacific region is strong and has been made persistently, in particular in the recurrent editions of the Asia Water Development Outlook. The fourth edition refines the definition of water security for the region, as a combination of five interdependent Key Dimensions (Fig. 9.1):

1. Satisfy rural household water and sanitation needs in all communities.
2. Support productive economies in agriculture, industry, and energy.
3. Develop vibrant, liveable cities, and towns.
4. Restore healthy rivers and ecosystems.

[2] All data and projections in this section refer to AWDO, 2020 (ADB, 2020). They were developed and compiled by the OECD (see OECD, 2021) building on a range of data sources, which are acknowledged in the figures. Further analyses and developments for this chapter remain the authors' own and may not reflect the opinion of the OECD Secretariat, OECD member states and the Asian Development Bank.

5. Build resilient communities, which can cope with water-related extreme events.

9.2.1 Drivers for investment in water security in Asia and the Pacific

The main drivers of water security in the Asia Pacific region are well known. Rapid urbanization continues: it is projected that 2.5 billion people—55% of the population—will be living in cities by 2030; the United Nations project that 1,050 million additional city dwellers in Asia-Pacific will need to be connected to water systems in 30 years (UNDESA, 2018). This dynamic drives needs (and creates opportunities) to provide city dwellers with safe water and improved sanitation services, and protect them from water-related risks. A compounding challenge is that 40% of urbanization expansion fuels informal settlements. At the same time, small farms remain an integral part of rural economies in many countries in the region, employing over 40% of people in South Asia and receiving on average 80% of Asia's freshwater.

The challenges to water security driven by population growth and rapid urbanization are exacerbated by a changing climate, which translates into high impact disasters: de Souza and Stumpf (2012) note that the Asia Pacific region is impacted by more than 40% of disasters worldwide and home to 84% of people affected globally. UNEP (2015) computes that weather-related disasters have amounted to USD 750 billion losses from 2003 to 2013 in the region, with Myanmar, the Philippines, Bangladesh, Viet Nam, and Thailand among the most affected countries.

Other drivers of future investment needs are factored in the analyses, but not detailed here. They include: operation, maintenance, and renewal of (expanding) assets; compliance with international commitments and national regulations; people, assets, and GDP exposed to flood risks; and food security and access to global markets for agriculture. In addition, a range of emerging issues in the region is not captured in the quantitative data synthesized below. Consider the improvement of individual and other appropriate sanitation systems, such as off-grid sanitation; stormwater management, in particular in cases of combined water drainage and sewerage systems; contaminants of emerging concern, such as pharmaceutical residues and microplastics in freshwater streams; and sludge management, which grows in volume and potential toxicity, as more stringent wastewater treatment standards are enforced.

These drivers add to the need to invest in water-related infrastructure and services to serve the unserved population[3], to protect exposed and vulnerable groups, assets, and ecosystems, and to develop water resources to support sustainable economic development and adaptation to climate change in the region. The projections and analyses below cover three water-related subsectors:

- Access to water supply and sanitation.
- Flood protection (riverine and coastal).
- Irrigation infrastructure (for both efficiency upgrades and expansion).

9.2.2 Projected investment needs

As regards water supply and sanitation, the OECD, building on World Bank figures—see Rozenberg and Fay (2019)—projects that total annual investment needs for the period 2015–30 to achieve universal access to safely managed water supply and sanitation services in the Asia Pacific region amount to USD 198 bn/yr (Fig. 9.2). This aggregate figure covers capital, maintenance, and operation costs, to connect those without access and to improve the level of service for those who have access but fail to meet the ambition of SDG6 in water and sanitation. This translates into expenditures that range between 1% and 2% of GDP for most countries in the region, with several outliers needing to spend up to 6% of GDP: Afghanistan, Nepal, Pakistan, and Timor Leste.

Investment needs to protect population and assets against flood risks are difficult to monetize, essentially because very few countries monitor current levels of expenditure in this domain. Moreover, possible responses to flood risks vary, with diverging consequences in terms of capital and operating costs: banning construction in flood plains can save huge amounts of investments, compared to building dikes to protect vulnerable assets or communities. Therefore, changes in the volume or share of population and GDP exposed to flood risks are used as a proxy to determine the magnitude of future investment needs. All projections below are imported from the World Resources Institute's Flood Analyser.

Investment needs for flood protection are concentrated in low- and middle-income countries. Bangladesh, Myanmar, Viet Nam, and Cambodia have the greatest percentage of the population exposed to flood risks (Fig. 9.3). The greatest increase in absolute numbers is in India (over 20 million additional people), Bangladesh (approximately 8 million additional people), and Indonesia, Pakistan and Viet Nam (each over 3 million

[3] 300 million people in the region still do not have access to safely managed or basic services of drinking water, 1.2 billion lack adequate sanitation; see UNICEF, 2019.

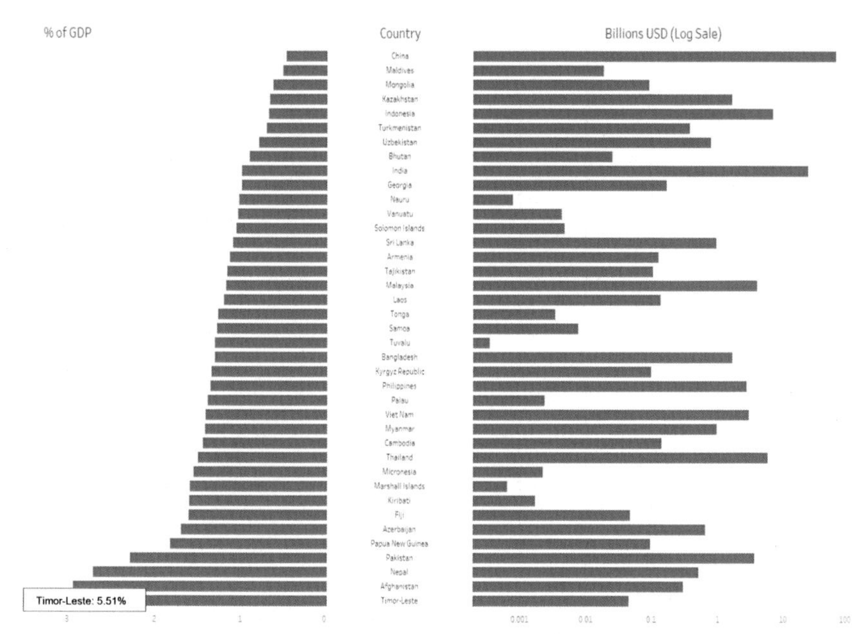

Figure 9.2 ***Projected annual expenditure needs for WSS 2015–30.*** (% of GDP - Billions USD/year) Note: No data for Australia, Singapore, New Zealand, Brunei, South Korea, Japan, Hong Kong (SAR China), Niue, Cook Islands. Scenarios: indirect pathway of basic connection first, and then safe managed connection; SSP2. Calculation for GDP over the period derived from actual GDP in 2015–18, forecast of GDP over the period 2019–24 and extrapolation of average growth rate until 2030. *(Source: Asian Development Bank, 2020, OECD (2021); OECD calculations based on cost of service delivery from Rozenberg and Fay (2019), 2015 dollars. GDP data from IMF.).*

additional people). When it comes to GDP, the exposure is most substantial in India (over USD 280 billion), China (USD 220 billion), and Indonesia (over USD 100 billion). As a percentage of GDP, Bangladesh, Cambodia, Afghanistan, Kyrgyz Republic, Tajikistan and Viet Nam are most exposed, with 6% or more of GDP in 2030 exposed to flood risks. Riverine flood risks are a greater risk to the Asian economy than coastal flood risks. However, coastal flood risks are projected to strongly affect the GDP of Bangladesh, the Solomon Islands, Vanuatu, and Viet Nam (Figs. 9.4–9.6).

Approximately 2.6 million km^2 of agricultural land in Asia is irrigated (about 70% of the world's total irrigated land) (Meier et al., 2018), with China, India, and Pakistan leading. Irrigated area in Asia is projected to expand further by 2050 (by 22% compared to 2010 levels) (Rosegrant et al., 2017). Such trends are driven by population growth, dietary changes, and consideration for food security.

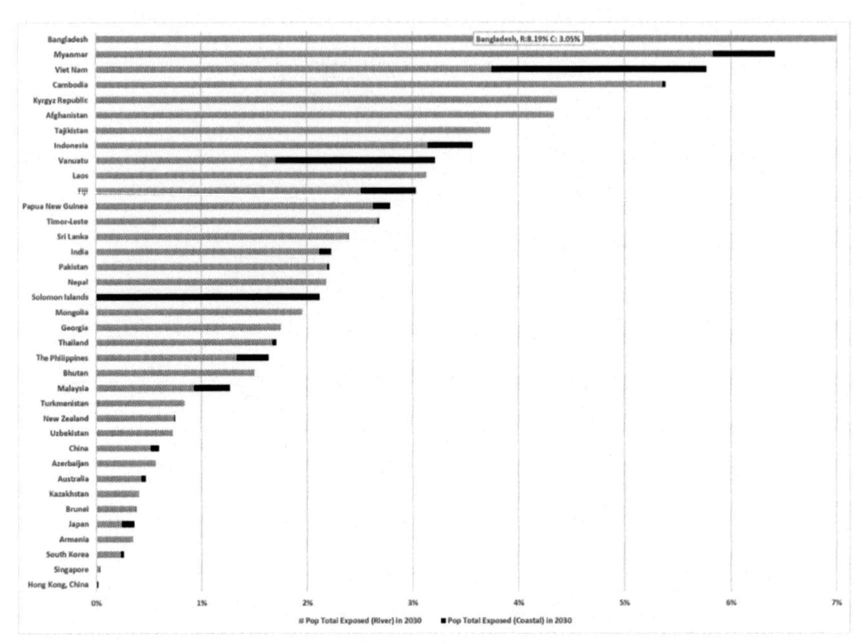

Figure 9.3 ***Projected share of the population exposed to flood risk, 2030.*** Flood risk as a percentage of the population in 2030. Note: Subsidence included in coastal flooding. *(Source: WRI data (accessed in 2020)).*

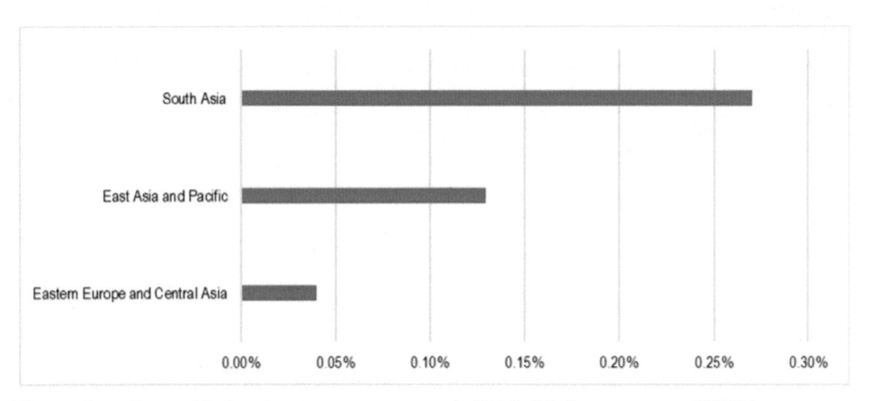

Figure 9.4 ***Annual irrigation investment needs 2015–30.*** Percentage of GDP/year. Note: EECA region includes 13 ADB countries, as well as 10 non-ADB countries. *(Source: Asian Development Bank, 2020, OECD (2021); data from Rozenberg and Fay (2019)).*

No reliable projection of the costs of irrigation exists at the country level. Rozenberg and Fay (2019) model financing needs for irrigation expansion and irrigation efficiency, taking account of socioeconomic and climate change (SSP2/RCP8.5). Projections assume farmers will cover the costs of

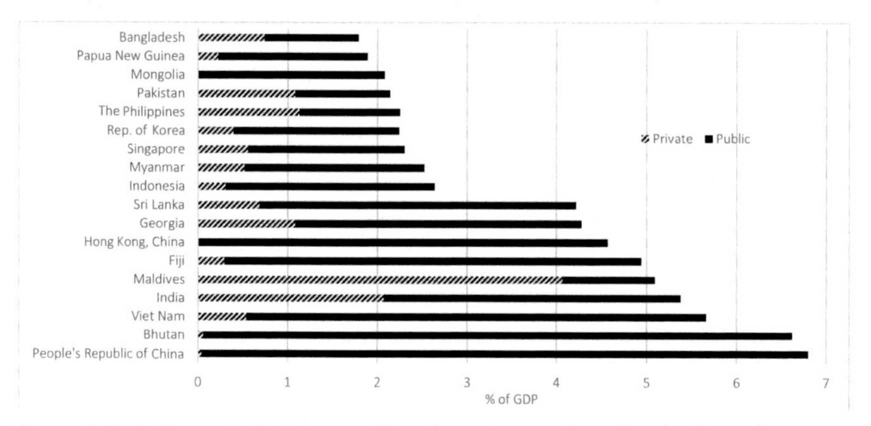

Figure 9.5 ***Public and private expenditure for water supply and sanitation, select economies, select years.*** Percentage of GDP. Note: Actual budget expenditure except Armenia, Bhutan, Georgia, Maldives, Myanmar, and Thailand, which are planned or estimated budget expenditure. Periods covered are 2010–13 average for Indonesia; 2010–14 average for the PRC, Fiji, and Malaysia; 2010, 2011, and 2014 average for Hong Kong (China); 2011 for Armenia, Bangladesh and Georgia; 2011–2012 average for Nepal; 2012–2013 average for India; 2011–13 average for Maldives; 2011, 2012, and 2014 average for Singapore; 2011–14 average for the Philippines, Sri Lanka, and Thailand and 2014 for Myanmar. *(Source: Asian Development Bank, 2020, OECD (2021); OECD calculations based on ADB data (2017). Original sources of country-level data: World Bank Private Sector Participation in Infrastructure (PPI) database, World Development Indicators, ADB estimates.).*

parts and materials for farm irrigation equipment and a water price, which reflects the relative scarcity of water due to increasing demand from other sectors. The figure below presents such estimates at the sub-regional level, as a share of GDP. In absolute costs, East Asia and the Pacific face the greatest investment costs of approximately USD 20 bn/year.

9.2.3 Prevailing sources of finance

Some data are available to characterize prevailing sources of finance for water-related investments in selected countries in the region. In the case of water supply and sanitation, public budgets are the dominant source of funding, and the level of effort can be higher than 5% of GDP, in Bhutan, China, India, the Maldives, or Viet Nam. Private investment can represent a significant share of total levels of expenditures in some countries (Bangladesh, India, the Maldives, Pakistan, the Philippines). AIIB (2019) reports that China has attracted the major share (49%) of private investment in water infrastructure in emerging economies in the Asia-Pacific region since 2000, followed by Malaysia (8%), Philippines (6%), and India (6%). This

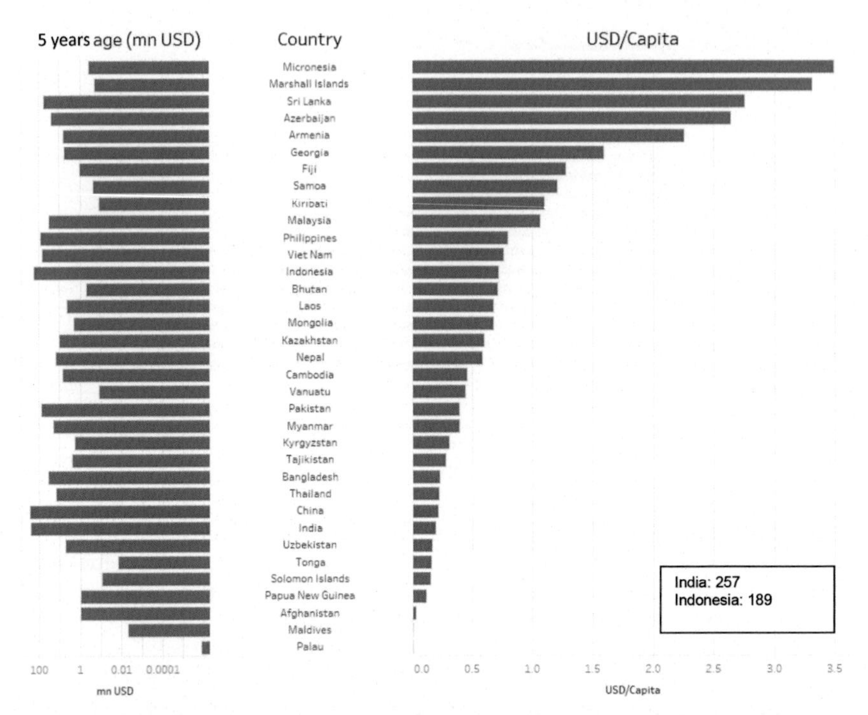

Figure 9.6 ***Annual ODA for water-related infrastructure in Asia-Pacific countries (average 2013-2017; Millions of USD)*** Notes: Includes ODA disbursements for the following water-related infrastructure: water supply and sanitation, water storage, flood protection, irrigation. Data are in millions of USD, gross disbursements, constant 2017 prices. 2017 population figures. Data issues for some countries prevented their inclusion in the graphic. Particularly: Brunei, Cook Islands, Nauru, Niue, Timor Leste, Turkmenistan, and Tuvalu. *(Source: Asian Development Bank, 2020, OECD (2021); data from OECD (2019)).*

reflects the size of the country, but also distinctive capacity to attract investors' attention and to accelerate investment in water-related infrastructure in the last two decades.[4]

The OECD Creditor Reporting System tracks official development assistance (ODA), per country of destination. It reports on selected policy areas. Water data reported below combine ODA flows for water supply and sanitation, water storage, flood protection, and irrigation. The total level of water-related ODA has remained stable over the last decade. This may reflect more reliance toward general budget support (i.e. ODA flows not earmarked for a particular policy domain). The figure below indicates that

[4] Recent trends in China are discussed in some details in the next chapter in this volume.

India and Indonesia have received comparatively high volumes and shares of water-related ODA in the region, while countries, which may have fewer alternatives, benefitted comparatively less (if at all) such as Afghanistan, Bangladesh, Myanmar, Papua New Guinea, and Timor-Leste.

The projections above signal that achieving water security in the Asia Pacific region—when in line with the ambition of the sustainable development agenda and in particular SDG6 on water and sanitation—will drive significant levels of investment and finance, in both developed and developing countries in the region. Obviously, countries are unevenly equipped and able to cover these needs. Tailored financing strategies will be required, which combine a range of initiatives, at local, subnational, national, and regional level. Regional collaboration can contribute to support and strengthen such strategies, most importantly by collecting data on water security, financing needs, and capacities; and coordinating policy responses and investment plans at the level of transboundary water basins. The next section flags a couple of areas, where progress can support further investment in water security in the region.

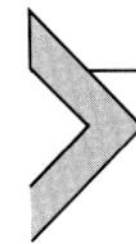

9.3 Market-based mechanisms and valuation tools to finance water-related investments in South-East Asia and China

Tailored financing strategies will combine a range of measures, at several geographical levels. The table below suggests that this is not merely about channeling more money to water-related investments. Policies and institutional frameworks can be designed to make sure that the best is being made of existing assets and financial resources, thereby postponing the need to additional investments. Similarly, significant progress in water security can be achieved by minimizing future investment needs, typically by avoiding building future liabilities: robust plans would be a giant step forward, that account for a changing climate and related uncertainties, that drive investment decisions in domains that affect water security (including agriculture, energy, urban development or land use) and that are supported by realistic financing strategies.

Obviously additional funding will be required. However, this is best seen as one element of robust financing strategies, not as a condition for action toward water security. Additional funding can originate in prevailing sources of finance. With regards to water supply and sanitation, available data suggests that additional revenues could be generated by adequately designed tariffs for water supply and sanitation services, while taking account and addressing affordability issues (see OECD, 2021, for a discussion). When it comes to flood protection, such fiscal instruments as land value taxes can capture some of the rent generated by increased water security that benefits property developers. None of these options are silver bullets. But they all deserve detailed consideration while devising financing strategies for water security that reflect local conditions and opportunities at local or national levels.

Leveraging commercial finance toward green infrastructure is a key aspect in supporting governmental recovery strategies in Asia Pacific and in South East Asia. The need for such leveraging is likely to only increase in a post-COVID world. In addition, Multilateral Development Banks and International Financial Institutions can utilize green financing instruments to guide national governments in the development of mechanisms and approaches toward leveraging commercial finance by focusing on context-specific risk allocation between relevant entities—as well as the development of underlying project pipelines and capacities (see also Chapter 4 in this volume).

Green financing addresses all three columns in Table 9.1 via capacity building, alignment of public financing with coherent, multi-disciplinary climate and development priorities, and the development of clear financing strategies for entry into private equity markets. The sub-sections below explore examples of market-based mechanisms and valuation tools that can potentially serve as models for addressing water-security investment needs in Southeast Asia and the People's Republic of China (PRC).

9.3.1 Shandong Green Development Fund (SGDF)

The SGDF pilots an innovative leveraging mechanism to fund climate-impacting projects in Shandong Province, PRC. Created with ADB support in 2019, the fund is financed by USD1.5 billion in investment from a blended range of private, institutional, and public sources and seeks to leverage public funding to increase private support for a portfolio of high-risk climate-relevant products contributing to climate mitigation and/or adaptation activities. In particular, it focuses primary on early project development and attempts to subsidize upfront project risks for a capped time period—with the goal of making the project more bankable for private funding beyond this initial capped period. In this way, it seeks to mitigate risks related to climate impacts and support an integrated public/private approach to climate finance. The SGDF was the first ADB and PRC Catalytic Green

Table 9.1 Options to address financing needs for water security in Asia-Pacific.

Make the best use of existing assets and financial resources	Minimize future investment needs	Harness additional sources of finance
Enhance the operational efficiency of water and sanitation service providers. Build capacity for economic regulation. Encourage connections, where central assets are available. Strengthen capacity to use available funds.	Develop plans to future-proof the water sector. This requires plans that set priorities and drive decision-making, manage uncertainties, and increase resilience. Support plans with realistic financing strategies. Encourage policy coherence across water policies and other policy domains. Manage water demand, and strengthen water resource allocation. Develop flood risk mitigation strategies. Exploit innovation in line with adaptive capacities.	Ensure tariffs for water services reflect the costs of service provision. Consider new sources of finance from users and beneficiaries. Leverage development and public finance to crowd-in commercial finance.

Source: adapted from OECD (2021).

Finance Facility to apply a financial intermediary approach. Due to rapid industrialization in recent decades, Shandong Province has experienced a rapid intensification of local green-house gas emissions. Therefore, key SGDF mitigation objectives include a commitment to fund projects capable of reducing carbon dioxide emissions within the province by 3.75 million tons per annum by 2027.

However, SGDF priorities extend beyond renewable power generation and also address water-relevant climate vulnerabilities. The demand for domestic and industrial water use in Shandong Province has increased rapidly in recent years and has not been met by an analogous increase in water resource use efficiency. Likewise, water pollution is a serious regional concern in the province—particularly the eutrophication of rivers and lakes due to industrial and domestic wastewater. If current trends continue, domestic water pollution will negatively impact the sustainable socio-economic development of the province. Reflecting this risk, the SGDF has a vigorous water sector component and includes priority project areas in water supply, sanitation, storm drainage, urban water footprints, and the water–food–energy nexus at the core of the province's vulnerability to climate variability.

To reflect the funds' initial ambition, all SGFD-funded projects must meet predefined SGDF green framework criteria regarding project financial eligibility, governance, implementation arrangements, environmental and social management systems, gender considerations, monitoring and evaluation, and verification, and be aligned with the Green Climate Fund investment framework. In addition, funded projects must respect the principles, terms, and conditions agreed upon by ADB and SGDF co-financiers. In this way, the fund attempts to better leverage private funding sources while simultaneously keeping a long-term (i.e., "future proof") focus on sustainability that is anchored by specific development goals.

9.3.2 ASEAN Catalytic Green Finance (ACGF) Facility

Green infrastructure finance needs in Southeast Asia alone amount to US$3.1 trillion or US$210 billion annually for the period of 2016 to 2030 for climate change-adjusted infrastructure investments. Given the US$102 billion per year financing gap for selected SE Asian countries in 2016–20, the need for private sector financing was acute even prior to the COVID-19 pandemic. The economic impact of the COVID-19 crisis has only increased the size of this financing gap. To achieve investment on the scale required, private capital sources must be secured within a competitive finance market. At the same time, investments must be realized in ways that minimize the environmental impact of infrastructure projects.

Based on US$1.4 billion in funding commitments extended by the ADB and multiple global development partners (e.g., the EIB, Agence Française de Développement (AFD), Kreditanstalt für Wiederaufbau (KfW), the Republic of Korea and the EU), the ASEAN Catalytic Green Finance Facility was established in 2019 under the ASEAN Infrastructure Fund (AIF) with the goal of catalyzing private investment and speeding up green infrastructure development within SE Asia. The facility, which is owned by the ASEAN member states and the ADB (which also administers it), supports ASEAN governments to prepare and source public and private financing for infrastructure projects that promote environmental sustainability and contribute toward climate change goals.

As with the SGDF, ACGF's financing approach is based on the concept of front-end investment as a tool for derisking projects not currently suitable for private investment and providing a bridge to capital, technologies, and management efficiencies available in the private sector. Specifically, using AIF equity funds, the ACGF applies stage-financing approach whereby projects are given access to low-interest loans for the first seven years—with the hope that this upfront investment can make projects suitable for (more extensive) private financing, at higher interest rates, from their eighth year onward. In this way, the ACGF attempts to meet its goal of leveraging US$3 in commercial support for every US$1 of public funding invested. In addition, all ACGF projects must be both sufficiently "green" to make a significant contribution to climate change mitigation/adaptation goals (or other key environmental sustainability issues in the region) as well as suitably "bankable" to present a credible roadmap for eventual private capital investment.

The ACGF focuses mainly on infrastructure projects within the energy, water, transport, and urban sectors. Examples of such projects include (1) renewable energy, (2) energy efficiency, (3) sustainable transport systems, (4) green cities, and (5) sustainable water supply and sanitation. In addition to projects, the ACGF also contains a strong building knowledge and awareness component that seeks to develop viable green project pipelines by leveraging the strengths of existing knowledge and policy-focused organizations operating in the region. In this way, it contributes to the "future-proofing" of the water sector, helps establish realistic long-term financing strategies, and develops multisectoral connections.

9.3.3 The rise of eco-compensation in the PRC

In 1997 and 1998, the PRC experienced two major water crises. The Yellow River dried up in 1997, failing to reach the sea for a record of 267 days, and causing a severe water shortage in the basin. In 1998, devastating floods in the Yangtze River killed over 3600 people, causing approximately US$26 billion economic damages (Zhang and Bennett, 2011). The central government identified deforestation and widespread soil erosion in the upper and middle reaches of the Yangtze and the Yellow rivers as the main contributing factors to both events. To address the crisis, the government launched two of the largest payment for ecosystem service schemes in the world: the Natural Forest Conservation Program (NFCP) and the Sloping Land Conversion Program (also known as Green-to-Grain Program - GTGP). The GTGP is a cropland conversion program, providing farmers with grain and cash subsidies to convert cropland on steep slopes to grassland and forests. From 2000 to 2013, the central government made the equivalent of US$55.5

billion in afforestation investments, which resulted in 31.8 million ha of new forest, reduction of soil erosion by 30%, and surface runoff dropping by approximately 20% in the Yangtze and the Yellow River programs. To date, over 32 million rural farmers participated in the program (Mandle et al., 2019). The peculiarity of the GTGP lays in its intertwined objectives of reducing soil erosion and environmental degradation, with alleviating poverty, and diversifying people's livelihoods. Following the success of the GTGP program, the PRC has been piloting numerous programs falling under the umbrella of ecological or eco-compensation.

Ecological or eco-compensation is defined by the National Development and Reform Commission (NDRC) as "a type of public system or institution bearing either public or private sector measures to adjust the relative benefits and costs of ecological service provision among the key stakeholders in order to realize the goals of protecting the environment" (Zhang and Bennet, 2011). They are aimed at promoting a more harmonious coexistence of people's and nature, duly factoring in the opportunity costs of foregone development and the value of ecological service provision. Wang et al. (2020) have summarized three main strands of eco-compensation for watershed services in the PRC, embedded in its public-private, state-market and central–local relationships. Firstly, although the economic value of water resources is recognized under the 2002 Water Law, water fees are still very low and participation of market forces in eco-compensation schemes is fairly limited. Secondly, a large institutional reform on water rights is ongoing in the PRC, which will allow water transfers from different water users and between industrial and agricultural sectors. De facto, formal water markets are still highly regulated, from planning down to implementation and monitoring phases, with the government being the major player. Thirdly, the suppliers and users of ecosystem services are upstream and downstream county-level governments, with payment made on a formula calculated at provincial or subprovincial level. Individuals, communities, and non-state actors are not effectively part of the program. Central, provincial, and county governments, driven by the state-centred conservation and poverty alleviation agenda, are still the main stakeholders. More recently, the government signaled its commitment to leave more manoeuvring space to market-based forces, through shifting its role from program implementer to 'rule setter'. The first national regulation on eco-compensation will be published in 2022 and will lay out the foundations for strengthening market-based and green finance mechanisms.

Eco-compensation is also used as an incentive mechanism for the water fund established in Qindao Lake, the largest artificial freshwater lake in the PRC, which supplies drinking water to approximately 10 million people across Zhejiang province. The lake's water quality is at critical risk due to land degradation, use of chemical fertilizers and pesticides, and livestock waste. Partnering up with the Alibaba Foundation, Minsheng Insurance Foundation and Wanxiang Trust, The Nature Conservancy is supporting local farmers in producing organic tea and implementing high-yield rice paddy agriculture. Digital technologies and cost-effective solutions are being deployed to reduce the dosage of fertilizer applications and nonpoint source pollution across 30,000 acres, which will improve 15,000 livelihoods.

9.3.4 Leveraging investments in water services through gross ecosystem product accounting

Valuing natural capital is the first step to assess the status of ecosystem services that nature provides to benefit people and the economy, leveraging investments in watershed services (IWS). IWS is an umbrella term, indicating the broad range of finance, policy, and governance mechanisms to fund watershed restoration or conservation and promote ecosystem-based co-benefits (Vogl et al., 2017; Bennet and Carrol, 2014). Watershed services provided by healthy landscapes generate both water quality and quantity benefits (i.e., filtration of nutrients and contaminants, flow regulation, water supply, and aquatic productivity). IWS programs build on basin-wide approaches and three main assumptions. First, tackling water quality pressures at the source can be more cost-effective than mitigating their negative impacts downstream. Second, protecting water sources upstream, through agricultural restoration and conservation practices can provide both ecological benefits and livelihood opportunities. Third, their design is tailored to natural and socio-economic conditions at the local level (Vogl et al., 2017).

To better align policy reforms with environmental protection and integrated water management targets, the PRC government has developed a national balance sheet to calculate the country's aggregated value of ecosystem services, known as gross ecosystem product (GEP). The GEP uses the United Nations' Millennium Ecosystem Assessment framework for categorizing ecosystem services. Categories include ecosystem goods (food, materials, energy, biodiversity), regulating services (carbon sequestration, soil retention, sandstorm prevention, water retention, flood mitigation), and cultural services (recreational, eco-tourism, cultural values). Analogously to gross domestic product (GDP), GEP multiplies the quantities of ecosystem goods and services by their prices, using market and nonmarket valuation methods. These are aggregated to obtain a traceable single measure to reflect the growth or decline of ecosystem services, including those provided by watersheds (namely, water retention and flood mitigation). Like GDP, GEP is an accounting and not an economic welfare measure. By measuring the performance of ecosystems, GEP is a powerful science-based tool to inform water security-driven decision making, policy reforms, and future IWS programs (Ouyang et al., 2020). GEP accounting is an effective tool to inform decision-making for ecological planning and conservation policies. GEP can be also widely applied for evaluating government policies and designing performance indicators for land-use and integrated natural resources management. Several actions can be taken to implement the GEP metrics, such as the creation of a natural capital accounting system[5]; the setup of a leading group of key ministries and agencies with multisector responsibilities[6]; the establishment of a scientific committee to incorporate ecological metrics into the national statistical reporting system for planning purposes. GEP has been supported by the Asian Development Bank with pilot sites in three provinces—Qinghai, Guizhou, Yunnan—and in the Qiandongnan Autonomous County. A recent study has comprehensively analyzed and compared ecosystem service values in the Yangtze and the Yellow River basins, offering input for regional eco-environmental planning and policymaking[7].

9.4 Conclusions

As a region, Asia is projected to be heavily affected by water-related risks, driven by a combination of population growth and urbanization,

[5] For examples, see the Stanford University Natural Capital Project's InVEST. *Integrated Valuation of Ecosystem Services and Tradeoffs;* and World Bank's WAVES. Wealth Accounting and Valuation of Ecosystem Services.

[6] In the case of the Yellow River, this may include the Ministry of Ecology and the Environment (MEE), the State Forestry and Grassland Administration of the Ministry of Natural Resources (MNR), the Ministry of Water Resources (MWR), the Ministry of Agriculture and Rural Affairs, the National Bureau of Statistics and the National Development Reform Commission (NDRC) and the Yellow River Conservancy Commission (YRCC).

[7] Assessment of Ecosystem Service Value and Its Differences in the Yellow River Basin and Yangtze River Basin. https://www.mdpi.com/2071-1050/13/7/3822.

economic growth, and climate change. Analytical work for the 4th edition of the Asia Water Development Outlook provides a contrasted picture of financing needs and capacities in the region, reflecting diverse exposure and vulnerability to water risks and capacities to respond across countries.

Lessons could be learned from Europe (see Chapter 11 in this volume) on the benefits of (sub)regional cooperation to set common levels of ambition, share information, and data and learn from good practices in neighboring countries. There is scope to intensify water-related cooperation in the region.

Prevailing financing mechanisms can be enriched by innovative ones, which are pioneered in select Asian countries. The chapter explores recent developments in relation to market-based mechanisms and valuation tools to finance water-related investments in South-East Asia and the People's Republic of China. These developments are well-aligned with the attention to ecosystems and their value to support water security, especially under shifting conditions.

Scaling up such mechanisms requires distinctive institutional and policy frameworks, which need to be characterized further, and tailored to national circumstances. There is scope for further research to better characterize such enabling environments and better understand the role national governments can play to facilitate their deployment.

References

AIIB, 2019. Asian Infrastructure Investment Bank Water Strategy: Water Sector Analysis. AIIB.

Asian Development Bank, 2007. Asia Water Development Outlook, 1st ed. ADB, Manila.

Asian Development Bank, 2020. Developing gross ecosystem product accounting for eco-compensation, https://www.adb.org/sites/default/files/project-documents/48469/48469-001-tcr-en.pdf.

Asian Development Bank, 2020. Asia Water Development Outlook, 4th ed. ADB, Manila ISBN 978-981-4136-06-8.

Bennett, G., Carroll, N., 2014. Gaining depth: state of watershed investment 2014. www.ecosystemmarketplace.com/reports/sowi2014.

de Souza, R., Stumpf, J., 2012. Humanitarian Logistics in Asia-Pacific. Haupt Verlag, Bern, Switzerland.

Mandle, L.A., Ouyang, Z., Salzman, J.E., Daily, G.C., 2019. Green growth that works: natural capital policy and finance mechanisms around the world. Island Press.

Meier, J., Zabel, F., Mauser, W., 2018. A global approach to estimate irrigated areas–a comparison between different data and statistics. Hydrol. Earth Syst. Sci. 22 (2), 1119–1133.

OECD, 2021. Financing water security for sustainable growth in the Asia-Pacific region. In: OECD Working Paper. OECD.

OECD, 2019. OECD creditor reporting system. https://stats.oecd.org/Index.aspx?DataSetCode=CRS1.

Ouyang, Z., Song, C., Zheng, H., Polasky, S., Bateman, I.J., Liu, J., Ruckelshaus, M., et al., 2020. Using gross ecosystem product (GEP) to value nature in decision making. In: Proceedings of the National Academy of Sciences of the United States of America.
Rosegrant, M.W., et al., 2017. Quantitative Foresight Modeling to Inform the CGIAR Research Portfolio, Project Report. International Food Policy Research Institute (IFPRI), Washington, DC.
Rozenberg, J, Fay, M., 2019. Beyond the Gap: How Countries Can Afford the Infrastructure They Need while Protecting the Planet. Sustainable Infrastructure. World Bank, Washington, DC https://openknowledge.worldbank.org/handle/10986/31291.
UNDESA, 2018. World Urbanization Prospects: the 2018 Revision. United Nations Department of Economic & Social Affairs, New York.
UNEP, 2015. Aligning the Financial Systems in the Asia Pacific Region to Sustainable Development. UNEP Inquiry: Design of a Sustainable Financial System. UNEP, Geneva.
UNICEF, WHO, 2019. Progress on household drinking water, sanitation and hygiene 2000-2017. Special focus on inequalities. UNICEF, WHO.
Vogl, A., Goldstein, J., Vira, B., Bremer, L., McDonald, R., Shemie, D., et al., 2017. Mainstreaming investments in watershed services to enhance water security: Barriers and opportunities. Environmental Science & Policy 75, 19–27. doi:10.1016/j.envsci.2017.05.007.
Wang, R.Y., Ng, C.N., Qi, X., 2020. The Chinese characteristics of payments for ecosystem services: a conceptual analysis of water eco-compensation mechanisms. Int. J. Water Res. Develop. 36 (4). doi:10.1080/07900627.2019.1605889.
Zhang, Q., Bennett, M.T., 2011. Eco-compensation for watershed services in the People's Republic of China. Asian Development Bank, Mandaluyong City. https://www.adb.org/sites/default/files/publication/29290/ecocompensation-watershed-prc.pdf.

CHAPTER TEN

Financing mechanisms for water treatment projects in China

Elaine Wu
J.P. Morgan

10.1 Introduction

China is the most populous nation but also one with a severe water shortage. Since the 1990s, the government has focused its efforts on treating wastewater and ensuring a high standard for drinking water. To do so, investments in water pipelines, wastewater treatment plants, and other-related infrastructures had to be made. The private sector, including state-owned enterprises, played an important role in funding and constructing these projects. During these years, China's water financing has evolved from the simple build-operate-transfer project mode to the more sophisticated public-private partnerships. More recently, the central government's announcement of real estate investment trusts (REITs) for infrastructure projects is again an example of using innovative financing instruments to fund water investments. The Chinese government will continue to lend support in this regard as needed. The COVID-19 pandemic has heightened global awareness of climate change and environmental issues. President Xi Jinping's announced target for China to reach carbon neutral by 2060 is a testament to that. The rise of environmental, social, and governance investing in recent years will support this effort because investors will pay more attention to furthering United Nations' Sustainable Development Goals. Two of the 17 goals, "clean water and sanitation" and "life below water" bode well for this development. This should attract more private capital to the China water sector.

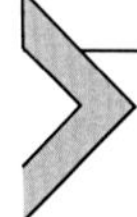

10.2 Background on China water situation and relevant government entities

China has one of the lowest water resources per capita globally. Its water resources have been stable, averaging 2.8 trillion cubic meters in the past

Financing Investment in Water Security: Recent Developments and Perspectives.
DOI: https://doi.org/10.1016/B978-0-12-822847-0.00010-7

decade while the population continues to grow (Source: Water Resources Bulletin, China Ministry of Water Resources; http://www.mwr.gov.cn/sj/tjgb/szygb/202008/t20200803_1430726.html. Link to bulletin's portal in Chinese: http://www.mwr.gov.cn/sj/#tjgb). The central government's strategy has been to reduce water pollution as a way to increase or maintain the amount of raw water (from rivers, lakes, underground) that could be treated to the level of potable water. It has also been increasing water tariffs to improve water conservation.

The central government has several entities overseeing these efforts. The National Development and Reform Commission (NDRC) issues five years plans on the construction of wastewater and recycle water with capacity and investment targets for each province. Toward the end of each 5-year period, local governments (provinces, counties, and municipalities) are held accountable for meeting these targets. Local government officials' own year-end evaluation could be affected if they fail to meet these targets.

From 1990 to 2010, a key focus for the government was ensuring proper treatment of wastewater and lifting treatment ratios. In 2011, the central government expanded on this with the issuance of a document called, "Decision from the Chinese Communist Party Central Committee and the State Council on Accelerating Water Conservancy Reform and Development." This statement, also known as the "No.1 Document," outlined the government's plan to increase investments in water conservancy projects in the next 10 years (Link to document in Chinese: http://www.scio.gov.cn/xwfbh/xwbfbh/wqfbh/2014/20140321/xgzc30632/Document/1366941/1366941.htm). It called for doubling annual investments in these projects from Rmb200 billion in 2010 to Rmb400 billion (USD31 billion to USD62 billion). The types of construction projects required included the construction of drainage canals, irrigation facilities, ancillary facilities, pipeline construction, and renovation of rivers. In 2015, the State Council followed up on that document with the "Action Plan on Prevention and Control of Water Pollution," commonly known as the "Water Ten" document in Chinese. It implemented new targets to lift tap water and wastewater treatment standards. (Link to document in Chinese: http://www.gov.cn/zhengce/content/2015-04/16/content_9613.htm).

In addition to the State Council, several other central government entities also set policies on the water sector. The Ministry of Housing and Urban-Rural Development (MOHURD) oversees policies related to the construction of water treatment facilities and infrastructure in urban and rural regions. The Ministry of Water Resources and Ministry of Ecology and Environment focus on maintaining the quality of water resources and setting standards for water treatment. The NDRC sets economic policies and also works with MOHURD to set national policies for the pricing of tap water and wastewater treatment tariffs. The Ministry of Finance regulates the financial instruments used for funding these infrastructure projects.

Investments for water projects are generally made at the local government level. In 2018, China invested Rmb660 billion in water projects, of which 2% of the amount, or Rmb12 billion, was contributed by central government while the remainder came from local governments. Of the 2018 national water investment amount, Rmb255 billion, or 39% was allocated to water resources projects, 33% to flood control projects, 17% to hydropower development and capacity building, and 11% to soil and water conservation and ecological restoration projects. (Source: *2018 Statistic Bulletin on China Water Activities*, Ministry of Water Resources; http://www.mwr.gov.cn/sj/tjgb/slfztjgb/201912/t20191210_1374268.html).

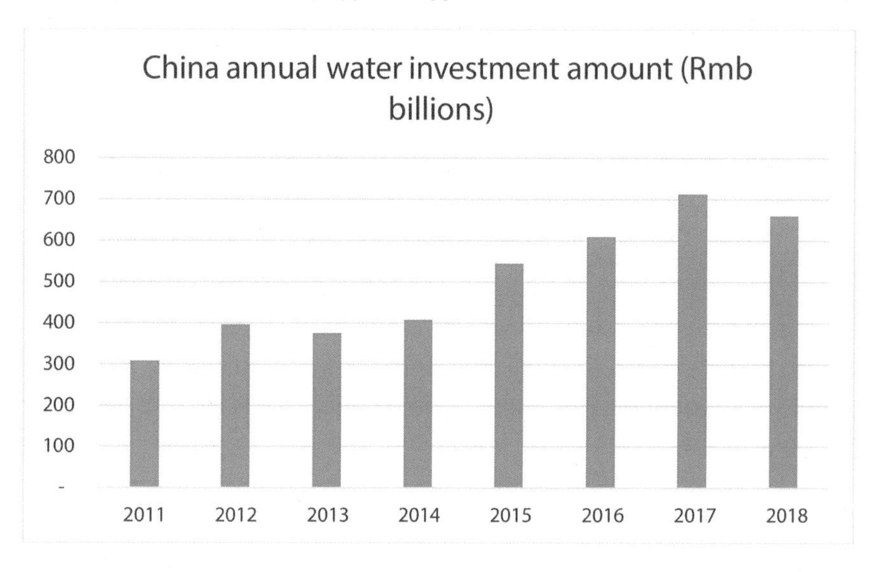

(Source: *2018 Statistic Bulletin on China Water Activities*, Ministry of Water Resources; http://www.mwr.gov.cn/sj/tjgb/slfztjgb/201912/t20191210_1374268.html).

10.3 Various types of water projects

There are five main types of water projects in China, starting from the source of the water to the treatment of water at various stages.

Raw water distribution projects transport water from the natural body of water, such as lakes, rivers, and reservoir, to downstream tap water utilities. These projects are typically owned by local government entities due to the sensitive nature of these projects and rarely require funding from third parties. One example of this is the Dongshen Water Supply Project operated by Guangdong Investment Ltd. This project takes water from Dongjiang, or Dong River, and transports it to the utilities of Hong Kong, Shenzhen, and Dongguan.

Tap water projects are water treatment facilities with networks of water distribution pipelines owned by local governments. They treat the water supplied by the raw water company and deliver potable tap water to residential, commercial, industrial, and public sector users. They are also in charge of collecting water and wastewater treatment tariffs from end-users. Private and "societal" capital—which includes state-owned enterprises and businesses owned by private individuals—is sometimes invited to invest in the pipelines or to take over the operation of these tap water projects to ensure continued investments are made. A common issue with tap water projects has been the leakage of water in the pipelines, which results in low operational efficiency and financial losses. However, in recent years, it has become less common for local governments to sell their stakes of these projects to third parties. This is likely because some of the low-performing projects have already been sold and because local governments prefer to keep their stakes in these projects if they see potential for profitability. The sensitive nature of these operations from the standpoint of national security may also be a considering factor.

Wastewater treatment projects are the most common type of projects which sought funding from banks and investors. This is because the central government had called for increasing the share of total wastewater treated during the 2000s. Most of these projects involved the treatment of urban sewage. Typically, a city could have a few sewage treatment plants. These projects were planned as the need arose. Local governments would solicit bids for tender. Domestic wastewater treatment operators would bid for the concession right to construct and operate the projects. The company with the best qualification and offers would win the project.

For drainage of rain and stormwater, the central government launched in 2015 a pilot plan to build "sponge cities," which includes the installation of permeable pavements, roof gardens, wetlands, and other facilities that would absorb rainwater to become underground water and released back into rivers and lakes. This is meant to be a solution to flooding.

Recycle water projects are less common but could be found as an additional facility at some wastewater treatment projects. Some municipalities and industrial parks construct recycle water facilities so they could reuse water by further treating wastewater that has been treated. Recycled water is typically used for industrial purposes or for watering urban landscape.

Desalination projects take seawater and remove the salt to produce tap water. These projects are located in coastal regions where there is a severe shortage of water. Because the cost of desalination is high, these projects are usually located in the north where the cost of water is higher. Like wastewater treatment projects, local governments typically solicit bids for rights to build and operate desalination projects. The winning firm would typically win the concession right to operate the project for 30 years.

10.4 Development of sewage treatment in China

Since the 1990s, the Chinese government has been focused on fighting water pollution by properly treating sewage and wastewater. It increased urban wastewater treatment ratio from 15% in 1991 to 95% in 2018. (Note: reference to urban regions includes all the major cities in China, but not the counties or townships.) The central government's 2020 goal was to reach wastewater treatment ratio of more than 85% for counties and 70% for townships. (Source: 13th Five Year Plan on National Urban and Township Wastewater Treatment and Recycle Water Facilities. Link to Chinese document: https://www.gov.cn/xinwen/2017-01/23/5162482/files/8f985361f1184a5189238f31facadcf7.pdf)

During the 1990s to 2018, there was a surge in investments in municipal sewage projects. The nature of sewage projects is less sensitive than tap water projects from the national security perspective, hence, businesses are allowed to wholly own these projects. This benefitted both the local governments and businesses. From the local governments' point of view, investment from the private operators allowed these jurisdictions to achieve the sewage treatment ratio targets set by the central government without investments from the government. From the private sector's point of view, this created growth in revenue from the construction of these sewage plants. Urban daily wastewater treatment capacity grew from only 3 million tons in 1991 to 170 million tons in 2017.

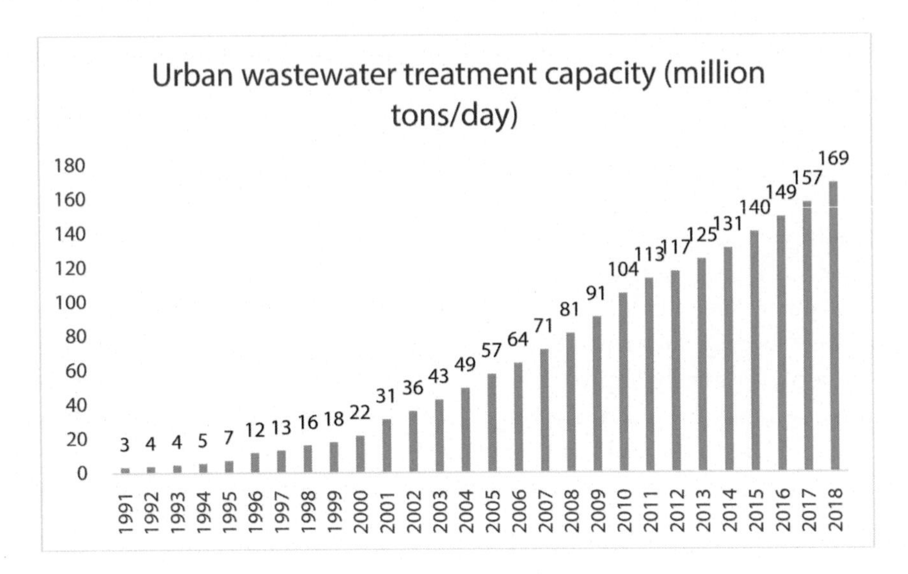

Source: China Ministry of Housing and Urban-Rural development.

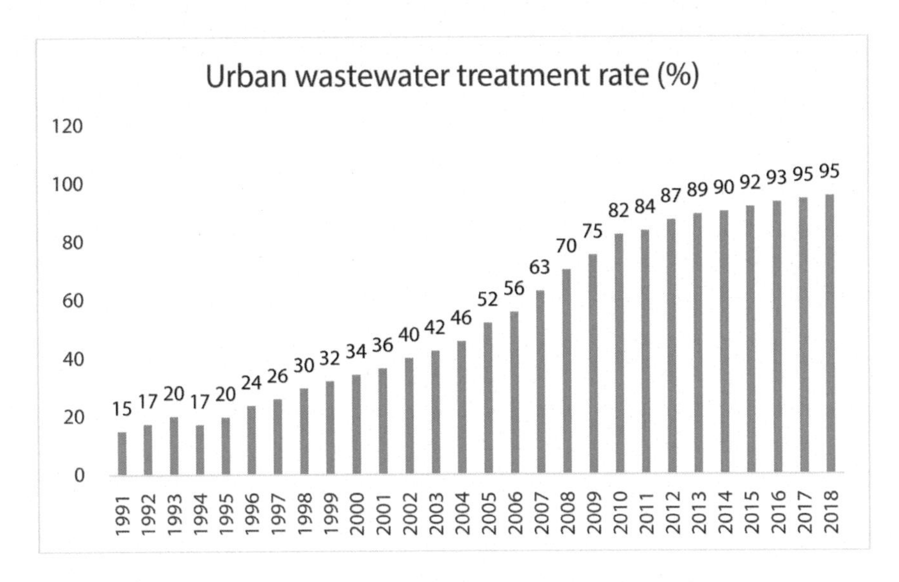

Source: China Ministry of Housing and Urban-Rural development.

The government's budgeted investments for the sewage treatment sector have been steadily growing from Rmb332 billion (USD51 billion) for the 11th Five Year period (2006–10) to Rmb564 billion (USD87 billion) under the current plan covering 2016–20. An aggregate investment of

Rmb1.3 trillion (USD200 billion) was budgeted for 2006–20. This created a pipeline of projects for water companies to invest in.

Details of China's budgeted investment in urban, counties, and townships wastewater treatment and recycle water facilities.

	Government target investment amount (Rmb in billions)			
Category	**2006–10 11th 5 years plan**	**2011–15 12th 5 years plan**	**2016–20 13th 5 years plan**	**Total**
New sewage pipelines	209	244	313	766
New wastewater treatment plants	81	104	151	336
Upgrade existing wastewater treatment plants	0	14	43	57
New sludge treatment facilities	32	35	29	96
New recycle water facilities	10	30	16	57
Facility management and others	0	3	12	15
Total (Rmb in billions)	332	430	564	1326
Total (USD in billions)	51	66	87	204

Source: NDRC's documents on 11th 5 years plan, 12th 5 years plan, and 13th 5 years plan on urban wastewater treatment and recycle water facility construction.

10.5 Funding for projects

Sewage treatment projects were funded using the "build-operate-transfer" method. These projects were awarded by local governments via a bidding process. Water treatment companies, comprising state-owned enterprises and companies owned by private individuals, would bid for these projects in various cities of China. The successful bidder would engage in a build-operate-transfer, or BOT, contract with the local government. This company would invest, construct and operate the sewage treatment plant. In exchange, the local government would award the company with a 25–30 year service concession contract, giving the company the right to collect sewage treatment tariffs from the local government. Typically, end-users (i.e., residential, commercial, and industrial users) pay their water bills, which include tap water and wastewater treatment tariffs, to the local tap water utility. This entity then pays the local government the wastewater treatment tariffs. The government then pays the wastewater treatment operator.

There is usually a minimal payment amount guaranteed from the government to the wastewater treatment operator to ensure it could earn a minimum amount of return. During the concession period, wastewater treatment tariff could be adjusted based on inflation or when there's been an increase in operating cost. These adjustments were typically made every 2–3 years. While there is not a lot of transparency in the wastewater tariff setting system, affordability of the local government is likely taken into consideration in the process. At the end of the concession period, the operator would transfer the asset to the local government at no cost. That would complete the BOT contract. Depending on the capacity, the investment size of an urban sewage project is typically Rmb50 million to Rmb200 million (USD7.7 million to USD31 million).

Investors and banks liked BOT projects because they have a steady stream of cash flow from the collection of sewage tariffs to fund debt payments. Usually, water treatment companies would fund 60% of the investment amount with project-level bank loans and 40% with equity, which usually comes from debt-financing or equity-financing at the listed company level. Since these are environmental projects supported by the government, the borrower can typically get a rate that is slightly lower than the standard lending rate for these projects.

10.6 Water conservancy projects using build-transfer modes

In 2011 when the central government issued The No. 1 Document and stepped up its efforts to improve water quality and conservation, a flurry of new water conservancy projects emerged. These projects involved the construction of pipelines, intercepting canals, and irrigation facilities. It also required the renovation of rivers, or cleaning up a river that has been polluted.

These water conservation projects were configured in the form of short-term build-transfer (BT) projects. This meant that the company would fully finance and build the infrastructure and then transfer the asset to the local government upon completion. These projects did not have recurring revenue streams as there was no "operation" component of these projects. The company would be responsible for the cost of the project. During the construction period, it would recognize construction revenue on a percentage-of-completion basis. The company would receive payment from the government after the project is completed. So, initial funding of the project is first borne by the contractor. After construction is completed and ownership of the asset is transferred to the government, the government would have to pay for the cost of the construction. This is unlike the BOT project mode, during which the government does not have to pay for the lump-sum cost of construction. The sewage treatment fee from end-users would fund the project. The difference is that, under the BT mode, the local government would gain ownership of the project sooner (usually in a few years) than under the BOT mode (usually 30 years).

In July 2011, the central government issued a document that called on local governments to use 10% of proceeds from land sales for the construction of water projects. Some local governments also gave land to water companies as collateral for payment of these projects. Upon completion of the project, if the local government did not have the cash to pay, the water company could sell the land to recover its payment.

10.7 Emergence of public–private partnerships

In 2014, public-private partnerships started to emerge as a funding method of choice for water projects. This was further emphasized with an opinion statement in May 2015 from the Ministry of Finance and Ministry of Environmental Protection on the "Implementation Opinions on Promoting the public–private partnership in the Water Pollution Prevention and Control Area."

Under the PPP structure, the local government is alleviated from the burden of fully funding an infrastructure project on its own, the way it did with the BT model as described in the previous section. In the past, local governments would hire a third-party contractor to construct infrastructure projects for them. They would then have to pay the contractor after the construction was done. With the PPP structure, local governments typically take a minority stake in a joint venture with the private partner (which could be a state-owned enterprise or other companies with no government ownership). The government could make the investment by injecting a government asset into the vehicle. The private partner would make the cash investment and take the lead in constructing and operating the project. Like with BT and BOT projects, there could still be a bidding process in place for PPP projects.

Unlike a single sewage project, which could cost Rmb50–200 million to build, the investment size of PPP water projects could range from Rmb1 billion to Rmb5 billion (USD153 million to USD769 million). One PPP project could involve a group of several related infrastructure projects, including construction of roads, pipelines, sewage treatment facility, and renovating a river. Construction of these facilities may take a few years, but the payment from the government could take 10–30 years. The local governments would pay interest payments to the water companies during this time. With the pure construction contracts, such as renovating of rivers where there is no revenue stream from services offered upon the completion of the projects, the water companies could run into higher accounts receivables risks if the local governments fail to pay them in the future. To incentivize private enterprises to take up these investments, the China Development Bank announced in 2015 it would provide long-term loans of up to 30 years for PPP projects.

There are various project modes under PPP depending on the nature of the project. They include BOT (Build-Operate-Transfer), which is often used for sewage treatment projects as they involve a 30-year operating concession of the plant upon completion. O&M (Operation and Management) is purely operating and managing a facility for the local government. This could be a local park, for instance, after the company finished building roads and landscaping for a neighborhood. BT could be used for the construction of pipelines, as there is no operational element attached to these projects.

10.8 Rooting out low-quality projects

The emergence of PPP projects created a problem. Some local governments allegedly used this as a vehicle to obtain off-balance-sheet funding. The central government found that some of the infrastructure projects created were not legitimate and were mainly used as a way to allow the local government to obtain funds. To crack down on this, in 2017, the central government announced measures to scrutinize the quality of PPP projects. Several government entities—comprising of Ministry of Finance, NDRC, Ministry of Justice, People's Bank of China, China Banking Regulatory Commission, and China Securities Regulatory Commission—issued the joint notice called Document No 50, which prohibited local governments from providing illicit guarantees on PPP projects. (Link to the notice in Chinese: http://www.gov.cn/xinwen/2017-05/03/content_5190675.htm) The MOF also issued Document No 87, "Notice on Preventing Local

Jurisdictions from Government Services Procurement to Illegally Secure Financing." This document prevented local governments from dressing up unqualified infrastructure in a PPP format. (Link to document in Chinese: http://www.gov.cn/xinwen/2017-06/03/content_5199529.htm) It wanted to stop local governments from improperly raising debt and overleveraging. It also targeted to root out equity investment made that were actually debt instruments. The way this was done was via special purpose vehicles for these PPP projects. Typically, banks would extend loans to the SPVs to fund up to 70% of the PPP projects. The remaining 30% were funded by equity investments from local governments and the contractor. Banks also participated in the equity investments via wealth management products and investment portfolios One of the exercises done was the Ministry of Finance taking the lead in reviewing the PPP projects in the government database to ensure they were legitimate projects. This was completed in March 2018.

10.9 Equity market for funding water projects

During the growth of sewage treatment project construction in the mid-2000s to 2015, many water companies sought listings in the capital market as a way to raise equity funding for their growth. The reason they needed the cash was because of the long-term payback period of these projects. Typically, a sizeable sewage treatment project would cost about Rmb100 million, or USD15 million, to construct. The water company would fund 60% of the investment from bank loans. The remaining 40% was from equity. The typical payback period of these projects is 7–10 years from the collection of sewage treatment tariff upon completion of the project. Hence, if the company wanted to invest in several sewage treatment projects simultaneously and find new projects to invest in every year, it would require continuous support from equity raising.

The China water sector is a fragmented market. The largest wastewater treatment company has only 5–10% of the total treatment capacity in the country. There are currently about 20 listed China water companies, mostly listed in the Hong Kong and Singapore markets. There were also some that listed in Shanghai. Valuation of these HK listed water companies surged from ~10x price-to-earnings ratio in 2008 to 25x P/E in early 2015 at the height of its earnings growth. The high valuation multiples allowed the companies to issue equity at an attractive price. Some also used equity issuance as a way to acquire some of its smaller peers with lower valuation multiples.

In the second half of 2015, as the China equity market started underperforming, China water utilities also took a hit. These highly leveraged companies with negative free cash flow were abandoned by investors as they sought companies with less risks. Some also begin questioning the BOT accounting methods, which recognize construction revenue before actual cash is received. Within one year, valuation of some of the HK-listed water companies fell from 25x P/E in 2015 to less than 15x. This was a problem for the listed companies because they were dependent on equity issuance to raise cash for the construction of new projects. With the stocks trading at a much lower valuation, many were reluctant to issue equity anymore. Instead, some of them sought alternative methods of funding, including the issuance of asset-backed securities, disposals of existing projects, and establishment of investment funds with third parties.

Water companies can issue asset-backed securities by securitizing the future payments they expect to receive from their sewage treatment projects. So, instead of waiting for 10 years to receive the sewage treatment fees from a group of projects, they could sell the right of those future payments to investors at a discount. The second alternative is to outright sell their ownership of their existing water treatment projects to third parties. Lastly, some companies also sought partnerships with insurance companies and other financiers to establish investment funds to invest in new water PPP projects. By doing so, the water companies would own a smaller stake in the project and can leverage at the investment fund level.

Despite the decline in valuation of these China water companies, the central government continues to see water infrastructure projects as an important part of its environmental protection campaign. State-owned enterprises will have to continue to play a role in supporting this sector. In 2019, state-owned owned China Three Gorges Corporation took a 5% stake in Beijing Enterprises Water Group Ltd by subscribing to a private placement of equity issuance, and it has increased the shareholding to about 15% in two years. Meanwhile, other strategy and policy banks such as China Development Bank and Asia Development Bank continue to support China water companies as financiers.

10.10 REITs to fund infrastructure projects

In April 2020, China Securities Regulatory Commission and National Development and Reform Commission jointly launched a pilot scheme on REITs for infrastructure projects. This program would allow water companies to sell water projects into REITs so that they could cash out of their investments. The REITs would retain the water companies to provide operating services on the projects. The benefits of this are that it would allow water companies to recover their initial investment sooner and for them to continue to earn revenue from providing operation services. These REITs would be publicly listed and will likely attract investors looking for stable dividend yield.

With the announcement of REITs for infrastructure projects, the central government is again using various financing instruments to allow for innovative funding options for water investments. China's evolution of water financing started with the simple build-operate-transfer project mode to the more sophisticated PPP cooperation between private and public sectors in recent years. Investments in wastewater treatment parallel the sophistication of the finance industry and financial markets in China. During this time, public money (from national and local governments to publicly-owned companies) played an important role in these water infrastructure investments. The viability of funding water projects in China will continue to depend on whether cash flow generated from the projects could support the initial investment. If not, then it is taxpayers and the government that will need to pick up the tab at the end of the day.

CHAPTER ELEVEN

Financing needs and capacities for the water supply and sanitation sector in the European Union

Nele-Frederike Rosenstock[a] and Xavier Leflaive[b]
[a]European Commisssion DG Environment
[b]OECD Environment Directorate

11.1 Introduction

This chapter focuses on financing needs and capacities in 27 European countries[1]. It highlights the foundational role of EU legislation on water, which drives investment needs and financing options. Building on recent data and projections[2], it sketches the financing challenges European member states face, now and in the future. The chapter further discusses emerging challenges, which can drive additional investment needs but, if properly managed, can also generate additional value and trigger opportunities for innovative financing options. These developments illustrate how a well-managed and financially robust water supply and sanitation sector contributes to climate-resilient and sustainable development in Europe.

11.2 The EU water sector—contextualizing investments in the EU

The European Union's water supply and sanitation sector are governed by a long-established EU water *acquis*, setting a common level of ambition

Disclaimer: The information and views set out in this chapter are those of the authors and do not necessarily reflect the official opinion of the European Commission, the OECD or their member countries.

[1] The chapter effectively covers 27 EU member states and the UK, as data collection and most analyses took place before Brexit.

[2] All data and projections refer to OECD (2020a), which was a joint endeavor by the OECD and the European Commission. Drafts of the report were presented, discussed and endorsed by EU member states and OECD countries. Further analyses and developments for this chapter remain the authors' own and may not reflect the opinion of the OECD Secretariat, OECD member states and the European Commission.

Financing Investment in Water Security: Recent Developments and Perspectives.
DOI: https://doi.org/10.1016/B978-0-12-822847-0.00003-X

across the 27 member states. The two main pieces of legislation are the Drinking Water Directive 98/83/EC (DWD), which aims at ensuring high-quality tap water across the EU, and the Urban Waste Water Treatment Directive 91/271/EEC (UWWTD)[3], protecting human health and the environment from wastewater pollution. Along with the much broader Water Framework Directive 2000/60/EC, and other associated laws, these directives set the legal context in which all water-related activities take place. From an environmental and economic point of view, it is important for ambitions across borders to be similar, to create a level-playing field for regional competitiveness and trade, and to avoid jeopardizing downstream action through upstream inaction.

The development of the *acquis* started with legislation on bathing water in the 1970s (i.e., the Bathing Water Directive 2006/7/EC). EU water legislation evolved over time, reflecting the need to address urgent pollution issues across the European community, but also member states' willingness to engage in further coherent protective measures. At the national level, member states, after transposing the legislation into national law, can go beyond EU ambition, adapt their legislation to their national, regional, and local situations, and strive for even better protection.

Fifty years after the first piece of water legislation was adopted, EU member states still stand at different implementation levels, a number of them failing to comply with EU law. The reasons for delays in reaching compliance vary across countries. Most countries cite the costs of compliance (new infrastructure to secure access to unserved populations, investments to reach more stringent quality standards) and lack of financing capacity as the main reasons for delays. Multiple factors contribute to the lack of investment.

As support to reach compliance, EU member states can receive different types of EU funding, and other support. Funding can come from a number of sources: the European Regional Development Fund, the European Social Fund, and the Cohesion Fund[4] (all together forming the Cohesion policy funds). In addition, the European Agricultural Fund for Rural Development operates under the Common Agricultural Policy (note that it is not clear how such funding will contribute to achieving the ambition of the WFD). From 2000 to 2020, about EUR 57.7 billion was allocated to the water

[3] Please note that the UWWTD covers only wastewater collection and treatment, not sanitation within houses such as toilets.

[4] Cowi et al. (2019), found that data from the European Social Fund (ESF) was not comparable across the three funding periods and thus ESF data is not included here.

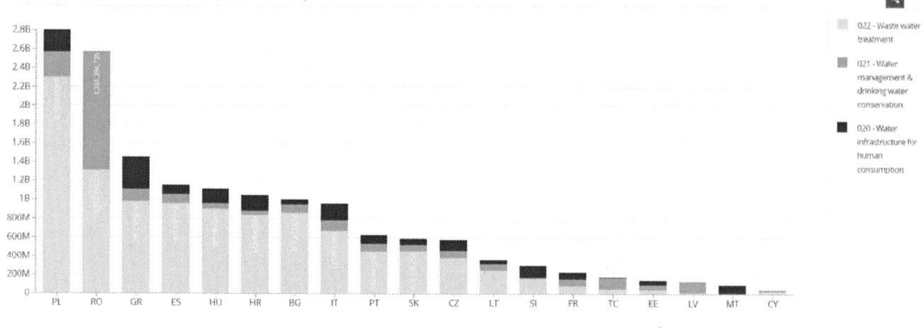

Figure 11.1 Planned EU funding for water projects, status quo 2019 (2014–2020 funding period). *Source*: Extracted from: EU Funding Data Portal (2020). 2014-2020—Water infrastructure investments—plans 2019 (ERDF/CF). Graph produced by the authors.

supply and sanitation sector to help Member States meet the requirements of the Directives (European Commission, 2019a) (Fig. 11.1).

To receive EU funding support, EU countries need to fulfill certain conditions (e.g., "ex-ante conditions" or "enabling conditions") and show how the funding will support them in reaching compliance; these conditions typically require thorough investigation and planning to ensure efficiency of expenditure programs; they also require some level of co-financing by domestic sources of finance. Other types of support are of more technical nature, such as exchanges among member states on technical matters or support through the EU's reform mechanism that was founded in reaction to the 2008 financial crisis. For instance, the Joint Assistance to Support Projects in European Regions (JASPERS, a joint initiative of the European Commission, the European Investment Bank, and the European Bank for Reconstruction and Development) assists countries with the technical and financial preparation of investment proposals. The Structural Reform Support Program supports and facilitates peer learning and sharing of good practices across member states.

Non-compliance with EU water legislation can lead to infringement procedures, launched at EU level. These procedures can lead to financial penalties, which can be either lump sum and/or penalty payments of several tens of thousands of EUR from the date of the judgment delivered to the date of compliance with the judgment. Regarding the Urban Waste Water Treatment Directive, in particular, slow and wrong implementation has led to infringement cases against many EU member states (European Commission, 2019a; European Commission, 2020a).

11.3 Assessing investment needs in the EU—methodological and data issues

Estimating investment needs can serve two objectives. First, at EU level it allows to understand which countries need further support and can thus trigger EU level funding prioritization to that country. Zooming into the investment needs, and the drivers behind the needs, can also help to understand if other EU support is needed (such as technical assistance or further policy guidance). Second, at national level these estimates can help prioritize funding for the water supply and sanitation sector. They can also support discussions among ministries and provide leverage to otherwise potentially overlooked areas. Whereas these estimates provide a notion of the order of magnitude of the investment gap, the planning of individual projects always needs to be supported by more detailed analysis. The main approaches to estimating investment needs and their relative utility for distinct purposes are discussed below.

Gauging investment needs in the water sector, specifically in the water supply and wastewater sector at EU or national level, can be done by considering the reported data and assessments by member states (e.g., as part of the River Basin Management Plans (RBMPs) under the Water Framework Directive[5] or information provided under Art. 17 of the UWWTD). In these assessments, member states report planned or ongoing projects and the investments required to complete them. In this way, they monitor progress toward compliance with EU water law, and provide the European Commission with the information to follow-up on implementation efforts where they are lagging behind. The assessments, however, vary in terms of comprehensiveness; they often lack comparability, as countries do not use the same methodology when assessing their investment needs.

Alternatively, the investment needs can be estimated by using available data and projecting the targeted expenditure levels to reach compliance with the legislation into the future. In addition to the reported data, Eurostat (the statistical office of the European Union) provides substantial amounts of comparable data on other aspects of water (Eurostat, n.d.). By using these data, assessing investment needs in a "top-down" manner is possible with limited resources. In contrast, a "bottom-up" approach for all member states is more difficult and requires more resources.

[5] Both, Art 17 UWWTD and the RBMPs, are planning obligations that EU member countries need to follow when implementing EU water legislation.

However, the "top-down" approach does not necessarily produce the same figures as national assessments. Top-down approaches can also differ from assessments done by other institutions. Good communication is key, depending on what the estimates would be used for, their data sources, their reliability, and the methodological choices. Relying on comparable data for modeling also means that some aspects cannot be well captured, especially if data are not collected through legally set processes (e.g. renewal of infrastructure, or use of individual sanitation systems).

Finally, investment needs assessment is possible for those parts of the water sector, where the solutions or the approach to implementation are fairly homogenous across countries, e.g., water supply and sanitation. Investment needs for other water-related subsectors, such as flood protection or scarcity management, can be more difficult to assess and project, as solutions are more time and location specific, and can depend on risk preferences. Typically, water scarcity can be mitigated through various combinations of water demand management (e.g., nudging water users, or water abstraction charges), reform of water allocation regimes, and supply augmentation (e.g., storage, water reuse, desalination, groundwater pumping): each combination triggers very different—and even divergent—investment needs.

11.4 Financing water supply and sanitation in Europe—the state of play[6]

Estimates of current levels of expenditures (with no distinction between CAPEX and OPEX) for water supply and sanitation build on a series of datasets from Eurostat, which provide a robust basis for cross-country comparison. An annual average expenditure of EUR 100 billion is spent across the 28 EU member states (including the UK at the time of computing the data), with the lion's share attributable to EU15 (Germany, France, United Kingdom, and Italy in particular); see Fig. 11.2 for a breakdown by country.

EU member states vary significantly as regards the financing sources of current levels of expenditure for water supply and sanitation (see Fig. 11.3 for a cross-country comparison). Some rely almost entirely on revenues

[6] This and the subsequent sections build on recent collaboration between the OECD Environment Directorate and the European Commission—DG Environment on the issue. For more detailed information on data, analyses, and method, please refer to OECD (2020a).

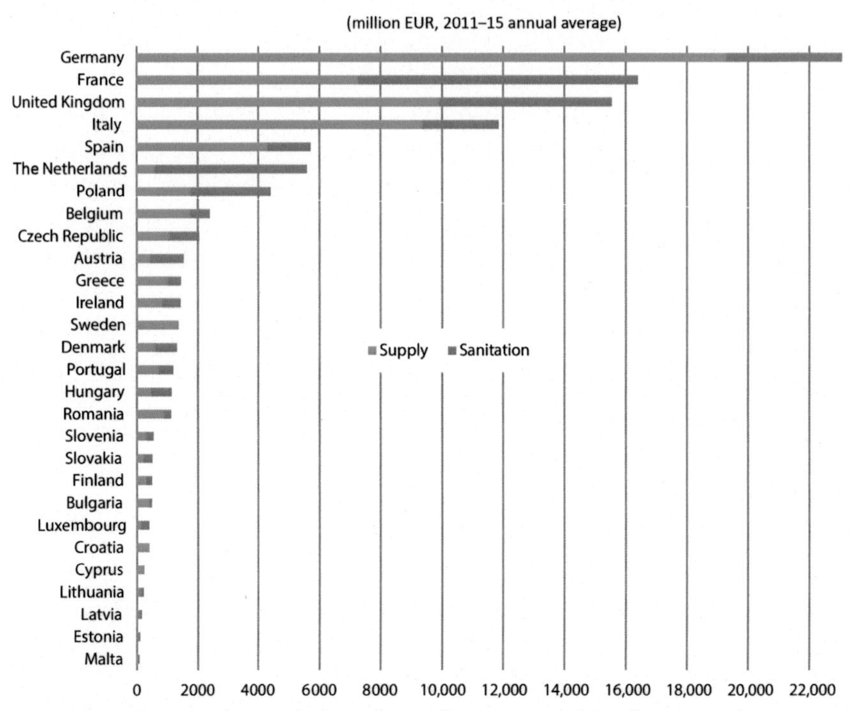

Figure 11.2 Estimated annual expenditures for water supply and sanitation per member state. *Note*: Likely overestimate of supply related expenditures (and corresponding underestimate of sanitation) in countries where waste water-related charged are included in the water bill (e.g., Cyprus). Total expenditure for Finland, Croatia, and Sweden are known to be underestimated due to data limitations. *Source*: OECD (2020a); data from EUROSTAT (General government expenditure by function, Final consumption expenditure on environmental protection services by institutional sector, final consumption expenditure of households by consumption purpose, mean consumption expenditure by detailed Classification of Individual Consumption according to purpose—COICOP level).

from water tariffs. Denmark is the country approaching most closely cost-recovery from such revenues. On the other hand of the spectrum, Ireland (almost) exclusively relies on public funding to finance the provision of water and sanitation services. There are debates about the respective strengths and limitations of these national strategies: while some argue that financing through revenues from tariffs is more transparent and promotes water-use efficiency, others claim that some elements of the infrastructure have a public-good dimension that justifies financing through taxpayers' money. A more detailed analysis of the source of budget finance (potentially earmarked

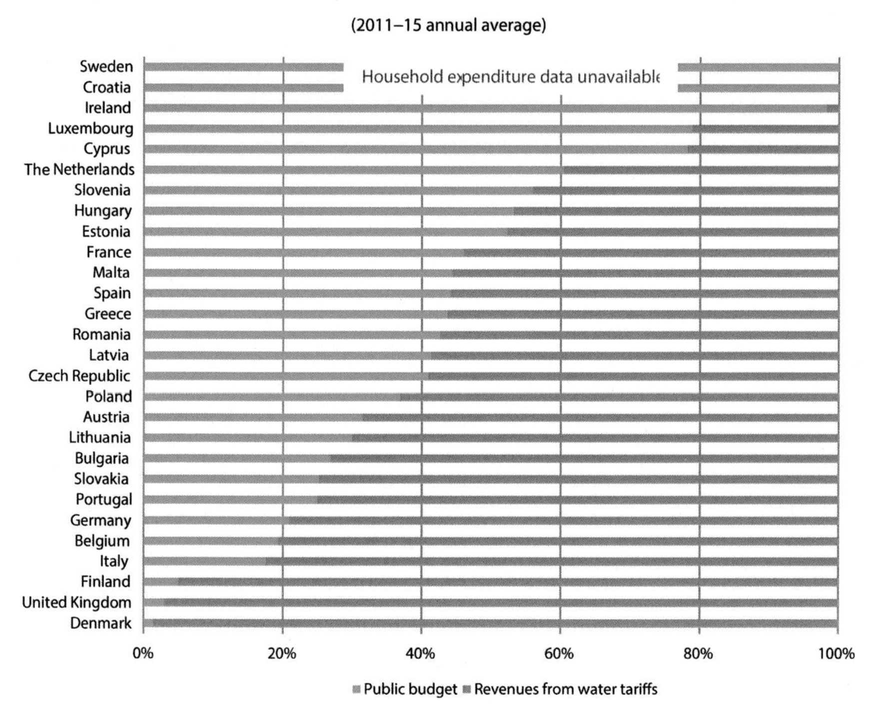

Figure 11.3 Sources of finance for water supply and sanitation services per member state. *Note*: Household expenditures missing for Croatia and Sweden. *Source*: OECD (2020a); data from EUROSTAT (General government expenditure by function, Final consumption expenditure on environmental protection services by institutional sector, final consumption expenditure of households by consumption purpose, mean consumption expenditure by detailed COICOP level).

for) water-related expenditure would add value to the discussion: different fiscal instruments have distinct properties (rate, basis, options for earmarking), which can inform financing strategies.

The source of public funding (domestic *versus* foreign) is an interesting variable, as several countries essentially rely on EU funding for their investments, meaning that they allocate a very small part of domestic public finance to the sector (be it to finance investment or to cover operation and maintenance costs) (see Fig. 11.4). This situation can be problematic, as EU funding for the sector is projected to be gradually phased out. The problem is particularly acute in countries that heavily rely on EU funding and have made little progress toward compliance. Additional analysis would be required to assess member states' ability to use EU funds effectively.

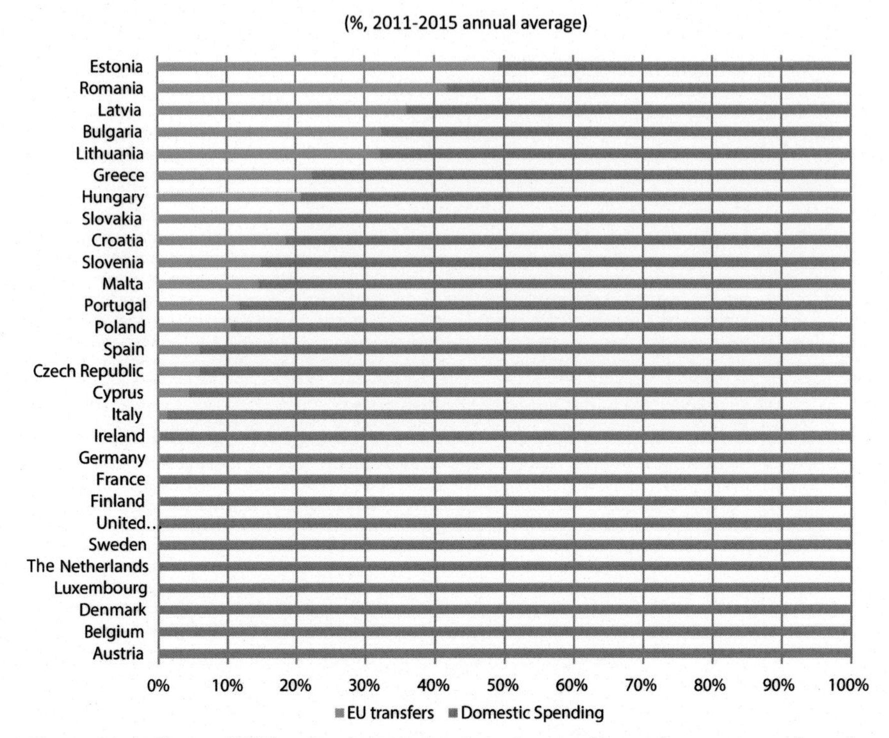

Figure 11.4 Share of EU funding in estimated total expenditures for water supply and sanitation per member state . *Note*: It is assumed that EU funding are always channeled through domestic budgets of each member state and that they are, therefore not additional to government expenditures presented in previous figures. *Source*: OECD (2020a); data from EUROSTAT (for past estimated expenditures), European Commission Directorate-General for Regional and Urban Policy (Open Data Portal for European Structural and Investment Funds).

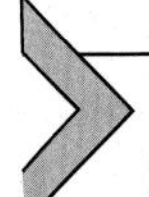

11.5 Financing water supply and sanitation in Europe—Projections to 2030

This section projects financing needs to be required to ensure—and maintain—compliance with the Directives on Drinking Water and Urban Waste Water Treatment by 2030. It discusses financing capacities and their uneven distribution across EU member states.

Projections on financing need consider several drivers. For water supply, urbanization (and the number of additional people to be connected to water supply systems) and compliance with the DWD are key. However, since most EU member states already are in close compliance with the DWD, two additional drivers were considered: the number of people who do not have access to water (essentially vulnerable and marginalized groups who need to be provided access to water in general; at places, this may require the adjustment of prevailing technologies and service provision) and additional investment to approximate the best performance in terms of water network efficiency (minimizing nonrevenue water or resource losses; see Alaerts, in this volume); the assumption here is that water conservation and energy efficiency are increasingly valued and are likely to drive further investments in the near future. For sanitation, projections are driven by urbanization (i.e., the number of additional people to be connected to sewage collection and wastewater treatment) and compliance with the UWWTD[7]. Projections discussed below do not account for the state of the existing assets and the potential backlog for investment. This can explain the at times significant discrepancies with projections made by national authorities.

The projections consider several scenarios, which reflect different combinations of these drivers. Business-as-usual is only driven by the number of city dwellers. A Compliance scenario also reflects the cost of compliance with the revised DWD (the new Directive entered into force in January 2021) and the UWWTD; it reflects the cost of connecting vulnerable groups; it captures the additional level of effort required to comply with the UWWTD[8] as well. The Efficiency scenario also covers the additional efforts to converge toward 10% non-revenue water for water supply services.

No particular assumption is made regarding climate change. Obviously, a changing climate will affect water demand, water-related risks (water scarcity and heavy rains), and the dilution capacity of water bodies, all contributing to further investment needs for the water supply and sanitation sector in Europe. The projections assume that (part of) these additional needs are already captured in current levels of investment. Fine-grained analyses are required to assess whether current trends and expenditure patterns are adequate and sufficiently robust and flexible to address greater variability and other future challenges. Such analyses can inform how distinct adaptation strategies translate into additional investment and expenditure needs. In

[7] Note that investment needs for small communities (not covered by the UWWTD) are driven by the Water Framework Directive; they are not considered in quantified projections here but are discussed qualitatively in the subsequent section.

[8] The distance to compliance is affected by countries' reliance on individual and other appropriate sanitation systems (IAS; for instance, sceptic tanks). When reporting on distance to compliance, countries assume that IAS comply with UWWTD requirements. Greece, Hungary, and Latvia feature among the countries which report the smallest distance to compliance. Additional country-specific research is required to assess whether IAS are properly designed, their performance is monitored, and compliance is enforced.

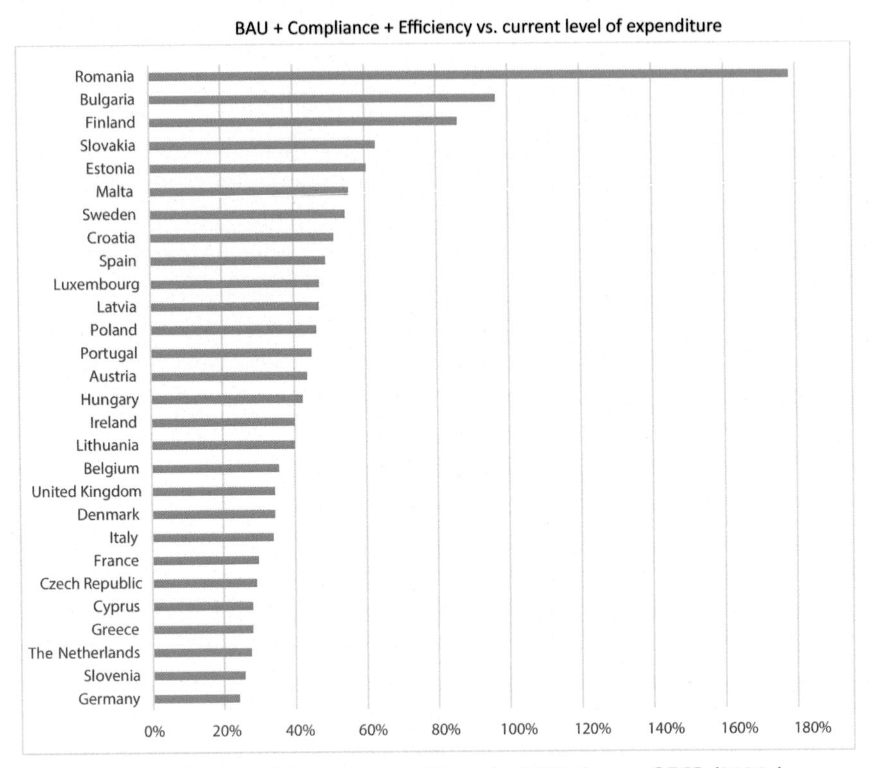

Figure 11.5 Per Annum additional expenditures by 2030. *Source*: OECD (2020a).

the absence of such systematic and comparable analyses across countries, discussions on how climate change will affect financing needs in the future can only remain qualitative and speculative.

The aggregate figure of the investment gap for the 27 member states and the UK amounts to EUR 289 billion by 2030. Sanitation represents the lion's share of the total additional expenditures. This is particularly the case in Italy, Romania, and Spain and—at lower levels—in Bulgaria, Croatia, Portugal, and Slovakia. In these countries, urban population growth plays a minor part (sometimes nil) in projected future expenditures for water supply and sanitation. Fig. 11.5 compares the projected level of effort with the current level of expenditure. It provides a clear indication of the additional level of effort required for countries to comply—and remain compliant—with the DWD and the UWWTD by 2030. The chart indicates that all countries will need to increase annual expenditures for water supply and sanitation by more than 25%. Romania and Bulgaria need to double (or more) their current levels of expenditure. At the lower end of the spectrum, Cyprus, the Czech Republic, France, Germany, the Netherlands, and Slovenia are projected to face comparatively minor needs for increase. This is likely to reflect different situations, including high levels of past expenditures and good anticipation of future needs, significant catch up in the recent decades (Czech Republic), or an underestimate of future needs, possibly driven by overreliance on individual and other appropriate sanitation systems (IAS; e.g., Slovenia).

These projections can be used to discuss how robust the financing strategies mentioned above are, up to 2030. This discussion makes no assumption on the availability of EU funding. It focuses on revenues from water tariffs and allocation of domestic public budgets

Revenues from tariffs for water supply and sanitation play a pivotal role in financing the sector: they provide for a (relatively) stable source of revenue (usually reflecting water uses) and contribute to the creditworthiness of service providers, thereby directing the capacity to attract commercial finance (typically bank loans). Fig. 11.6 suggests that there is room to increase revenues from tariffs in prevailing financing strategies across EU member states. While exploring this option further, affordability of water bills for households needs to be factored in. A thorough analysis of affordability[9] is required to assess rooms for maneuver, identify the share and size of the population in need of support, and appropriate, targeted social measures to address the needs of these communities. The figure suggests that in most EU member states (and the UK), 95% of the population can afford to pay more for the service they receive, should this be required. Bulgaria and Romania are exceptions, where targeted social measures would need to cover more than 10% of the population, suggesting that carefully designed tariffs may be a valid option. The assessment of affordability probably needs a thorough revision, to reflect the consequences of the economic and financial crisis driven by governments' responses to COVID-19: it is likely that such responses will increase the size of vulnerable groups at risk of poverty. Moreover, early social measures in several countries result in waiving sanctions for nonpayments of water bills. This can result in short-term liquidity crisis for service suppliers, which, should the situation prolong, could translate into long-term solvency issues, undermining the financial sustainability of the sector and limiting future investments.

[9] Affordability here is crudely measured as the share of household disposable income allocated to water bills. Chapter 7 in this volume discusses more sophisticated ways to characterize affordability issues.

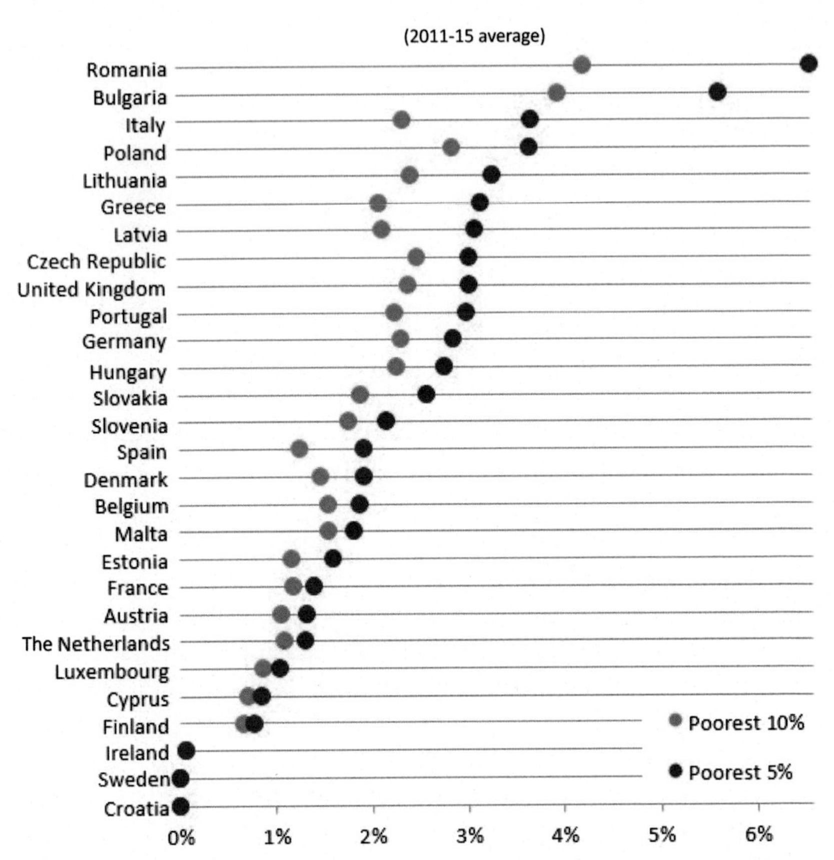

Figure 11.6 Share of water supply and sanitation expenditures in households' disposable income. *Note*: Lack of household expenditure data for Croatia and Sweden. *Source*: OECD (2020a); data from EUROSTAT (household expenditures and income data).

EU member states (and the UK) also vary in their ability to allocate more public finance to water supply and sanitation. Fig. 11.7 plots countries according to the share of public finance in GDP and the share of public debt to GDP. Countries in the upper right corner have less room for maneuver than countries in the opposite corner. Here again, the economic consequences of COVID-19 are likely to intensify pressure on public finance, hindering the likelihood that more public budget can be earmarked for water services.

The analyses above confirm that financing for water supply and sanitation is not merely a developing country issue. Each EU member state (and the UK) faces similar—albeit distinct—challenges and capacities to address them. Robust and tailored financing strategies are required.

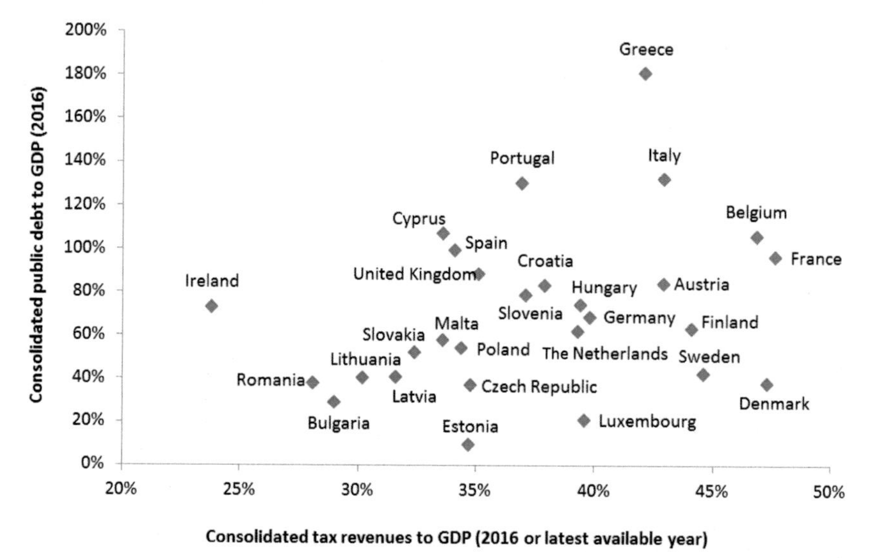

Figure 11.7 Ability to increase public spending based on raising taxes or borrowing. *Source*: OECD (2020a); 2016 data from EUROSTAT.

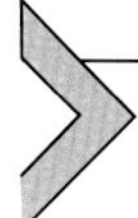

11.6 Options to close the financing gap for water supply and sanitation in Europe

Analysis of the state of play and current and future challenges points to a series of options to close the financing gap of the water supply and sanitation sector in Europe. A range of conversations with national authorities in EU and OECD fora suggests that this is not merely about additional funding: much can be done with the available assets, water resources, and finance. Three avenues can be explored more systematically to minimize future financing needs for water supply and sanitation in European cities.

First, thorough and cost-effective maintenance of assets can prolong the operational life of existing infrastructure and postpone the need to renew or rebuild it. More generally, the operational effectiveness of utilities is a major feature of the financial sustainability of water services in Europe. From that perspective, the introduction of independent economic regulation can support building a resilient sector—to benchmark the performance of service providers, and incentivize efficiency gains. While OFWAT—the economic regulator for England and Wales—is often referred to as a role model, the Water Industry Commission in Scotland and ERSAR—the water and waste services regulation authority—in Portugal also have a lot to share; in Estonia, the creation of the Competition Authority, under the

Ministry of Justice, paves the way to improved regulation and performance for water services.

Second, urban planning that factors in water-related needs and risks can minimize the need to build additional pipes to collect and treat rainwater: green roofs, permeable pavements, and sustainable urban drainage can be cost-effective options to manage rainwater. Such options can minimize future liabilities and related investments to cope with the harmful consequences of decisions made today. No doubt the transition in European cities is challenging, but proper planning and sequencing of developments and investments can minimize future costs (see OECD, 2015) for a more detailed discussion).

Third, anecdotal evidence suggests that there is room to improve the efficiency of expenditure programs in water supply, sanitation, and related domains. Sadoff et al. (2015) document investment pathways for urban water security: the case studies illustrate how investment planning and sequencing can cost-effectively enhance benefits in terms of access to water services and water security more broadly. Here again, independent economic regulation has a role to play. The institutions listed above illustrate how an economic regulator can be tasked with reviewing the efficiency of expenditure programs, thereby creating an environment that is conducive to efficient investments, both public and private. As suggested in the introduction to this volume, this avenue points at some limitations of a project-based approach and the mere reliance on more bankable projects: while individual projects may be appropriately designed and able to attract finance, they may not be justified or the best option to contribute to resilient urban environments at the least cost. In a non-European context, The Dutch initiative Water as Leverage illustrates how cost-effective water resilience can be more difficult to finance than a suite of costly—but bankable—projects (for more information, see https://wateraslеverage.org/).

Of course, the three avenues sketched above do not exclude that additional finance will be required for water supply and sanitation in Europe, to connect unserved populations (e.g., vulnerable and marginalized groups), to comply with more stringent health and environmental standards, and to adapt to climate change (e.g., to adjust treatment capacities to dilution capacities of water bodies, or to cope with heavy rains). Appropriate options vary according to the distinctive challenges and capacities of each EU member state. Well-designed and sustainable financing arrangements between the water sector, governments, and potentially other financing institutions are required. As a first step, these arrangements would build on a good understanding of investment needs—supplementing the projections above with a thorough bottom-up assessment of the needs, informed by a reliable review of the state of assets, of options to adapt to climate change and a fine-grained analysis of affordability of water bills—and the development of a strategic vision for the future.

Robust financing options will build on the strategic use of tariffs for water services and public funds: these tools can be used as sources of finance and as levers to harness others, namely the range of commercial finance and private capital that is available across European countries.

As discussed above, raising water tariffs will be necessary to secure revenues and enhance the creditworthiness of service providers. Revenue from tariffs should at least cover the operational cost of the provision of services, and ideally move toward full cost recovery. Ring-fencing tariffs can help to ensure that revenues actually benefit the water sector and water users (Bender, 2017). However, raising tariffs should not be done in a blanket approach, but rather by targeting those who can afford to pay more (OECD, 2020b). This would also enhance compliance with EU legislation, especially Article 9 of the Water Framework Directive, which demands that the environmental and resource costs of water policies and water services are fully recovered through a combination of revenues from tariffs and budget transfers (European Commission, 2019b). Raising tariffs for water services is politically challenging, and should therefore come with activities to improve service quality and cost-efficiency of service provision (see the role of economic regulation, discussed above), and to raise the awareness of what water users pay for. An economic regulator could play a crucial role in this regard, ensuring tariff levels, structure, and tariff setting processes are based on robust and transparent data and methods, and appropriate consultation with water users.

While transitioning away from the dependence on EU or other concessional funding, blended finance can be an intermediate solution (OECD, 2019a). Blended finance is here defined as mobilizing (domestic or external) public finance to lower the risk of water-related investments and crowd-in commercial finance. Using domestic and international public funding as a way to crowd in additional commercial finance can help build a financially viable sector in EU member countries that still need to catch up significantly in terms of compliance with the EU *acquis*. Although not yet much used in the water sector in the EU (OECD, 2020a), it could play a role in those EU member countries that still depend highly on EU funds or other external funding sources. The capacity to blend multiple sources of finance could be enhanced by the development of intermediaries, which can minimize the transaction costs of water-related investments and provide a bridge between project owners and a heterogeneous financing industry. Part 3 in this volume documents the role and operation of diverse intermediaries, in a non-European context. Existing ones in Europe deserve further attention, such as the *Agences de l'eau* in France or the *Nederlandse Waterschapsbank* (NWB Bank) in the Netherlands. While such instruments can hardly be replicated in other countries, the functions they deliver and their mode of operation can inspire governments and non-state actors in other European countries.

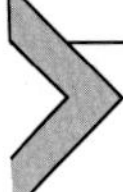

11.7 Looking ahead: renewed ambition for water supply and sanitation in Europe

In the future, the water supply and sanitation sector will need to consider improving its business case to make it attractive for investments from various sources. Water supply and sanitation services will be better able to cope with changing circumstances such as climate, demographics and urbanization and additional investment needs if they contribute to a range of policy objectives and create additional value on which innovative financing mechanisms can build.

From an environmental and financial point of view, a better cross-sectoral integration of the water sector in general, and the water supply and sanitation sector in particular, is key. Putting water (more specifically investment in water supply and sanitation, in rainwater collection and in adapting urban environments to a changing climate) on the radar of urban development (e.g. city planners, property developers), public health, or energy supply can change the game in many ways: as discussed below, it can avoid building future liabilities (e.g., when the expansion of impervious surfaces increases risks of flash floods) and channel new sources of finance to robust and resilient water systems in Europe.

The cross-sectoral integration of the sector in the broader policy agenda of the EU can be structured around the five following topics: (1) maintaining high levels of water quality, (2) managing water quantity, (3) integrating the sector in the circular economy, (4) contribution to climate change mitigation; and (5) digitization of the sector. Whilst some of these options go beyond the current business case of the water sector, further integration and expansion of the scope can create additional value to society and new investment opportunities, or reduction of investment needs, as well as potentially new revenue streams. Although for most of these challenges, limited large-scale evidence is currently available.

11.7.1 Maintaining high levels of water quality

In Europe, society at large has benefitted from generally high drinking water quality for decades. The water supply sector as described above has been and is key for ensuring that this quality is maintained. While some of the pollution is well dealt with by conventional treatment practices in place, removal of micro-pollutants is usually not the focus. A consumerist and aging society is likely to lead to the presence of an increasingly diverse group of micro-pollutants in water that can be harmful to the aquatic environment, and eventually to human health. Pharmaceutical residues are one of the many examples that attract increasing attention by the public and policymakers (OECD, 2019b). Some of the contaminants of emerging concern can pass through the wastewater system and end up in water bodies. If they are removed they are often transferred into the sewage sludge, which in some parts of Europe is reused in agriculture as a fertilizer (Pistocchi et al., 2019). The required treatments can be rather costly, and new financing schemes will need to be explored. In Switzerland, a tax was levied on citizens connected to the sewer system to finance the upgrade of about 100 wastewater treatment plants (Micropoll, n.d.). However, there is also an opportunity for pro-active integration with policy agendas of other sectors. To deal with new types of pollutants cooperation with upstream sectors such as the pharmaceutical industry could be explored (see OECD (2019b) for a discussion of policy options along the life cycle of the industry). A number of European countries have already started to develop such an integration, to the benefit of all, by setting up multi-stakeholder dialogues (e.g., France, Germany, The Netherlands). An additional route that can be contemplated is extended producer responsibility, already employed in the EU's solid waste sector. Next to burden sharing, this could potentially incentivize the research and development of "green chemicals."

11.7.2 Managing water quantity

In terms of water quantity, in the EU and the UK, 52 million people already live in water-scarce regions. In contrast, some central and northern EU countries will have a rise in water availability due to climate change (European Commission, 2020c). In cases of too much water, cities will need to deal with increased occurrences of stormwater overflows and flash floods.

These can endanger properties, pollute bathing sites used for tourism and generally harm the environment.

The water supply and sanitation sector will have several options to play a central role when it comes to dealing with too much and too little water: in urban contexts, the sector will need to ensure that stormwater and urban flooding do not harm the environment and properties through close cooperation with those in charge of urban planning. In urban settings, rainwater harvesting might become an important water efficiency measure. Some EU cities have adopted integrated approaches to deal with these situations, such as Copenhagen and Rotterdam, by including in their planning appropriate storage capacities and nature-based solutions (Gawlik et al., 2017; OECD, 2015). In water-scarce regions, the sector can contribute through providing alternative water sources. Water reuse will need to play a role in the agricultural sector (European Commission, 2020b), and potentially in other sectors too, e.g., industry and aquifer recharge. Water reuse for different purposes will reduce pressures on water bodies and open up new revenue streams for treatment plant operators.

As much as for water quality it will make sense to enhance cooperation across sectors to create value for society: given that urban runoff contains new kinds of micro-pollutants, dealing with them "upstream" likely decreases water treatment costs for water suppliers. Preparing for floods and dealing with storm water will be costly, as in some cases, it might require rethinking urban planning (e.g., shifting toward sustainable urban drainage). However, it also offers unique opportunities in terms of bringing back nature to urban environments through the use of nature-based solutions. As regards floods, insurance schemes and land value capture mechanisms (such as local taxes on property value) can be considered. Insurance schemes could also be used as incentives not to build in flood-prone areas.

AnglianWater "Love every drop" strategy to protect water bodies and secure water supplies

As part of its "love every drop" strategy, AnglianWater employs catchment advisors. These advisors help to identify pollution sources and advise on solutions. One task of the advisors is to work with farmers. In some regions, farmers use metaldehyde containing pesticides that are harmful to the aquatic environment and are difficult to remove through treatment. By working closely with the farmers, alternative fertilizers were identified that achieve similar results on the crops and are less harmful. Through the cooperation, raw water quality improves and thus drinking water treatment costs decrease. Other cities in Europe are taking major initiatives in catchment protection (*Eau de Paris* in France and *Münchner Stadtentwässerung* in Germany are well-documented); these and other examples of cooperation can inspire other communities, where competition remains a challenge across water uses and policy agendas.

Source: *AnglianWater (2019); EurEau (2019)*

Water reuse for industrial purposes

An example for the reuse of waste water is the Tarragona site in Spain, where secondary effluent is used for cooling and as process water in a nearby petrochemical park. Using reclaimed water for industrial processes in the petrochemical park reduces pressure on the Ebro river, which now serves primarily as a drinking water source.

Source: *Sanz et al. (2015).*

11.7.3 Water supply and sanitation and the circular economy

Another option to create value and to support the broader policy agenda of the EU is the integration with the circular economy approach of the EU. Wastewater contains raw materials that can be exploited and that could support the sector's integration in the circular economy. The main resources that can be recovered are nutrients, reclaimed water, and energy. Technology allows for the extraction of other materials such as struvite (a magnesium ammonium phosphate), phosphorus-biofertilizer, cellulose, biopolymers, biogas, and biomass fuel (see for example, SMART Plant, an EU Horizon 2020 sponsored project). In addition, treatment plants could be consistently upgraded to recover nitrogen as a concentrated ammonia product (when and where it makes sense), which can be used as a fertilizer. However, the cost/benefit ratio, as well as energy use requirements, necessitate further research (Beckinghausen et al., 2020).

Equipping wastewater treatment plants to be part of the circular economy requires initial investments, but could create new revenue streams for water utilities in the future. For this to work, public acceptance, the absence of unwanted contaminants and a stable production in terms of composition will need to be ensured by governments and the water sector (Hidalgo et al., 2021). In addition, stable markets for recovered materials are a condition for revenue generation. They depend on robust regulation in adjacent policy areas, in particular reflecting the Polluter Pays Principle and the opportunity cost of using scarce resources. It is likely that the EU's new Circular Economy

Action Plan (European Commission, 2020d) will drive further research and regulatory action in this regard.

11.7.4 Water supply and sanitation and climate change mitigation

In the EU, driven by the political ambitions set in the European Green Deal (European Commission, 2019c), there will be a push to reduce energy use and greenhouse gas emissions in the water supply and wastewater sector. The sector consumes about 3% of electricity in the EU, which is projected to increase (IEA, 2017). The costs are estimated at around EUR 2.9 and EUR 3.9 billion a year (Magagna et al., 2019). As one of the main energy uses in the water supply sector is the pumping of water, part of this money is lost due to the high leakage rates across the EU (leakage rates in the EU vary between 7% and 60%) (European Commission, 2017).

The EU's sanitation sector uses 1% of all energy consumed in the EU (Ganora et al., 2019), and the energy costs can make up to 20% of a municipality's bill (Powerstep, 2019). Research shows that the wastewater sector can reduce its energy use by about 5500 GWh if all treatment plants shift to the average use of energy; or reduce it by as much as 13,500 GWh per year if all plants reach the performance of the best-performing ones (Ganora et al., 2019). Shifting from high-energy use to being energy neutral, or even energy production, requires initial investments and the rethinking of established processes (Powerstep, 2019). It offers the sanitation sector an opportunity to contribute to climate change mitigation, thereby changing its role from an energy consumer to a (clean) energy producer. It remains to be seen if these initial investments can be offset by the subsequent savings or revenue streams that could be created by supplying energy. Again, the answer partly depends on cogent policy initiatives in other domains (typically reflecting the negative externality of greenhouse gas emissions).

Energy recovery from waste water

Waste water discharge is a source of heat with effluents being at above 10°C at all times of the year. This heat can be used for the operation of heat pumps. In Sweden, the waste water treatment plant Hammarbyverket in Stockholm (Sweden), has seven heat pumps with a total capacity of 225 MW. The heat pumps recover 1,235 GWh of heat per year, which is then used to warm up

95,000 residential buildings. In addition to the heat, the treatment plant also generates cooling energy for the district cooling network.

Source: ***Gawlik et al. (2017)***

11.7.5 Digitization

Lastly, as a horizontal topic, digitization is both a challenge and an opportunity for the sector. Through the possibility of gathering, consolidating, and comparing large amounts of data, digitization further allows for better targeting investments, as regards investment planning and asset management. It can help reduce cost by enabling more efficient management. Traditionally, a water company covers four main activities: (1) planning and design, (2) maintenance, (3) operations and control, and (4) customer service. All of these can benefit from digitization and thus efficiency gains that translate in lower operation costs, enhanced operational efficiency and (as mentioned earlier) potentially postponing investments to renew or rebuild existing assets.

As regards water supply, digitization has multiple applications: it can support leak detection, improve water demand modeling to ensure efficient energy use, and it can support improved and continuous monitoring of pollution levels to better target treatment and thus reduce energy and treatment costs. In the sanitation sector, real-time monitoring of overflows can help protect touristic sites.

11.8 Conclusions

Member states of the European Union benefit from a unique position for cross-country comparative analysis as regards current and future investment needs for water supply and sanitation (and other water-related investments not covered in this chapter). First, they enjoy a common level of ambition, in addition to global goals such as SDG6: European legislation sets standards for levels of service that drive decisions and convergence across Europe. Second, they benefit from coordinated and robust data and processes to monitor progress toward set levels of ambition, and share experience on good practices. Third, the European Commission provides a range of supporting tools to facilitate investment and compliance, including (but not limited to) significant financing instruments.

Over the last decades, such a framework has expedited investment and convergence toward high standards of water services and water security in most countries. It is now tested against a range of challenges. First, for countries lagging behind, the "last mile" toward compliance (e.g., securing access to unserved communities) may be the most costly (per capita) and difficult (politically) to achieve. Moreover, whenever EU legislation is revised to be fit for purpose and to ensure that health and the environment are appropriately protected, there will be (initial) additional costs of compliance. This new context calls for innovative responses, which can deliver compliance at least cost (e.g. through enhanced operational efficiency of utilities, innovative architecture of water services, and best use of new data).

Second, whilst the use of EU funding was important for many EU countries to increase compliance when they initially joined the EU, with this funding likely to decline, member states now need to explore other sources of finance, e.g., commercial finance. Whilst EU funding still is available, there is an urgent need to make a more strategic use of the funds to fund cost-effective expenditure programs that contribute to compliance on the ground and to crowd in new sources of finance. The same reasoning applies as regards domestic finance, as, in a post-COVID-19 environment, public spending may become scarcer and earmarked for such distinctive objectives as public health, resilience to multiple crises, and labor.

Third, climate change is changing the landscape for urban development, water supply, and sanitation services in the coming decades. A thorough assessment of how resilient current developments and infrastructures are to a changing climate remains to be done at the European level. It is a prerequisite to more specifically explore appropriate financing needs and strategies.

This chapter argues that a range of options is accessible to European countries to adjust to this new context as regards financing water supply and sanitation services and related issues. They coalesce around the following three axes: independent economic regulation, valuing water, the role of intermediaries.

This chapter foremost argues that independent economic regulation can contribute in many ways to a financially sustainable water sector in Europe: it can drive the efficiency of service providers; it can check and stimulate the effectiveness and efficiency of public expenditure programs that affect access to water services and the cost of water-resilient settlements, and it can contribute to robust tariff policies for water services. In addition to prevailing approaches, economic regulation should cover the opportunity of individual projects and situate them in a sequence of investment: this may be an adequate scale to assess expenditure programs and the relevance of projects (in addition to bankability).

Moreover, this chapter suggests that the financial sustainability of water services in Europe will depend on the capacity of the sector to demonstrate added value and contribute to other policy agendas. At EU level, the level of ambition is set out in the European Green Deal (European Commission, 2019c), which continues to be the guideline throughout the recovery from the sanitary and economic crisis evoked through the outbreak of the COVID-19 virus. In a post-COVID-19 environment, resilience to concomitant crises (health, climate, or else), efficiency of public expenditure, labor, livable cities come to mind. Robust economic analysis can help make the case for such added value. Innovation is required to capture part of that value and transform it in a stream of revenues that can be the basis of financing models. Such analysis can pave the way to a wider use and dissemination of existing mechanisms in Europe, such as insurance, land value capture or extended producers' responsibility.

A cross-cutting theme of this Volume, innovation, is likely to be triggered by intermediaries, which have the capacity to minimize transaction costs for water-related investments, and to connect project owners and a diverse financing community. More work is needed to characterize the financing community in Europe, its experience with and appetite for a range of financing mechanisms, thus defining the role of intermediaries. While European countries have a lot to share on that front, they can also be inspired by innovations in other regions.

In addition to local and national authorities, the European Commission has the capacity to explore further, incentivize and facilitate such a transition toward a financially sustainable and resilient water sector in Europe. Support can come in the form of policy development—both on environmental issues as well as the financing industry, regional programs in support of a green recovery as well as reform support to member states.

References

AnglianWater, 2019. Love every drop—vision and plan. <https://www.anglianwater.co.uk/about-us/who-we-are/sustainability/love-every-drop/> (accessed April 17, 2021).

Beckinghausen, A., Odlare, M., Thorin, E., Schwede, S., 2020. From removal to recovery: an evaluation of nitrogen recovery techniques from wastewater. Appl. Energy 263.

Bender, K., 2017. Introducing commercial finance into the water sector in developing countries. Guidance Note, World Bank, Washington, DC.

Cowi et al., 2019. Integration of environmental concerns in Cohesion Policy Funds (European Regional Development Fund, European Structural Fund, Cohesion Fund). Publications Office of the EU, Luxembourg.

EurEau, 2019. Cooperation projects between water operators and farmers. <https://www.eureau.org/resources/briefing-notes> (accessed April 17, 2021).
European Commission, 2017. Drinking water directive—impact assessment. SWD 2017, 451 final.
European Commission, 2019a. Evaluation of the urban waste water treatment directive. SWD 2019, 701 final.
European Commission, 2019b. Water fitness check. SWD 2019, 440 final.
European Commission, 2019c. European Green Deal. COM/2019/640 final.
European Commission, 2020a. Tenth report on the implementation status and programmes for implementation (as required by Article 17 of Council Directive 91/271/EEC, concerning urban waste water treatment). COM/2020/492 final.
European Commission, 2020b. Water reuse. <https://ec.europa.eu/environment/water/reuse.htm> (accessed April 17, 2021).
European Commission, 2020c. Climate change impacts and adaptation in Europe (PESETA IV). EU Science Hub, Italy.
European Commission, 2020d. Circular economy action plan. COM/2020/98 final.
Eurostat (n.d.). Water statistics. <https://ec.europa.eu/eurostat/web/environment/water> (accessed April 17, 2021).
Ganora, D., et al., 2019. Opportunities to improve energy use in urban wastewater treatment: a European-scale analysis. Environ. Res. Lett. 14, 044028.
Gawlik, B.M., Easton, P., Koop, S., Van Leeuwen, K., Elelman, R., 2017. Urban Water Atlas for Europe. European Commission, Publications Office of the European Union, Luxembourg.
Hidalgo, D., Corona, F., Martín-Marroquín, J.M., 2021. Nutrient recycling: from waste to crop. Biomass Conv. Bioref 11, 207–217. https://doi.org/10.1007/s13399-019-00590-3/.
International Energy Agency, 2017. Water-Energy Nexus—world energy outlook special report. https://www.iea.org/reports/water-energy-nexus (accessed April 17, 2021).
Magagna D., Hidalgo González I., Bidoglio G., Peteves S., Adamovic M., Bisselink B., DeFelice M., De Roo A., Dorati C., Ganora D., Medarac H., Pistocchi A., Van De Bund W., Vanham D., 2019. Water energy nexus in Europe. Publications Office of the European Union, Luxembourg, 2019, ISBN 978-92-76-03385-1, doi: 10.2760/968197.
Micropoll, n.d. Verband Schweizer Abwasser- und Gewässerschutzfachleute. <https://www.micropoll.ch/faq/> (accessed April 17, 2021).
OECD, 2015. Water and Cities: Ensuring Sustainable Futures, OECD Studies on Water. OECD Publishing, Paris.
OECD, 2019a. Making blended finance work for water and sanitation. OECD Publishing, Paris.
OECD, 2019b. Pharmaceuticals in fresh water. OECD Publishing, Paris.
OECD, 2020a. Financing water supply, sanitation and flood protection – challenges in EU member states and policy options. OECD Publishing, Paris.
OECD, 2020b. Addressing the social consequences of tariffs for water supply and sanitation - Environment Working Paper No. 166. OECD Publishing, Paris.
Pistocchi, A., Dorati, C., Grizzetti, B., Udias, A., Vigiak, O., Zanni, M., 2019. Water quality in Europe: effects of the Urban Wastewater Treatment Directive.
POWERSTEP, 2019. Policy brief. <http://powerstep.eu/system/files/generated/files/resource/policy-brief.pdf> (accessed April 17, 2021).
Sadoff, C.W., Hall, J.W., Grey, D., Aerts, J.C.J.H., Ait-Kadi, M., Brown, C., Cox, A., Dadson, S., Garrick, D., Kelman, J., McCornick, P., Ringler, C., Rosegrant, M., Whittington, D., Wiberg, D., 2015. SecuringWater, Sustaining Growth: Report of the GWP/OECD Task Force on Water Security and Sustainable Growth. University of Oxford, UK, 180pp.
Sanz, Joan, Suescun, J., Molist, J., Rubio, F., Mujeriego, R., Salgado, Blanca., 2015. Reclaimed water for the tarragona petrochemical park. Water Sci. Technol.: Water Supply 15, 308.

PART III

Financing models in practice: Case studies

CHAPTER TWELVE

Introduction to the water financing landscape: Select proven and emerging approaches

Kathleen Dominique[a,1] and Alex Money[b]
[a]OECD Environment Directorate
[b]Smith School for Enterprise and the Environment, University of Oxford

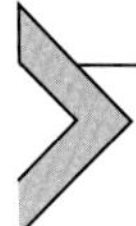

12.1 The water financing landscape: a brief introduction

Investments that contribute to water security and sustainable growth comprise a very heterogeneous range of projects and hence require a range of financing approaches. Broadly defined, "water-related investments" refer to investments that contribute to water security[2] through the delivery of water and sanitation services, the management of water resources, and water-related risks ("too much," "too little," and "too polluted").

Investments contributing to water security often deliver multiple benefits for other sectors and policy agendas, notably agriculture, energy, urban development, public health, biodiversity and ecosystems, education, and among others. By way of illustration, De Bièvre and Coronel (this volume), document the range of cobenefits delivered by water funds (a well-established funding approach for source water protection). Biodiversity is likely the most significant cobenefit and also the most visible one in the startup or early years of many such initiatives. Further, where these initiatives contribute to conserve and restore ecosystems, carbon is captured, generating demonstrable benefits in terms of avoided greenhouse gas emissions. In addition to this contribution to climate change mitigation, water funds also clearly contribute to climate change adaptation, for example when restoring hydrological regulation in wetlands, while this same regulation is being lost

[1] The views expressed in the chapter are the author's own and do not necessarily reflect the views of OECD member countries.

[2] See OECD (2013) water security for further discussion and definitions related to.

Financing Investment in Water Security: Recent Developments and Perspectives.
DOI: https://doi.org/10.1016/B978-0-12-822847-0.00016-8

due to glacier melt in the same catchment (De Bièvre and Coronel, this volume).

12.1.1 A diversity of investment types and financiers

The diversity of investments span a wide range, each investment with a unique risk-return profile and distinct project attributes. Investments can include a range of infrastructure types (including conventional "grey," and nature-based solutions, or a combination thereof) as well as large, centralized infrastructures and small-scale, decentralized systems. This can also include investments designed for other purposes, which also contribute to water management (e.g., green roofs or permeable surfaces that limit rainwater runoff).

At the same time, the range of financiers is also very diverse: with different mandates, investment objectives, risk appetites, and liquidity needs (OECD, 2022). This diversity creates complexity, but also opportunities, particularly where actors have unique access to critical assets. For example, projects that require land or other fixed capital assets may benefit from including the local government as an investment partner. There is also potential for land value capture as asset prices rise in response to improvements in water infrastructure. Other actors such as corporate entities may also have unique assets such as supply chain distribution networks that can help reduce transactional and operational costs associated with project development. Incorporating "in-kind" benefits of this nature into financial term sheets is not without complexity, but can help to catalyze investment flows.

Additional classifiers for water investments include scale (from watershed to household), function (water supply, wastewater management, flood protection, etc.), and operating environment (ownership, governance, and regulation) (Money, 2017). Table 12.1. lists categories of risks, returns, and project attributes relevant for financing investments that contribute to water security.

12.1.2 A diversity of financing approaches

Heterogeneity can be a catalyst for innovation, and this holds true in financing water-related investments. As noted above, every project has its own attributes, which several case studies in this part of the book explore in more detail. Here it suffices to recognize that optimizing the financial structure of a project (often involving the innovative use of instruments)

Table 12.1 Categories of attributes of investments that contribute to water security.

Feature	Description
Risk	
Macroeconomic and business risks	Operational and performance risks (e.g., weak performance of utilities or projects; overestimation of demand for water resources or services; poor governance or managerial quality of utilities or water agencies) Credit risk (incapacity of counterparty to honor contractual arrangements) Currency risk arising from a mismatch between the currency in which revenues are generated and currency of debt service Difficulty to secure collateral in terms of assets or revenue streams
Regulatory and political risks	Political risk (potential for political interference in the tariff setting process; nationalization of utilities) Economic or environmental regulation may be weak, absent or subject to change over the course of the investment lifecycle Regulatory regimes may preclude the possibility of including debt service in the costs that can be covered by the tariff, thereby limiting the feasibility of access to commercial finance
Technical risks	Performance risks may arise or become more acute over the lifetime of the investment due to aging infrastructures, leakage, obsolescence of technologies, especially in the context of long-lived and capital-intensive infrastructures and under-investment in maintenance
Environmental/ social risk	Environmental risks (e.g., increasing water scarcity can lead to increase of cost of bulk water supply as a result from the variability of rainfall and increasing uncertainty about future conditions in the context of climate change) Social risks (e.g., particularly for low-income households, relative to tariff increases as a result of major new capital investments)
Return	
Cash-flow generation	Utilities collect tariffs and other payments (e.g., connection fees) from customers. Tariffs may not fully cover operational and maintenance costs and rarely cover capital expenditure Improvements in operational efficiency can create more cash flow to invest in service expansion and increase the customer base and revenues

(continued on next page)

Table 12.1 Categories of attributes of investments that contribute to water security—cont'd

Feature	Description
Developmental return	Improved access to water and sanitation services produce a range of valuable benefits for individuals, communities, and the environment, including a reduction in adverse health outcomes, increased educational attainment (especially for girls), enhanced labor productivity and improved environmental conditions
Project attributes	
Greenfield versus brownfield	Greenfield projects face additional business or technical risk due to the construction or new facilities, lack of performance track record and/or new technologies
Scalability	Some projects and financing structures could be scaled and replicated, with adaptation to local contexts and institutional structures. Other models present limitations to replication due to specific contextual circumstances
Size	Depends on whether the service is provided to an urban or rural area. The population density of the service area is a critical factor
Transaction costs	Vary depending on the complexity of the project, institutional arrangements, enabling conditions, and the parties involved in the financing In cases where there is weak capacity of service providers to maintain an asset registry and sufficient financial and accounting record, transaction costs can be very high
Tenor/Longevity	Minimum average of 15 years of debt financing contributes to a sustainable debt service over the lifetime of the asset

Source: Authors, adapted from Dominique and Bartz-Zuccala (2018).

is usually necessary in order to align the attributes of a project with the requirements of an appropriate investor (Gietema, this volume).

One of the ways in which the water sector could access a greater share of private sector capital is by improving how the "supply side" (e.g., project developers) makes its investment case to the "demand side" (e.g., project investors). In order for this investment case to be made effectively, it is necessary to reconcile this heterogeneity of water infrastructure with the need to align projects within an investable "asset class." Water infrastructure is sometimes mischaracterized in the policy-facing literature as a homogenous "thing"; often in an attempt to reductively frame the financing challenge to a common set of problems that simply require a generic solution. In reality, the opportunities and challenges of financing water infrastructure need to be understood from the project level, as that is the unit of account for investment. Without a project-level typology, knowledge asymmetries between the supply side and the demand side are inevitable. This is where understanding the distinctive risk-return profile and specific project attributes (as described in Table 12.1) is valuable.

These asymmetries can create a form of market failure where unsuitable or inappropriate sources of finance are pursued to fund projects, while investors whose objectives are better aligned with the projects are either not identified or not approached. We suggest that by aligning projects with their most appropriate funding sources, it could be possible to reduce some of the frictional costs associated with project financing. This should help to accelerate the pace at which projects are funded; which in turn could increase the probability of a broader spectrum of water infrastructure projects finding appropriate funding.

12.1.3 A diversity of investor types

Research by The New Climate Economy, a think-tank, estimates that of the US$2.5 to US$3 trillion spent on infrastructure each year, around 30%–40 % comes from the private sector; principally institutional investors and corporations. Investment in infrastructure is usually through debt (loans and bonds), equity (shares), or a combination thereof (Bielenberg et al., 2016). Bank loans are the most significant source of private sector finance to infrastructure projects followed by loans funded by investment companies, insurance companies, and public pension funds (Bielenberg et al., 2016). Bank loans are often ultimately funded by the capital markets through bond placements. These private-sector sources share the objective of achieving yields that either match their liabilities (such as insurance claims or pension payments), or the returns that investors require. Infrastructure developers and private equity funds often invest at an earlier stage of the project, where risks are higher. Private equity investors generally participate in middle-income countries where risks are perceived to be lower. They typically take an equity position and target a higher return on investment to compensate for the additional risk. Depending on the country context, sensitivities about private investment, private ownership, and private operation (and in some cases, outright prohibitions) may limit private investment for certain subsectors. For example, as noted in Wu's (this volume) analysis of financing water in China, private operators are welcome in wastewater and industrial bulk water, but are barred from entry into water supply services.

What is sometimes overlooked in commentary about infrastructure finance is that over US$1 trillion of private sector investment in infrastructure is accounted for by the corporate sector. Companies, including large multinational organizations, invest in infrastructure principally for directly operational reasons. Most investments are deployed within their "fence line," that is, on production facilities that they own or control. In recent years, however, new initiatives such as corporate water stewardship have increased the volume of corporate spending on infrastructure beyond the fence line, although these investments are still typically concentrated within a vertically integrated supply chain.

Typologies that include corporations as investors in water infrastructure are relatively undeveloped, perhaps because the data is difficult to come by. Unlike most institutional investors, corporations have few obligations to disclose what investments they make in infrastructure, particularly when they are investing in their own production capacity. However, corporations are likely to become more significant as investors in the next few years, as the implications of the infrastructure gap start being felt on company revenues and profits.

Corporations reap benefits from infrastructure investments. Without adequate infrastructure to make and sell products to consumers in local markets, businesses will not be able to fully tap latent demand. They are also well-positioned to mobilize finance and bring their internal expertise to bear on project evaluation and development. For example, large corporations that are listed on stock exchanges often have access to the international capital markets and can borrow money at much lower rates of interest than state-owned utilities in many developing countries. There is even the potential for corporations to take advantage of variations in the cost of capital by borrowing money at low rates on the international markets, and then investing in the debt of local infrastructure projects.

In addition, corporations may also be able to leverage their internal resources and knowledge networks to evaluate risk and return more adeptly than many institutional investors. Finally, corporations with a local presence in multiple markets are well-positioned to avoid "home bias," which institutional investors tend to exhibit to the extent that they prefer to invest in the region that they are most familiar with.

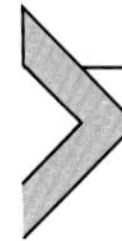

12.2 Select proven models and emerging approaches: insights on prerequisites for replication

12.2.1 Proven models and key prerequisites

Water-related investments deliver a range of public benefits and public finance has traditionally played a central role, with concessional finance playing an important role in developing countries. Public authorities (government agencies, municipalities, publicly owned utilities) often have the mandate for the delivery of water supply and sanitation services as well as water resources management. Ensuring such authorities have access to financing, notably via capital markets (where feasible) to finance long-tenor debt for long-term investments is key. Hence, the role of public funding, access to capital markets, and the capacity to leverage commercial finance with existing public funding are vital elements to address the financing challenge.

The case studies documented in this part of the book provide robust illustrations of several proven financing models for water-related investments. Drawing on the US experience, Gebhardt et al. (this volume) discuss the history of the public funding provided to water and sanitation infrastructure illustrating the US government programs established to address national water management goals. The authors detail the two dominant US federal loan programs—notably the Clean Water and Drinking Water State Revolving Funds managed by the states and the Water Infrastructure Finance Innovation Act (WIFIA), which target large-scale investments in water infrastructure of national importance. Revolving funds have also been successfully deployed for water and sanitation infrastructure investment in the Philippines and Cabo Verde (OECD, 2019).

Also drawing on the experience of the United States, Davis and Johnson (this volume) focus on mobilizing private capital for large-scale ecological restoration and conservation. They detail how the Clean Water Act created a market for restored wetland and streams via mitigation banking. This market now attracts significant private capital investment. Such policies have provided an attractive incentive for private investment in public goods (in the form of improved environmental outcomes).

Another investment approach with a proven track record is the water fund, as thoroughly described by De Bièvre and Coronel (this volume), drawing in particular on the experience of FONAG, the water fund in Quito, Ecuador, the first of its kind established over 20 years ago. Water funds are long-term financial mechanisms that allow multiple stakeholders to join efforts for a common purpose, mainly source water protection. This is an illustration of pooling funding from several beneficiaries—such as urban dwellers using the municipal water supply, along with corporations and hydropower producers reliant on secure water supply. These financial mechanisms benefit from structured and secure legal and institutional arrangements that ensure funding flows over long periods of time.

12.2.2 Emerging approaches

Much has been written over the years, not least by both authors of this chapter, regarding the opportunities of blended finance as a way to mobilize private sector investment; but also the particular challenges of achieving this objective within the water sector. The intention is not to rehearse these arguments here, although interested readers are encouraged to refer to the existing literature (see WWC, 2018; OECD 2019). An example of a blended finance vehicle provided in this part of the book is the Kenya Pooled Water Fund (van Oppenraaij et al., this volume). The WIFIA direct financing program managed by the US Environmental Protection Agency is also an example of using public instruments to de-risk investments, as WIFIA relies on US Treasury funding and lends at US Treasury rates (Gehbardt et al., this volume).

These examples, useful as they are, also highlight the relative paucity of blended finance structures that operate at scale in the water sector (several examples are documented in OECD (2019) along with the challenges for replication and scaling). This remains an area for ongoing innovation, particularly given the urgency of universally achieving SDG 6 as a first-order improvement in sustainable development outcomes.

As an example of this innovation, in January 2022, a new investment platform was launched, seeded with a US$ 750 million commitment from the Dutch pension fund provider, APG. Called the ILX Fund 1, its approach is to coinvest with multilateral development banks and development finance institutions, in order to give its investors exposure to emerging market opportunities. The rationale of investing in syndicated loans originated and structured by DFIs is to mitigate the investment risk, on the assumption that the development banks have completed some due diligence, and would likely have funded through technical assistance some of the basic improvements necessary to make a project investable. ILX has highlighted four investment themes as current priorities: energy access, infrastructure, food security, and inclusive finance. While water does not feature as a named theme, it is directly in scope under infrastructure, and indirectly in scope under energy access and food security. Indeed, Muruven (this volume) describes "landscape based" approaches to financing water, which is potentially well-aligned with the impact orientation of the ILX fund. In particular, it facilitates investments of a sufficiently material ticket size to justify the transaction costs typically associated with a syndicated DFI loan.

12.2.3 Key enabling conditions for scaling and adaptation to new contexts

When reviewing case studies of a range of financing approaches for investments that contribute to water security, a fundamental and persistent question arises: if and how can such examples be adapted and replicated for new contexts, notably for different types of water-related investments and for different political, regulatory, and institutional contexts? Exploring this question requires understanding the key enabling conditions that need to be in place. Each of the case studies in this part of the book devotes some discussion to the enabling conditions required for the specific financing approach to work well. This section summarizes some commonalities among the key conditions identified.

Policies, legal and regulatory elements

- Laws establish governing, contractual, and enforcement parameters for sustainable operating models.
- Financial contracts are supported by statutory authority and contract law precedents.
- Regulatory regime that defines an explicit goal for a defined environmental resource (such as the US Clean Water Act's "no net loss" of aquatic resources). Goals can be forward-looking or can account for prior environmental harm requiring remediation.
- In the case of ecological restoration, contractual means to procure ecological credits to provide an incentive for investment.
- Unwavering implementation of the regulated and agreed tariff adjustments (as well as annual indexation) is mandatory for sustainable, revenue-based, long-tenor debt financing.

Governance arrangements and political support

- Qualified entities that are empowered to administer programs at national and subnational levels.
- Clearly defined roles and responsibilities for WSS service delivery and for water resources management across the institutional landscape.

- Political support at the national and local levels, in particular in developing countries.

Market access and financial support

- Viable local capital markets with established securities laws and regulations are tested and resilient.
- Secondary market trading is well established. Securities firms are subject to standards of integrity established by law and accreditation.
- Federal and or state government investment quality enables market access at reasonable cost.
- A dedicated funding stream can be secured for investment or security support (i.e., guarantee facilities).

Capacity and resources for quality project development and selection

- Project development resources can be secured and sustained.
- Project selection criteria is established, publicly vetted, and reflected in the published project prioritization list.
- There is an emerging critical mass of projects in development that can support aggregating models and private investor support.
- Ensuring responsiveness and capacities of local utilities to the demands of the project preparation phase.

Secure revenue streams and verifiable performance

- For WSS service delivery:
  - Creditworthy borrowers.
  - Revenue streams are established and supported by high collections.
  - Cost management and investments that reduce nonrevenue water loss.
  - Track record of overcoming operational challenges.
- For water resources management and ecological performance:
  - Defining the basic principle of a credit founded on science-based criteria and a financial mechanism for long-term monitoring and maintenance.
  - A metric of ecological success that reflects scientific understanding of desired physical, biological, and chemical outcomes, applied in a predictable, consistent manner for a given resource type.
  - Monitoring and information generation to allow for adequate decision-making, effective implementation, as well as adaptive management and institutional learning. This includes impact monitoring to demonstrate the long-term impact and financial returns, drawing on rigorous data collection in collaboration with constituents and scientific partners.

- The principle that private investment in restoration must provide results *before* sales can occur and a profit obtained.

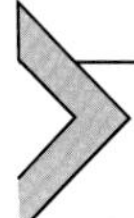

12.3 Reflections on future developments: a role for decision-ready data to inform financing

Looking forward, beyond what can be learned from experience, adapted, and scaled up, there is significant potential for technological innovation to stimulate investments that contribute to water security in the coming years. One area that is likely to receive increasing attention is the use of geospatial data and in particular its potential for improving the allocative efficiency of capital investments for water security.

To illustrate, we can look at the example of water stress. Water is an intrinsically local resource. Water stress can be defined as where the demand for water exceeds the sustainable supply, at that location. Of the many sobering facts regarding current and future water stress, few are more arresting than the reality that today, 1 billion people living in urban areas face water stress. Moreover, recent peer-reviewed research suggests that number could double by 2050 (He et al., 2021). The prospect of two billion people living in cities in towns where they cannot rely on accessing clean, freshwater when they open a tap is a sobering one. The risk to economic, societal, and financial assets as a consequence of water stress is becoming progressively more evident. For example, when Cape Town approached "Day Zero" in 2018, its rich and poor citizens alike queued daily at standpipes for their water ration. Crisis was only just averted by the serendipitous arrival of rain.

This is a challenge that can be addressed with current knowledge and technologies. Salt can be removed from seawater. Reservoirs can be built. Leaks can be repaired. Demand can be managed. Water can be transferred across basins, nationally, and internationally. Water embedded in products such as food and clothing can be traded, and so forth. But each of these options has their own profile of costs and benefits, that depend on many demand and supply drivers over the short, medium, and long term. Where many of these drivers have been relatively predictable historically, dynamics such as urban migration, rapid changes in socioeconomic conditions, and volatility associated with climate change, have made projections more challenging and the future more uncertain.

Further, water stress is hyperlocal. If water stress is to be managed at least cost to communities, decision-makers need ready access to reliable, good quality, forward-looking data at the relevant spatial scale. Because of the complexity of producing such data, they are scarce. Arguably, it is the lack of this hyperlocal, accurate, decision-ready data that is one of the biggest impediments to financing water infrastructure at scale—and cutting-edge technologies offer the prospect of changing the status quo.

These data would be valuable to financiers and investors in water infrastructure and could radically improve the efficiency of capital allocation. Where a utility is choosing between spending on a leakage reduction program, or a desalination plant, or a new reservoir—the decisions made have significant socioeconomic, environmental, and financial implications. The potential for over-investment, underinvestment, or poorly targeted investment is high. Access to good data that can underpin and more precisely target investment decisions can be a critical factor in enabling the finance to flow to where it generates the most value.

Earth observation data—derived from satellite imagery—now makes it possible to measure and track changes in surface water extents at very high resolution. With a historical catalog of imagery already going back some 40 years, it is possible to track inter- and intra-annual variations in water availability at the scale of individual rivers, reservoirs, and streams. Measuring changing water extents simultaneously across a networked basin system allows unique insights into flux dynamics for both surfaces and even subsurface flows. With high-resolution data, it becomes possible to use an observational record to project future water stress, even without the benefit of knowing local conditioning variables such as land cover, soil type, and geology. On the demand side, earth observation—when combined with other novel datasets—unlocks higher-resolution insights than those afforded by current models.

For example, consider a micro basin where a number of mines operate. Mining is typically a water-intensive activity. Combining geospatial insight with asset-level data, the physical production output of the mines can now be tracked, providing granular visibility on demand for the water resources, and how that is changing over time. Similar insight can be gained from tracking local agricultural production, or industrial activity. The innovations offer a step-change in how water stress is measured and managed; which by extension has clear implications for financing water infrastructure.

Investments in water security are a first-line adaptive response to climate change. Much as climate models today are improving our ability to project changes in local meteorological patterns, integrating these data in the future with a better understanding of hyperlocal water stress will enable decision-makers to optimize the infrastructure and investment choices that need to be made for the coming decades. Climate models project how rainfall, temperature, and evapotranspiration patterns—all fundamental drivers of water stress—will change under different scenarios. The water sector provides the best possible use case for how technology can be integrated with climate science to improve the efficacy of investment for adaptation. Challenges notwithstanding, these are important and exciting times to be engaged in financing for water security.

References

Bielenberg, A., Kerlin, M., Oppenheim, J., Roberts, M., 2016. Financing change: how to mobilize private sector financing for sustainable infrastructure — working papers. The New Climate Economy.

Dominique, K., Bartz-Zuccala, W., 2018. Blended finance for water investment: background paper. In: 3rd meeting of the Roundtable on Financing Water. OECD (accessed November 9, 2021).

He, C., Liu, Z., Wu, J., et al., 2021. Future global urban water scarcity and potential solutions. Nat Commun **12**, 4667. https://doi.org/10.1038/s41467-021-25026-3.

Money, A. (2017). *Projects, investors, risks and returns.* https://www.oecd.org/env/resources/Money%20(2017)%20Projects,%20investors,%20risks%20and%20returns.pdf (accessed July 27, 2021).

OECD, 2013. Water Security for Better Lives. OECD Publishing, Paris. https://doi.org/10.1787/9789264202405-en.

OECD, 2019. Making Blended Finance Work for Water and Sanitation: Unlocking Commercial Finance for SDG 6. OECD Publishing, Paris. https://dx.doi.org/10.1787/5efc8950-en.

OECD, 2022. Financing a Water Secure Future. OECD Publishing, Paris. https://doi.org/10.1787/a2ecb261-en.

WWC, 2018. Hybridity and Blended Finance. World Water Council www.worldwatercouncil.org.

CHAPTER THIRTEEN

Water infrastructure financing: the experience of the United States

James Gebhardt, CFA[a], Rick C. Ziegler[b] and Alex Mourant, CFA[a]
[a]Office of Water, US Environmental Protection Agency, Washington, D.C.
[b]Office of International and tribal Affairs (OITA), US Environmental Protection Agency, Washington, D.C.

13.1 Introduction

In support of national clean water and safe drinking water goals, the US Congress has created a series of national-level water-focused financing programs, accessible by state and local governments and other qualified project sponsors to drive water sector investment. These programs, in part, are geared toward delivering capital cost subsidies that serve to incentivize action by project owners and accelerate project completion timetables. The core incentives consist of the accessibility of capital at below-market terms that are made available in accordance with federal law to both creditworthy and credit-challenged project owners. There are currently five major programs: the Clean Water and Drinking Water State Revolving Funds (collectively, the CWSRF, DWSRF, or SRFs) which are federally sponsored and state-administered; the US Environmental Protection Agency administered direct financing program established by the 2014 Water Infrastructure and Innovation Finance Act (WIFIA); the Community Development Block Grant (CDBG) program administered by the United States Department of Housing and Urban Development which is used in part to provide grants for long term public infrastructure needs; and, the US Department of Agriculture Water Environment Program, which generally provides financial assistance to rural communities with populations of 10,000 or less. Each of these programs was established to help project sponsors and owners minimize project costs and to provide the needed financial incentives to compel investment action. Today, these programs operate alongside traditional sources of investment capital dominated by state and local government issuance of tax-exempt bonds accessed via the US Capital Markets. US local governments include all governing jurisdictions with responsibility for water, wastewater service delivery, and stormwater management.

Financing Investment in Water Security: Recent Developments and Perspectives.
DOI: https://doi.org/10.1016/B978-0-12-822847-0.00014-4

This chapter will discuss the value of the USEPA sponsored programs (i.e., why they are needed and what is their relative value in relation to each other and to capital market alternatives), how they operate and how they are juxtaposed against, and leverage, the US Capital Market. We will also discuss the potential for adapting elements of the SRF and WIFIA models (referred to together as US Water Finance Models) to support international development blended finance strategies, formalized in recent years by the Organization for Economic Cooperation and Development (OECD) (OECD, 2018) and endorsed by the G7 (G7, 2018), the G20 (G20, 2019), and bilateral and multilateral finance development institutions.[1,2,3]

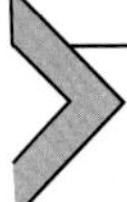

13.2 Federal policy support for local water infrastructure improvements

13.2.1 Early history

In the United States, water and wastewater infrastructure facilities are locally owned and managed. From the earliest days of the Republic, management of water resources and wastewater flows consisted of harnessing potable water for domestic and industrial use and disposing of used water often thru open conveyances directly back to ground sources and local waterways. By the mid-19th century, as populations grew and became more concentrated in urban centers, recognizing related public health dangers, cities embarked on the development of sewer systems to manage wastewater. These systems were largely paid for from general tax dollars levied by local governments. During the 20th century, the federal government began to play a role in the protection of the nation's waterways. The federal interest was precipitated by concerns about the integrity of the nation's navigable waters, which included refuse matter largely attributable to untreated wastewater flows. As early as 1911, the US Army Corps of Engineers, the entity then solely responsible for the nation's waterways, recommended that wastewater treatment facilities be mandated by the federal government to address the fouling of waterways from untreated sewage disposal (Folsom, 1958).

[1] OECD Development Assistance Committee, "Blended Finance Principles," http://www.oecd.org/dac/financing-sustainable-development/blended-finances-principles/#:~:text=The%20OECD%20DAC%20Blended%20Finance,philanthropies%20and%20other%20concerned%20stakeholders.

[2] G7 2018. *"Charlevoix Commitment on Innovative Financing for Development."* Charlevoix, Quebec. https://www.international.gc.ca/world-monde/assets/pdfs/international_relations-relations_internationales/g7/2018-06-09-innovative_financing-financement_novateur-en.pdf.

[3] G20 Osaka Declaration, https://www.mofa.go.jp/policy/economy/g20_summit/osaka19/en/documents/final_g20_osaka_leaders_declaration.html.

Federal efforts to address water pollution control needs began in earnest in 1948 when Congress passed the Federal Water Pollution Control Act (the FWPCA). The FWPCA created a comprehensive set of water quality programs and provided grant funding for state and local governments that was tied to population and income levels. Enforcement was limited to interstate waters. The US Public Health Service was assigned responsibility for financial and technical assistance. In 1965, Congress amended the FWPCA to require states to establish interstate water quality standards. The FWPCA also established the Federal Water Pollution Control Administration to set standards where states failed to do so. The federal government later determined that these federal standards did not produce the water quality gains needed. Consequently, major amendments were enacted in 1972. It was at this time that the FWPCA took on the moniker, the Clean Water Act (referred to henceforth as the CWA). These amendments established the objective to restore and maintain the chemical, physical and biological integrity of the nation's waters backed by a federally administered regulatory regime and a direct Municipal Construction Grants Program managed by the newly created US Environmental Protection Agency (US EPA or the Agency).[4] The direct federal grants program was intended to address a backlog of municipal wastewater treatment needs that was the primary cause of the nation's poor water quality.

The 1972 amendments established a direct linkage between federal enforcement and federal funding as a critical component of the nation's strategy to address the poor quality of the nation's waters. Access to the nation's public capital markets was not a factor that impacted eligibility for grant funding. Municipalities with investment-grade credit profiles had just as much access to federal grant dollars as those that did not. The core public policy consideration was pairing the newly created enforcement regime with publicly weighted capital incentives (Copeland, 2012). The following considerations were key:

- A federal funding commitment to local governments that would bear the investment burden of compliance was critical to gaining Congressional support for direct federal enforcement that could be applied nationally and uniformly.
- There was high political concern regarding the ability of local governments to establish and /or raise user rates for sewer systems at levels

[4] The USEPA is an independent operating agency of the federal governments. The Administrator is appointed by the president and approved by the US Senate. Its annual operating budget, which includes infrastructure funding programs, is funded through Congressional appropriations.

needed to support the levels of wastewater investment urgently needed to achieve water quality treatment standards and do it affordably.

When enacted the federal cost-share was set at 75% of eligible project costs. State and local governments were responsible for splitting the remaining costs which were funded from revenues or debt issuance. The federal share was reduced to 55% in 1981. According to the Congressional Research Service, between 1972 and 1984 federal appropriations for the Municipal Construction Grant Program totaled USD 41 billion. At the time this level of federal support represented the largest public works commitment since the Federal Interstate Highway System. As a consequence of fiscal constraints that emerged in the 1980s, the federal government sought to further shift the level of federal commitment. This shift included a focus on how future federal commitments would be delivered, including a role for the nation's capital markets.

13.2.2 Clean Water Act amendments jump-starts state revolving fund models

In 1987, reauthorization of the CWA ended the Municipal Construction Grants Program in favor of the creation of the Clean Water SRF. Justification for the change centered on: (1) the original intent of the program to address the backlog of sewage treatment needs having been met; (2) most of the remaining projects were believed to pose little environmental threat and were not appropriate federal responsibilities; and (3) state and local governments were fully capable of running construction programs with a clear responsibility to construct treatment capacity to meet environmental objectives. States and localities also had a view that favored greater state and local control that could lighten the burden of federal rules, regulations, and oversight (Copeland, 2012).

Under the CWSRF Program, states became the designated recipients of federal grant dollars and would be provided seed money to capitalize state-administered financial assistance programs to build sewage treatment plants and other water quality projects.[5] Repayments of obligations to the states by eligible recipients of SRF financial assistance would provide for buildup of a renewable source of capital for future investments. The new program gave the states flexibility to set priorities and administer funding. Federal appropriations were scheduled to end after FY1994

[5] The 1987 Amendments greatly expanded the types of water quality projects eligible for assistance. However, it limited financial assistance for point sources to publicly owned facilities. The Amendments also designated as eligible nonpoint source projects which could be either publicly or privately owned.

Table 13.1 US federal water infrastructure funding (Millions of U.S. Dollars Appropriated).

Reporting years	Construction grants	Clean Water SRF	Drinking Water SRF	Totals
1951–60	187			187
1961–70	1,881			1,881
1971–1980	36,636			36,636
1981–90	20,397	2,074		22,471
1991–95	141	6,896		7,037
1996–00	106	5,705	2,732	15,580
2001–05	70	6,501	4,033	10,604
2006–10	39	8,085	7,574	26,302
2011–15	66	10,328	3,611	14,005
2016–20	0	7,237	4,622	11,859
Totals	59,523	46,826	22,122	128,921

Source: United States Environmental Protection Agency, CW and DWSRF National Information Management Survey (NIMS). Reporting Years are July 1 to June 30.

(Copeland, 2012). Contrary to the initial plan, Congress has continued to provide national appropriations to the CWSRF each year since 1989. Continued bipartisan congressional support is due to the consistent stream of positive feedback provided to Congress by stakeholders, specifically state and local governments.

In 1996, federal support for local infrastructure investment was expanded to include drinking water facilities. Inclusion of language in the 1996 amendments to the Safe Drinking Water Act (the SDWA or together with the CWA, "the Acts") (P.L. 104-182) authorized the creation of a Drinking Water SRF, modeled on the Clean Water SRF, based on the following:

- Growing populations had altered land use sufficiently to elevate risks of nutrient contamination of drinking water supplies.
- Growth in the number of regulated contaminants required large investments in treatment technology to meet regulatory requirements.
- Policy concerns that many of the nation's small community water systems, comprising approximately 92% of the national total, were likely to lack the financial capacity to meet the rising costs of SDWA compliance.

Table 13.1 provides a history of federal support for pollution control and drinking water infrastructure investment. It shows both the construction grant funding that was made available for projects and the grant dollars made available to the states to capitalize the Clean Water and Drinking Water SRFs.

To date, the state programs have successfully revolved the USD 68.9 billion in federal appropriations and USD 13.5 billion in matching funds to close more than 56,000 financial assistance agreements and deliver almost USD 180 billion in financial assistance. The average loan size is USD 3.4 and USD 2.7 million for the CW and DWSRF programs, respectively. These small loan sizes reflect the fragmentation of water and wastewater service providers in the United States. This fragmentation is both jurisdictional and geographic. In many states, especially those dominated by smaller rural systems, small average loan size is the result of a deliberate prioritization of small project clientele, where project investment need and affordability challenges are greatest. In some instances, SRF programs impose hard limits on loan size.

13.2.3 National direct lending vehicle added to cover financing gaps, accelerate private capital investment

In 2014, Congress added on an additional financing program directed at water infrastructure when it enacted the Water Infrastructure Finance Innovation Act (WIFIA). This new financing mechanism provided statutory authority directing the US EPA to develop a direct loan program that would be administered by the Agency and funded by the US Treasury Department. As a federal credit program, WIFIA's funding is different than the SRF programs which are capitalized from a combination of annual federal and state appropriations. The WIFIA lending program establishes a US Treasury lending window from which monies are drawn to fund WIFIA loan draws and to which loan repayments are made. Appropriated dollars fund WIFIA program administration and capitalize a loss reserve that is sized based on WIFIA loan risk assessments made at the time each loan is underwritten. While the WIFIA program exercises proper due diligence to evaluate the credit risk of its borrowers for estimating the risk of each loan, the WIFIA program on-lends at the US Government borrowing rate, regardless of risk profile. Congress created the WIFIA program to address funding needs that were not being addressed by the SRFs due to state-specific project eligibility, loan capacity, or state-imposed loan size constraints.

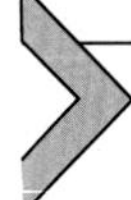

13.3 State revolving fund models: capabilities and designs

13.3.1 Underwritten with federal and state capital contributions

Federal appropriation made to states for the SRF programs are subject to states providing matching funds equal to 20% of each state's annual allocation.

Grant awards made by the US EPA are subject to states demonstrating that they will provide for the requisite state match. States have elected to provide required matching funds through legislative appropriations or from proceeds of publicly issued state match bonds.

State allocations of the Clean Water SRF national appropriation are based on a percentage allocation formula that is provided in the CWA.[6] Each state's percentage allocation for the CWSRF range from the statutory minimum allocation of 0.5%–11.2%. These allocations were established when the CWSRF was authorized in 1987. The Safe Drinking Water Act sets state allocations of DWSRF appropriations on the basis of a periodic needs survey required to be administered by US EPA at 4-year intervals. Per the Act, each state receives no less than 1% of the annual Congressional appropriation.

The programs are designed to be a sustainable source of funds. Sustainable fund operations are assured by the stipulation contained in the federal Acts that federal and state equity contributions, and program earnings, be held in the SRFs in-perpetuity and used solely for the purposes prescribed by the Acts. The longer that federal appropriations and state match dollars are made available the more financially resilient and less dependent the SRFs are on future appropriations to sustain robust support to eligible financial assistance recipients. The result has been an ever-rising level of funding certainty that benefits all SRF stakeholders, including contract project developers and equipment vendors. Consistent funding has become a program bellwether that has produced tangible long-term benefits by enabling SRF administrators to largely match financial and technical assistance needs year in and year out. The beneficial result has been a persistently strong signal to the water infrastructure investment sector that planning and project development efforts will be rewarded with funding at better than market terms.

The move away from grants as the primary funding method for municipal wastewater treatment shifted resources to the states. This shift retained the state match requirement which was a feature of the Municipal Construction Grants Program and set it at 20% for the SRFs. At 20%, it was deemed small enough to persuade states to commit the match. Congress also included a security requirement that would set a minimum standard of protection for the federal investment. This requirement subjects each SRF financial assistance commitment to a recipient pledge of dedicated revenue. Together these program requirements serve to align the interests of the federal

[6] Changes in the allocation formula requires an amendment to the Clean Water Act.

and state governments in supporting the long-term sustainability of the funds.

13.3.2 Financial assistance parameters

When created the Acts provided the states the authority to use SRF funds to provide a number of different forms of financial assistance, including:

- Loans or the purchase of local government debt obligations, including requirements to start repayment within one year of construction completion and maturity limits, generally 30 years.
- The use of program equity to over collateralize bonds issued by the SRF for the purpose of funding financial assistance obligations.
- The guarantee of obligations where proceeds fund eligible projects or the provision of loan guarantees for revolving fund programs established by local governments or intergovernmental agencies.
- The investment of funds to grow the equity of the program.

The Acts also require that SRF assistance be provided at or below market rates. In 2009, the American Recovery and Reinvestment Act (ARRA) allowed a percentage of the federal investment to be used as a source of principal forgiveness, negative interest loans, or grants, generally referred as "additional subsidization." This change deepened the value of financial assistance to recipients and consistent with the autonomy vested in states under the Acts, the SRFs generally target their allocations of additional subsidization for specific project solutions and/or economically stressed communities. The additional subsidization feature has now become a staple of the programs (see below).

13.3.3 Lending to disadvantaged communities

Many states have established criteria by which eligible recipients of financial assistance qualify to receive added financial benefits which generally consist of deeper than target interest subsidies that could be as much as 100% resulting in 0%, no-interest loans.

The 1996 SDWA amendments, creating the DWSRF, expanded financial assistance options for disadvantaged communities by authorizing states to implement a discretionary subsidy program for "disadvantaged communities" as defined by each state based on its' publicly vetted annual Intended Use Plan (IUP) and approved by EPA (see Box 13.1). Under this provision, states could provide a subsidy to those communities of up to 30% of the state's federal grant in the form of negative interest rates or principal forgiveness.

BOX 13.1 The Acts Require SRFs to be Administered with Intended Use Plans

Both the Clean Water and Safe Drinking Water Acts require states to prepare an annual plan identifying the intended uses of the funds in their respective SRFs and describe how those uses support the goals of the SRFs. A final IUP must meet all requirements of the Acts and regulations in order to meet annual federal grant award requirements. According to recent US EPA Guidance, a final IUP must contain the following:

List of Projects

For the CWSRF, the CWA requires that the IUP contain a list of publicly owned treatment works projects on the State's project priority list (PPL), that are eligible for CWSRF construction assistance. This list must include: the name of the community; permit number or other applicable enforceable requirement, if available; the type of financial assistance, and; the projected amount of eligible assistance. The IUP must also contain a list of the non-point source and national estuary protection activities that the state expects to fund from its CWSRF. For the DWSRF, the SDWA requires that the IUP contain a list of projects in the state that are eligible for DWSRF assistance and are scheduled to be assisted under the plan. This list must include: the name of the public water system, a description of the project, the priority assigned to the project, the expected terms of financial assistance, and the size of the community served.

The fundable list included in the CW and DWSRF IUPs must contain eligible projects for which the total cost of assistance requested is at least equal to the amount of the grant being applied for before a grant can be awarded.

Additional Requirements

The final CWSRF IUP must contain a description of the financial status of the state loan fund including sources and uses; contemplated loan terms and interest rates; the short-term and long-term goals of the state loan fund; a description of the means by which the state will choose those projects that are ready to proceed to construction; and a declaration that the state has or will have by a date certain the authority to provide assistance and has a process and appropriate criteria which it will use to determine how it will provide this assistance to applicants. The final DWSRF IUP must contain the criteria and methods established for the distribution of funds; a description of the financial status of the state loan fund; the short-term and long-term goals of the state loan fund; a work plan for any monetary set-asides permitted to be taken under the SDWA; a description of the means by which the state will choose those projects that are ready to proceed to construction; and a declaration that the state has or

(*continued*)

BOX 13.1 The Acts Require SRFs to be Administered with Intended Use Plans — cont'd

will have by a date certain the authority to provide assistance and has a process and appropriate criteria which it will use to determine how it will provide this assistance. See: U.S. Code of Federal Regulations, https://www.ecfr.gov/current/title-40/chapter-I/subchapter-B/part-35/subpart-K and https://www.ecfr.gov/current/title-40/chapter-I/subchapter-B/part-35/subpart-L (US Code of Federal Regulations, 2021)

The project scoring criteria developed for the New York CWSRF in accord with statutory requirements determines how that state prioritizes projects for SRF financial assistance. The criteria establish scoring requirements for the following areas:

A. Existing Water Source
B. Water Quality Improvement
C. Consistency with State Management Plans
D. Intergovernmental Needs
E. Financial Needs
F. Economic Needs

The total numerical score for the project being scored is the sum of the applicable scores for the six criteria elements. The project scores are computed based on information in the approved or approvable facilities plan, engineering report, or other equivalent documents. The state accommodates projects without approved or approvable facilities plans or engineering reports by accepting information from other sources that is later adjusted when a facilities plan or engineering report is approved or determined to be approvable. Projects must be adequately supported by technical documentation, data, reports, etc. Scoring determines where projects will be listed in the PPL. See: https://www.efc.ny.gov/2021-CW-IUP. (New York State Environmental Facilities Corporation, 2021)

States may also provide loan-term extensions to these communities.[7,8]

Criteria for disadvantaged communities are made available in each state's annual CW and DW IUPs. Today, both SRFs are well equipped to provide funding packages that can meet the needs of entities that qualify as disadvantaged. Qualified entities may receive assistance that includes 0%

[7] USEPA, Utilization of Additional Subsidization Authority in the Clean Water and Drinking Water State Revolving Fund Programs, Report to Congress, April 2014. See: https://www.epa.gov/sites/production/files/2015-04/documents/additional_subsidization_report_to_congress_.pdf.

[8] America's Water Infrastructure Act (AWIA) of 2018 changed the additional subsidization provision to set both a floor and a ceiling to this authority. For capitalization grants awarded after AWIA's passage, states were authorized to use up to 35% of DWSRF cap grant as additional subsidization and no less than 6% subject to there being sufficient applications from disadvantaged communities. See: https://www.congress.gov/115/bills/s3021/BILLS-115s3021enr.pdf.

interest and additional subsidization. States that, as a policy matter, limit principal repayment to 20 years from project completion, may offer loans to disadvantaged communities that can be repaid over 30 years. In the DWSRF, the SDWA allows repayment up to 40 years for disadvantaged communities.

13.3.4 SRF model adaptability: delivery vehicle for economic stimulus and disaster recovery

In response to the deep recession caused by the 2008–2009 global financial crisis, Congress passed ARRA which included a USD 6 billion appropriation for the SRFs—USD 4 billion for the CWSRF and USD 2 billion for the DWSRF. *The existence of the federally sponsored and state administered SRFs provided a ready platform to quickly deliver funds to assure infrastructure project development timetables would not be impacted by adverse economic developments.* State CW and DW IUPs were a ready source of shovel-ready projects which could absorb the funds which had to be committed within one year of ARRA's enactment (American Recovery and Reinvestment Act, 2009). For ARRA, Congress waived the 20% state match requirement. In 2013, the SRFs were looked to again as a vehicle for delivering funds—this time for infrastructure recovery. Special appropriations totaling USD 570 million were made to the states of New York and New Jersey to address extensive damage caused to water and wastewater treatment systems by Hurricane Sandy. The SRFs were used again in 2019 to deliver disaster recovery assistance to various states in the aftermath of hurricanes and wildfires.

ARRA introduced new requirements to both the CWSRF and DWSRF communities. Of the USD 6 billion appropriated to both programs, not less than 50% of a state's ARRA allotment had to be provided in the form of grants, principal forgiveness, or negative interest rate loans. These options have the effect of reducing the repayment amount or absolving the recipient obligation altogether. ARRA also introduced a Green Project Reserve to which states were required to commit 20% of federal resources. Projects eligible in this category included green infrastructure, and water and energy efficiency. Appropriations had to be contracted within a 1-year time frame (USEPA, 2014). ARRA ultimately committed the full USD 6 billion and funded 2,487 and 1,254 CW and DW projects, respectively. In response to Hurricane Sandy, the appropriations made to New York and New Jersey came with project resiliency requirements. In support of these requirements, the additional subsidization provisions made available in ARRA were also made available to New York and New Jersey.

In the years following ARRA, Congress used annual appropriation bills to authorize the continued use of additional subsidization by providing for a portion of a state's SRF appropriation to be allocated to additional subsidization. This benefit was no longer limited to disadvantaged communities. When Congress passed the Water Resource Recovery and Development Act of 2014 (WRRDA), it became a permanent part of the CWSRF provisions of the Clean Water Act. In November 2021, The U.S. Congress enacted, and President Biden Signed into law, the Infrastructure Investment and Jobs Act (IIJA). Like ARRA, IIJA utilized the existing SRFs to boost federal water infrastructure investment and economic growth (Infrastructure Investment & Jobs Act, 2021).

13.3.4.1 Technical assistance

Technical assistance is also a standard operating feature of the DWSRF. Federal DWSRF appropriations include specific funding allocations, known as "set-asides" for prescribed purposes that are geared toward supporting capacity-building efforts that are critical to sustainable operations and financial viability, including workforce development. The DWSRF also allows states to use permitted set-asides for investment in source water protection.

13.3.4.2 Administrative assistance

Administrative assistance is also provided by the Acts in the form of a "set-aside" to provide for SRF administration. The CWA and SDWA allow states to use the greater of 4% of all federal grant awards, USD 400,000 per year or $1/5^{th}$ percent per year of the current valuation of the fund. In addition, states may charge and collect administrative fees to support program operations. The administrative set-asides and fees are used to support all fund-related operations including all program-related overhead, staff salaries and benefits.

13.3.5 Federal-State Revolving Fund governing partnership

Under the federal-state governance framework responsibilities are shared. States have day-to-day management responsibilities for:

- The IUPs which establish, for stakeholders, the basis for the setting of project priorities and the criteria and methodologies for ranking projects.
- The processes by which loan applications are solicited, vetted, approved, and funded.

[9] More information can be sourced at https://www.epa.gov/infrastructure/water-infrastructure-investments.

- Assurance that with respect to eligible projects federal program requirements are followed (e.g., wage rate and material purchases limited to American Iron and Steel).
- Providing the oversight and technical assistance to assure each project's technical viability, inclusive of alternatives analyses, and financial sustainability as evidenced by the dedicated revenue pledge and credit analyses of the financial assistance beneficiary.
- Setting the financial terms that they make available in accord with federal and state laws.

  EPA responsibilities are delegated as follows:

  EPA Region Offices:
- Award and oversee administration of federal capitalization grants made to the states.
- Review states' IUPs and project priority lists for federal compliance.
- Conduct detailed reviews of state program performance.
- Communicate national policies to states.
- Enforce grant requirements.

  EPA National Office:
- Develops and disseminates national program policy.
- Oversee regions and state programs.
- Partner with Regions on program issues.
- Acts as clearinghouse and gateway for on-going national discussion about programs.
- Engages with the Administration, Congress, and stakeholders on issues of national importance. In this capacity EPA often is tasked with evaluating the benefits of new features that result from the national discussion and are under consideration by the Administration or Congress. The decision to integrate the SRFs in the national policy responses to the 2008-09 global economic crisis and for disaster recovery (discussed above) are prime examples of EPA's active role in working with the Administration and Congress to shape program enhancements.

13.3.6 State operational frameworks

Setting up the SRF programs at the state level required each state to adopt enabling legislation that established the state's institutional framework—such as identifying the federal grant recipients, the state entity(s) responsible for running the programs and, consistent with the federal language, the operational authority granted by the state. The states have developed two basic operating models for the program. The first model is represented by programs operated by the state environmental agency with responsibility for any debt issuance and non-program fund investment residing with the state treasurer. This model is dominant. The second model is represented by programs where oversight is provided by the state environmental agency, as grant recipient, and program operational responsibilities are shared between the agency and an independent state authority and its governing board.

Massachusetts is an example of the first model. The Massachusetts Department of Environmental Protection is responsible for its IUP and project selection but relies on the Office of the State Treasurer for financial and investment management services. The legislature created the Massachusetts Water Pollution Abatement Trust (Massachusetts Water Pollution Abatement Trust, 2021) which houses fund assets and is the obligor on SRF bonds issued to capitalize the Trust. The Office of the State Treasurer is responsible for all aspects of financial and investment management of the Trust while the State Office of Finance and Administration is responsible for budget and oversight.[10]

New York State ("NYS") is an example of the second model. In 1989, the NYS legislature amended its environmental conservation and public authority laws to accommodate the Clean Water SRF. The NYS Department of Environmental Conservation was designated as the federal grant recipient and was given authority to sign off on the State's Annual IUP. The NYS Environmental Facilities Corporation (EFC), created years before to serve as a financing conduit for private industry to access tax-exempt financing for infrastructure projects, was authorized to manage all financial aspects of the program. Principal responsibilities include IUP preparation, loan development and origination, loan servicing, bond issuance, managing the investments of the fund, and all budget and financial accounting functions. In 1995, the state amended its public health laws and further amended its public authority laws governing EFC to incorporate the DWSRF program into state operations.

13.3.7 Models adopted; structures and public capital market relationships

States have opted for a variety of models for the delivery of financial assistance to program-eligible recipients. How these models have been selected

[10] https://www.mass.gov/orgs/the-massachusetts-clean-water-trust.

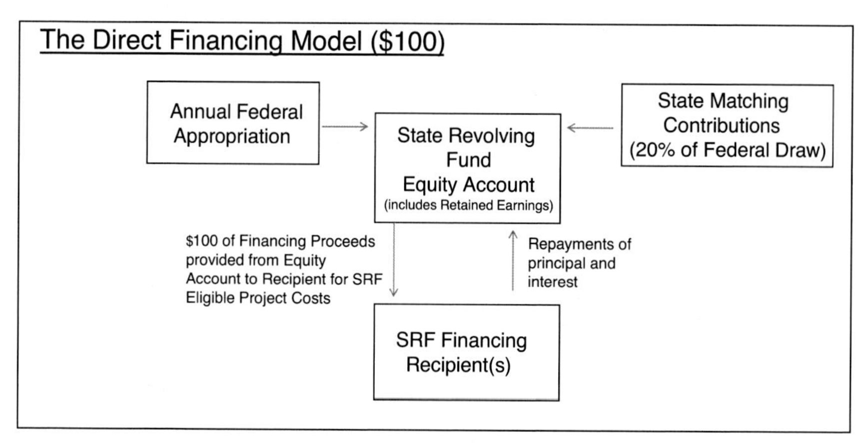

Figure 13.1 The direct financing model (USD 100). *(Source: Authors).*

and deployed reflects a number of factors. Principal among them have been the state-by-state distribution of available resources. States receiving a larger percentage of the national appropriation benefited from the economies of scale that came with the allocation. The Acts initially allowed up to 4% of the federal allocation to be drawn to pay for administrative expenses. For states with larger allocations this presented meaningful advantages in setting up their financial assistance delivery models. States with more limited resources opted initially to limit loan originations to direct financings funded solely from federal and state investment. States with higher allocations tended to be more aggressive making early entries into the bond market on behalf of eligible recipients. States that access the bond market to supplement their resources are leveraging the federal allocations and matching state contributions, using these resources to secure SRF bonds, thereby expanding their financial capacity to serve more projects.

Generally speaking, it is the smaller rural states that do not rely on leveraging, limiting financial assistance to direct loan programs where federal and state dollars are lent directly to eligible recipients (Fig. 13.1). Repayments of principal and interest are simply recycled and are available for additional lending. Through 2019, 23 states and Puerto Rico have limited their SRF financial assistance activity to direct lending.

The remaining 29 states have used federal and state dollars to secure additional funds from the US tax-exempt bond markets. Historically, states have leveraged program equity with the sale of SRF bonds on behalf of single or multiple recipients. Basic security for SRF bond issues consists of pledges of loans funded from bond proceeds and equity that is pledged to secure the transaction (commonly known as the Blend Rate and Cashflow Models) or from equity invested in a dedicated reserve (the Reserve Fund Model) that is invested in securities that meet rating agency requirements for the target bond rating.

Leveraging factors vary by state and are generally informed by the below-market rate targeted by the state. Because federal tax law limits return on pledged equity to the cost of tax-exempt funds it is common practice for SRFs to over-collateralize their bond issues in proportion to the interest subsidy target. This means that a 50% interest subsidy target translates into a 50% equity allocation for a given project (a 2:1 project-to-equity leveraging factor). This results in a trade-off between the extent of the subsidy and the amount of funds made available through leveraging. The deeper the subsidy target the greater the excess coverage and lower the leverage factor will be. Both the Blend Rate and Reserve Fund Models originate long-term SRF loans at the closing of the bond sale. The Cashflow Model originates loans from program equity (direct loans) and periodically enters the bond market to expand loan capacity by pledging all or a portion of the direct loans. The Cashflow Model is ideal for states that operate with more limited resources and cannot smoothly time multiple loan closings with a simultaneous bond close. The mistiming of loan and bond closings present interest rate basis risk for such state programs (i.e., the risk that interest rate fluctuations can adversely impact capital requirements). This means that if market interest rates have moved higher between loan origination and the bond sale, more loan assets need to be pledged to assure sufficient interest payments are collected to cover bond interest. If market interest rates move lower, less loan assets are required. In effect, due to basis risk, a Cashflow Model's transaction-specific leverage factor may fluctuate above or below the leverage target based on the loan/bond interest rate differential at the time of the bond sale.

13.3.8 Use of master financing indentures

In the early years of the SRF program, it was recognized by leveraging states that a master financing indenture[11] would be the best delivery vehicle

[11] For bonds issued to fund loans to large standalone borrowers, ratings remain heavily dependent on the credit quality of the underlying borrower. The actual rating assigned is a function of the credit strength of the borrower and pledge of the oversized SRF reserve. Rating services refer to this as a bottom up analysis. For the over-collateralized multi-borrower pooled financings, the ratings are driven largely by program factors with underlying borrower factors being of secondary importance. Rating services refer to this as a top down analysis.

SRF Equity and Master Trust(s) – Clean Water and Drinking Water

Assumption: $100 of Equity Secures a $200 Bond Issue

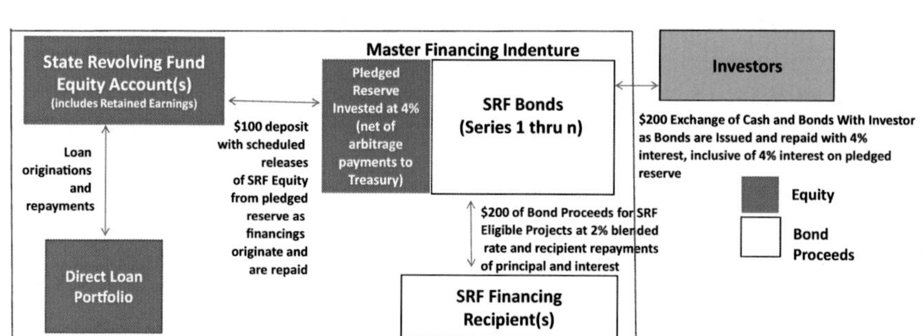

Note: The Clean Water and Drinking Water Financing Program can operate out of the same financing indenture and cross-pledge program assets to maximize credit value.

Figure 13.2 SRF reserve model. *(Source: Authors).*

for a revolving fund program. A master financing indenture[12] establishes all of the critical contract terms that protect the security interests of investors. SRF bonds are issued in accord with the master indenture and specific series supplemental indentures that establish the contract terms for each specific bond issued under the master (see SRF Bonds -Series 1 thru n in Figs. 13.1 and 13.2). Periodic issuance of bonds, in association with pledges of program equity, and the funding of loans from bond proceeds to eligible recipients produced, over a fairly short period, a diverse loan portfolio. The combination of a diverse and growing loan portfolio, over-collateralized with program equity, resulted in triple-A rating designations from each of the major rating services—Moody's, S&P, and Fitch Investors—for most state SRF programs by 1995.[13]

Consistent with program objectives—to assure market access to eligible recipients and minimize financing costs - achieving triple-A ratings provided states and program participants with access to least-cost funds which in many states sets the basis for the interest subsidy benefit. When the DWSRF was added in 1996, many states managed the needs of both programs thru the then-existing CWSRF master financing indenture. Although separate programs, for security purposes the respective statutes allow SRF administrators to cross-collateralize all assets pledged under the indentures.

[12] A master financing indenture is an overarching legal agreement between a bond issuer and a trustee that protects bondholders by (a) defining the rights and requirements of each of the parties to the agreement and (b) establishing threshold security parameters that bondholders can rely on in relation to any series of bonds issued and secured thereunder. Supplemental or series indentures are entered into in relation to specific series of bond issuances that fit under the master indenture. Supplemental indentures can add additional bondholder protections that are series specific but cannot weaken any provisions established under the Master Trust.

[13] https://www.transportation.gov/buildamerica/financing/tifia.

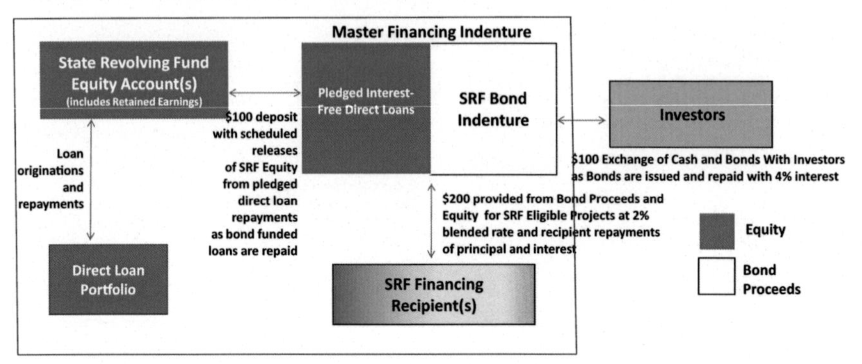

Figure 13.3 SRF blended rate model. *Note*: The Clean Water and Drinking Water Financing Program can operate out of the same financing indenture and cross-pledge program assets to maximize credit value. *(Source: Authors).*

Figs. 13.2 and 13.3 provide schematics representations of SRF Reserve and Blend Rate Models. The reader should note that they also include the basic fund flows for direct loans which are a part of every SRF program.

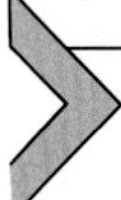

13.4 Water Infrastructure Finance Innovation Act ("WIFIA") National Direct Lending Model

13.4.1 Creation under federal authority

In 2014, the Water Infrastructure Finance Innovation Act ("WIFIA") established the statutory authority to create a second USEPA water finance model to complement the SRF model. WIFIA was modeled based on the US Department of Transportation's Transportation Finance and Innovation Act ("TIFIA") financing program (U.S. Department of Transportation, 2021).[14] WIFIA, TIFIA and other federal credit programs are governed by the 1990 Federal Credit Reform Act ("FCRA") (Federal Credit Reform Act, 1990).[15]

[14] For statutory language: https://fiscal.treasury.gov/files/ussgl/fcra.pdf. For more information: https://www.cbo.gov/sites/default/files/102nd-congress-1991-1992/reports/7916.pdf.

[15] The concept of "back-weighting" or "back-loading" debt involves skewing principal amortization to later maturities than what is normally expected under a level debt service payment structure where principal amortization is structured so that annual payments are roughly equal. This skewing may even include full deferrals that pushes the first principal amortization date to a later year. The effect is to extend the weighted average annual life if the debt instrument.

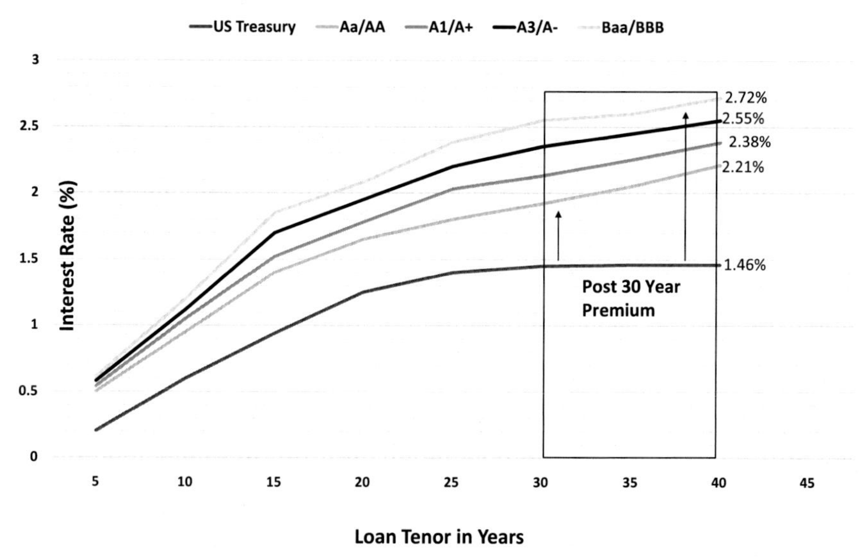

Figure 13.4 US treasury bond rates versus muni rates. *(Source: Bloomberg, July 2020).*

13.4.2 Model design

The WIFIA Model leverages the US Government's cost of funds (US Treasury Bond Rates) which is shaped by its credit ratings (Aaa:Moody's/AA+:S&P/AAA:Fitch) and by the US Dollar's position as the dominant international currency. The key value is the statutory authority to on-lend at the US Government's cost of funds (the on-lending rate is based on the loan's weighted average life matched against the comparable US Treasury rate). By securing funds at the US Treasury rate, loan recipients avoid higher borrowing costs typically imposed by investors for credit risk—referred to as credit premium. The US Government absorbs the premium, takes the credit risk and passes the interest cost savings on to the WIFIA loan recipients. The result is effectively the equivalent of a below-market concessionary rate that benefits critical water infrastructure and ratepayers within the US. The interest savings benefits vary depending on the credit quality of the underlying WIFIA loan obligor, the spread between US Treasury rates and tax-exempt municipal bond rates and the loan's term structure. Fig. 13.4 shows the credit premium spread differentials in the US Capital Markets versus US Treasuries as of July 2020.

The most notable attribute of this model is its capacity to leverage US Government credit and market liquidity to obtain and on-lend capital at prevailing US Treasury rates at a minimal direct cost to the loan obligor.

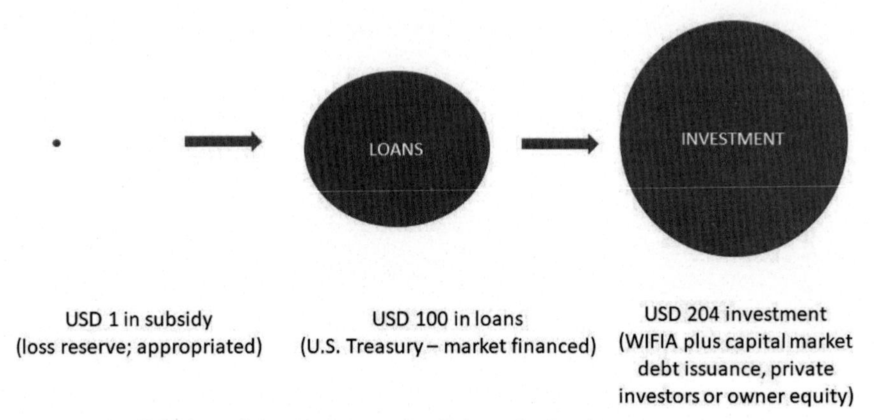

Figure 13.5 WIFIA model capital structure. *(Source: Authors).*

Capital allocated to the program is limited to that required to administer the program and capitalize the loss reserve. Loss reserve rules applicable to federal credit programs were established under FCRA to assure the budget impact of all federal lending and guarantee programs are evaluated fairly. The FCRA derived loss reserve estimates for the WIFIA Program currently allow it to operate at a risk-to-capital ratio of greater than 100:1. Based on risk modeling that reflects water sector historical performance, the US Governments expects a loss of USD 1 or less for every USD 100 in loans underwritten. This means that each dollar of appropriated loss reserves can support USD 100 of WIFIA project lending. The risk assessment results in only the loss reserve being expensed as a government outlay when loans are funded. The impact on the federal balance sheet is minimized by the balance of each loan that collateralizes the federal borrowing that funds the loan. Fig. 13.5 provides a synopsis of the capital structure that supports the WIFIA lending model.

13.4.3 Unique benefits

The WIFIA program is designed to provide low-cost funding for domestic water projects but is deliberately limited to no more than 49% of total eligible project costs. At 49%, the capped least cost financing terms are intended to improve project affordability while attracting co-funding in the form of private capital provided directly by private investors or indirectly thru the capital markets (i.e., blended finance). Life to date WIFIA Program performance is provided in Table 13.2.

Table 13.2 WIFIA program performance.

Program metrics	Year					Total program
	2017	2018	2019	2020	2021	
Invited	13	40	40	61	38	192
Applied	12	27	26	28	0	93
Applications Pending	0	0	7	25	38	70
Not Applying	1	12	7	8	0	28
Withdrawn	0	1	0	0	0	1
In Millions						
Funds Requested	USD 2,520	USD 4,587	USD 5,792	USD 3,258	USD 0	USD 16,156
Funds Obligated	USD 2,433	USD 4,032	USD 5,158	USD 1,335	USD 0	USD 12,959
Funds Disbursed	USD 502	USD 407	USD 101	USD 3	USD 0	USD 1,013
Loan Authority	USD 3,049	USD 6,710	USD 7,310	USD 11,500	USD 13,000	USD 41,569
Subsidy Appropriation	USD 25	USD 55	USD 60	USD 55	USD 60	USD 255
Subsidy Obligated	USD 19	USD 24	USD 42	USD 14	USD 0	USD 99
Closed Loans						
Number	12	24	22	12		70
Avg Loan	USD 203	USD 168	USD 234	USD 111		USD 185
Min Loan	USD 21	USD 16	USD 18	USD 14		USD 14
Max Loan	USD 699	USD 388	USD 727	USD 424		USD 727

Source: US Environmental Protection Agency:https://www.epa.gov/wifia.

Other program benefits include:

- Debt amortization which can be back weighted[16] allowing higher cost debt to be amortized on an accelerated basis further improving project economics and affordability.
- Loan interest that is only charged against loan disbursements eliminating negative arbitrage and re-investment risk on borrowed proceeds.
- The ability to close loans well in advance of loan draws allowing a forward rate lock at no additional cost to borrowers.
- An option to capitalize interest during construction.
- The ability to prepay at any time without penalty.

13.5 Context for the successful adoption of US water finance models

Investment in the United States benefits from three key factors; well-established and vetted legal and governance frameworks, generally strong operational capacity and financial viability of water service providers and a mature public finance capital market. The operational success of US EPA's Water Finance Models stems from these preexisting enabling conditions.

13.5.1 Legal and governance frameworks

The United States Constitution and underlying state constitutions define the fundamental terms of governance, including taxing powers and the authority to create subnational jurisdictions which in turn are empowered by state laws, enacted by legislatures, that grant powers to provide services and to collect the revenues necessary to deliver such services, including the contracting of debt obligations necessary to raise investment capital needed to standup producing assets. These contractual obligations, commonly referred to as bond indentures or loan agreements, are protected by a judiciary system that is empowered to interpret statutes and the powers derived therefrom. These protections extend to legal precedents built up from the long history of western jurisprudence that is foundational to contract law. The protections

[16] Operational and financial viability refers to a water utility's capacity to self-fund operations thru rate setting and as a consequence have sufficient financial strength to attract investors to support long term investment. In the US, smaller communities may back stop user revenues with ad valorem tax pledges making rate setting somewhat less critical to capital market access. For a detailed discussion of the critical links between viability and capital access see "Water Utility Turnaround Framework: A Guide for Improving Performance", World Bank Group Water Practice at https://documents.worldbank.org/en/publication/documents-reports/documentdetail/515931542315166330/water-utility-turnaround-framework-a-guide-for-improving-performance.

provided by governance and legal frameworks are core to the success of finance models that rely on the use of publicly issued debt instruments.

13.5.2 Utility operational capacity/financial viability

In the US the operational and financial viability challenges are largely limited to smaller service providers. Most of the population is served by large utilities that are operationally and financially viable.[17] Larger utilities are likely to be rated for credit strength by a nationally recognized rating service. Of the 1,578 water and wastewater utilities rated by Standard & Poor's, less than 1% are rated below investment grade.[18] The SRF model's set-asides from federal appropriations are designed to provide resources that can be invested in developing the managerial, technical and financial capacity of the smaller non-rated, and less creditworthy systems. This is critically important for small water systems, which EPA defines as serving less than 10,000 population and account for more than 80% of service providers.[19]

13.5.3 US public finance markets

The third leg is the US Public Finance Market. The introduction of US sponsored water finance models was made easier by ready-made access to mature financial markets that can easily accommodate debt issuance by the national government and state and local governments. In 2019, USD 426 billion was sold to investors through the US municipal debt marketplace.[20] Operating within and alongside this marketplace, the SRF finance model leverages program equity to attract private capital investment that it uses to expand lending capacity and deliver better than market terms for eligible borrowers that cannot otherwise secure financing on comparable terms.

[17] Investment grade ratings refers to the perceived strength of an obligor's credit and its ability to repay its obligations, as determined by credit rating agencies. A bond is considered investment grade if it has received a rating of BBB- or above from S&P or Fitch Ratings or Baa3 or above from Moody's, the three largest credit rating agencies.

[18] USEPA Safe Drinking Water Information System. https://www.epa.gov/ground-water-and-drinking-water/safe-drinking-water-information-system-sdwis-federal-reporting.

[19] Securities Industry & Financial Markets Association. https://www.sifma.org/resources/research/us-municipal-issuance/.

[20] In 2011, S&P downgraded the U.S. Government's credit rating. In taking this action, S&P excluded certain credit vehicles that it deemed unaffected by the action. These included all state revolving fund credit vehicles which were retained at AAA due to the inviolate nature of the resources held in trust in SRF accounts. See: Research Update: United States of America Long-Term Rating Lowered To 'AA+' On Political Risks, August 5, 2011 and Rising Debt Burden; Outlook Negative and Special report: U.S. Sovereign Downgrade's Global Effects, August 8, 2011 at: https://www.standardandpoors.com/en_US/web/guest/article/-/view/sourceId/6802837.

WIFIA also serves to improve overall financing terms by reducing the market access needs of eligible borrowers.

13.6 International context

The adoption of USEPA Water Finance Models, in whole or in part, depends on local conditions and their potential alignment with the benefits that these models can deliver.

13.6.1 Value proposition

With respect to the SRF model, a key-value has been the superior credit strength of the SRF credit model versus a state's underlying credit. It is this factor that underlies the SRF model's promise for international application—where sufficient financial resources can be dedicated to irrevocable trusts and concentrated to produce stable highly rated credit mechanisms that can operate and offer favorable market or below market terms independently of the sovereign's credit strength.[21] A related factor is a multiyear commitment of resources that, if secured, can serve to both strengthen the credit of the financing vehicle and its' prospects for sustainable funding operations. How much capital can be committed on a lock-box basis will affect the design of financial assistance offerings which can range from guarantees, which can leverage a smaller concentration of resources to serve market access at least cost (see paragraph below), to below-market rate loans which require resource concentrations sufficient to produce returns that can be allocated to loan subsidies.

In contrast to the SRF model, the WIFIA model is designed to leverage the sovereign's lowest in-country borrowing cost. Where developing nations have relatively good market access and sufficient debt capacity, a WIFIA-like program could offer significant value. Countries most likely to benefit are those with investment-grade ratings that can offer on-lending rates that improve financing terms for subnational governments or related entities with identifiable revenue streams. Countries that are also actively seeking to reduce subnational dependence on the nation's resources may also find a WIFIA model to be a useful tool to transition subnationals to revenue-supported lending arrangements, including nations that are actively engaged

[21] See: "Financing Options for the 2030 Water Agenda", November 2016; https://openknowledge.worldbank.org/bitstream/handle/10986/25495/W16011.pdf?sequence=4&isAllowed=y

in the development of domestic finance markets. A key variable that will drive adoption will be the capital required for loss reserves. In the United States, the federal government determines the risk-to-capital requirement. For targeted developing nations, the general credit quality of the water sector, its capital needs relative to national GNP, and the nation's credit stability will be key determinants of loss reserve requirements.

13.6.2 Credit enhancement vehicle and least cost capital provider

The core attributes of these models, their credit-enhancing properties, and the accompanying incentives that can stimulate project development, have attracted the attention of bilateral and multilateral finance development institutions. The few credit-enhancement vehicles that have been developed have leveraged finance development institutional products, primarily guarantees. Their intended purpose has been to construct local market access vehicles that can raise capital for water service providers on more affordable financing terms including loan tenors that are better aligned with the useful life of the projects financed.

In recent years successful credit enhancement vehicles have been created in the Philippines and in India. The Government of the Netherlands has been actively supporting a credit enhancement vehicle to support water infrastructure project financing in Kenya (van Oppenraaij et al., This volume).

13.6.3 International adoption and barriers

The adoption of finance models that can support market access and meaningfully improve financing terms are achievable. However, success hinges less on model adoption and more on the challenges that impede the development of active, viable project pipelines that can be complemented by active, viable local capital markets. The development of viable pipelines also depends on national and state-level commitments to governing and regulatory frameworks that can have a forcing effect on local government investment priorities and credit formation. Standing behind these frameworks are financial incentives that are essential for sustaining strong regulatory regimes. In the United States, initially financial incentives were weighted heavily toward grant support but later migrated toward lending facilities. Effective regulatory regimes supported by financial incentives, which can include grant and lending facilities, may be threshold requirements for sustained efforts that can meaningfully improve WASH coverages worldwide.

In addition to establishing the governance and regulatory frameworks, coordinated federal and state action must also address operational and financial viability challenges within the WASH sector. In 2016, The World Bank Group identified three critical barriers to investment. These were:

- System governance that includes enforcement frameworks.
- Operational efficiency.
- Financial viability is dependent on overcoming operational challenges.

The Bank also concluded that successful capacity building depended on an action sequence that included establishing a baseline assessment, cleaning up finances by increasing revenues and lowering costs, setting clearly defined objectives, and performance targets, making investments in information systems, and focusing on improving human resources.[22]

13.6.4 US water model replication versus "Bespoke" adaptation strategies

The US Water Finance Models described in this chapter were designed and adopted in alignment with the prevailing governing frameworks that dictate federal, state, and local actions. Their development also reflected (1) a federal commitment to boost investment within the sector as a core public policy objective and (2) the federal government's ability to allocate financial resources to backstop strong regulatory action as outlined by the Clean Water and the Safe Drinking Water Acts. The existence of state and local governments that could act as effective partners was also consistent with the enabling conditions critical to the success of these models.

The international adoption of US Water Finance Models will depend on both the similarities and differences in local conditions when compared to the US experience. One clear impediment to replication of the capital intensive SRF model is the general lack of capital to be dedicated to revolving funds on an annual or reliable periodic basis around which project planning can occur, project funding can be assured and upon which credit structures can be established that can cost-effectively leverage private investment capital. In the alternative, developing country leaders have engaged multilateral and bilateral agencies in efforts to pool Official Development Assistance ("ODA") that would normally have a project-specific target and

[22] FINDETER, Philippine Water Revolving Fund and TNUDF Case Studies, https://www.worldbank.org/en/programs/global-water-security-sanitation-partnership#4.

instead make capital and/or guarantee commitments that could provide the basis for a multi-project lending program that attracts private capital investment. ODA capital commitments supported the Philippine Water Revolving Fund ("PWRF") (2012) and Columbia's FINDETER, a credit enhanced lending vehicle, which was also supported by a Government of Columbia capital contribution. Multilateral and bilateral guarantees were included in the PWRF and the Tamil Nadu Urban Development Fund's (TNUDF) Pooled Financing (2003).[23]

Adoption of the WIFIA model by other countries would largely depend on:

- The credit strength of the sovereign relative to the credit strength of the WASH sector which will determine the loss reserve requirements needed to protect the sovereign's credit rating.
- The relative savings that the sovereign can offer water service providers by on-lending at cost.

[23] FINDETER, Philippine Water Revolving Fund and TNUDF Case Studies, https://www.worldbank.org/en/programs/global-water-security-sanitation-partnership#4.

Table 13.3 The US water finance model adoption checklist.

Enabling condition	Minimum standard	Remedial action (Examples)
Federal and state		
Governing frameworks	Qualified entities that are empowered to administer programs at federal and state levels	Assessment of - existing powers and need for new legal authority to support debt placement, investor interests and regulatory regimes; - institutional capacity development needs
Legal authority	Laws establish governing, contractual and enforcement parameters for sustainable operating models	Identify and enact necessary changes in law to assure debt powers, investor and environmental protections can be enforced

(continued on next page)

Table 13.3 The US water finance model adoption checklist—cont'd

Enabling condition	Minimum standard	Remedial action (Examples)
Judicial framework	Financial contracts are supported by statutory authority and contract law precedents	Investors may seek court validation for new legal authority
Public resources are available for project development and financial program support	Project development resources can be secured and sustained A dedicated funding stream can be secured for investment or security support (i.e., guarantee facilities)	Identify source funds that can be committed over multiple years. Can include commitments from DFIs.
Credit Quality/ Market Access	Federal and or state level government investment quality enables market access at reasonable cost	Assessment of sovereign investment quality and determination of credit enhancing structures that can leverage available resources
Project selection framework	Project selection criteria is established, publicly vetted and reflected in published project prioritization list	Develop the legal construct thru legislation and/or regulatory action that establishes sustainable project selection framework
WASH service sector		
Revenue models	Revenue streams are established and supported by high collections	Determine alignment between services and revenue models
Operational efficiency	Cost management and investments that reduce non-revenue water loss	Develop operational improvement strategies that attract performance investors
Financial Viability	Track record of overcoming operational challenges	Depends on operational improvements

(*continued on next page*)

Table 13.3 The US water finance model adoption checklist—cont'd

Enabling condition	Minimum standard	Remedial action (Examples)
Project Pipelines exist and can be sustained	There is an emerging critical mass of projects in development that can support aggregating models and private investor support	Coordination among sovereign, states and utility representatives to commit and attract matching project development capital
Capital markets		
Local markets attract buyers, security issuers	Established securities laws and regulations are tested and resilient	Barriers to market entry are low
Secondary market trading is well established	Securities firms are subject to standards of integrity established by law and accreditation	Market integrity is tied to adequate regulatory capital requirements

13.7 A US water model adoption checklist

Establishing workable in-country water finance models requires that (1) certain standards be satisfied at the federal, state and local government/utility levels and (2) there is capacity to leverage local capital markets. The US Water Model Adoption Checklist, provided in Table 13.3, attempts to identify these enabling conditions, related standards, and examples of remedial actions, where necessary.

References

American Recovery and Reinvestment Act (P.L.111-5), 2009. Summary and legislative history. Congressional Res. Service. https://crsreports.congress.gov/product/pdf/R/R40537.

America's Water Infrastructure Act (2018), https://www.congress.gov/115/bills/s3021/BILLS-115s3021enr.pdf.

Copeland, C., (2012), "Water infrastructure financing: history of EPA appropriations", Congressional Research Service, Washington, DC

Federal Credit Reform Act (1990), U.S. department of treasury, https://fiscal.treasury.gov/files/ussgl/fcra.pdf.

Folsom, MB., (1958), Water pollution control magazine. The Military Engineer.

G7 (2018), "Charlevoix commitment on innovative financing for development", Charlevoix, Quebec. https://www.international.gc.ca/world-monde/assets/pdfs/international_relations-relations_internationales/g7/2018-06-09-innovative_financing-financement_novateur-en.pdf.

G20 (2019), "Osaka declaration", https://www.mofa.go.jp/policy/economy/g20_summit/osaka19/en/documents/final_g20_osaka_leades_declaration.html.

Infrastructure Investment & Jobs Act, Water infrastructure investments (2021) https://www.epa.gov/infrastructure/water-infrastructure-investments.

Massachusetts Water Pollution Abatement Trust (2021), https://www.mass.gov/orgs/the-massachusetts-clean-water-trust.

New York State Environmental Facilities Corporation (2021), "Clean Water SRF Intended Use Plan", https://www.efc.ny.gov/2021-CW-IUP.

OECD (2018), "Blended finance principles", http://www.oecd.org/dac/financing-sustainable-development/blended-finances-principles/#:~:text=The%20OECD%20DAC%20Blended%20Finance,philanthropies%20and%20other%20concerned%20stakeholders.vv

US Code of Federal Regulations (2021), "Subpart K State Water Pollution Control Revolving Funds" and "Subpart L – Drinking Water State Revolving Funds", https://www.ecfr.gov/current/title-40/chapter-I/subchapter-B/part-35/subpart-K and https://www.ecfr.gov/current/title-40/chapter-I/subchapter-B/part-35/subpart-L.

US Environmental Protection Agency (US EPA) (2014), "Utilization of additional subsidization authority in the clean water and drinking water state revolving fund programs", Report to Congress, https://www.epa.gov/sites/production/files/2015-04/documents/additional_subsidization_report_to_congress_pdf.

U.S. Department of Transportation (2021), Transportation Infrastructure Finance Innovation Act (TIFIA), https://www.transportation.gov/buildamerica/financing/tifia.

CHAPTER FOURTEEN

Mobilizing private capital for large-scale ecological restoration and conservation: Insights from the US Experience

Adam Davis[a] **and Sara L. Johnson**[b]
[a]Ecosystem Investment Partners, Baltimore, Maryland
[b]Ecological Restoration Business Association, McLean, Virginia

14.1 Introduction

The Clean Water Act Compensatory Mitigation program is the largest environmental restoration program in the United States. Each year, government agencies and private companies spend about US$3.5 billion to restore, enhance, and protect hundreds of miles of streams and thousands of acres of wetlands in order to satisfy compensatory mitigation requirements under Section 404 of the 1972 Clean Water Act. The program has restored close to 700,000 acres (2,830 km^2) of private land with private dollars since program inception.[1]

To assist with the successful implementation of this work, the US Army Corps of Engineers has developed and refined a mitigation banking program over the last 25 years through which private companies invest capital to evaluate, propose, construct, protect, monitor, and maintain government-vetted environmental restoration projects. Through this performance-based process, private companies earn a return on investment through the sale of environmental credits after project success criteria set by the government are achieved. Pricing is set between the buyer and the seller of credits. By specializing in restoration work and operating at scale, the seller can make a return on investment while saving both time and money for buyers needing Clean Water Act compliance.

This chapter describes important lessons learned from this market for outsourced compliance with Clean Water Act requirements and how

[1] Based on analysis of RIBITS August 30, 2021 and author correspondence with consultant Steve Martin on August 30, 2021.

Financing Investment in Water Security: Recent Developments and Perspectives.
DOI: https://doi.org/10.1016/B978-0-12-822847-0.00013-2

this industry now attracts significant private capital investment. Mitigation banking policies have improved environmental outcomes while providing efficient compliance options for development activity and simultaneously incentivizing private investment in public goods.

The chapter is organized into five sections: (1) the history of policy development supporting private investment, (2) the credits used to measure durable restoration results, (3) how the use of credits incentivizes investment, (4) case studies, and (5) lessons learned.

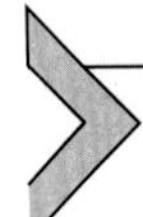

14.2 The history of mitigation banking policy development

A 2015 PLOS One study found the ecological restoration industry in the United States generates USD 25 billion in annual economic output and creates 225,000 jobs (Bendor et. al., 2015). How did we get here and what kind of opportunity lies in the future?

Following decades of increasing concern about water quality in the United States, a turning point was reached when the badly polluted Cuyahoga River in Ohio caught fire in 1969 and catalyzed a new environmental policy. Three years later, ongoing public outcry and resulting bipartisan support led to the passage of the Clean Water Act (CWA) of 1972, a bedrock statute for US environmental law. Among many objectives, and in recognition of the fact that the country had drained or paved nearly half of all its original existing wetlands, the CWA policy framework established both the goal of "no net loss" for aquatic resources and the framework for the US Army Corps of Engineers (the "Corps") and US Environmental Protection Agency ("EPA") to enforce the related rules.

In the early days of the "no net loss" implementation, permittees would conduct mitigation themselves on-site, or develop their own off-site restoration projects to make up for the impacts of their development projects. These efforts were expensive and time-consuming, as they required land control for the same amount and type of resource impacted as well as technical mastery of hydrology and ecological science. Even if restoration was entirely successful, which was not always the case, these restoration projects had to be maintained by the permittee, creating a long-term liability. As the Corps and EPA developed and refined mitigation guidelines and regulations, permittee demand grew for reliable mitigation options, and consequently, so did interest from specialized professionals with the skills to develop mitigation.

In the 1990s, the first mitigation banks were set up to protect and restore properties that could later serve to compensate for impacts of multiple permittees. This allowed restoration companies to concentrate on siting larger projects that would have more significant regional ecological benefits than individual "permittee responsible" projects developed in an ad hoc manner. These early mitigation bank development companies, however, were under-capitalized. Very often they were created by farmers or ranchers who had lower-lying property that was too wet to farm profitably, or by developers who had previously needed to buy mitigation and therefore realized the potential for selling mitigation from restoration activity.

Specialized restoration consulting companies began to form to bring additional resources to mitigation bank development and the advances in construction techniques required by strict requirements for ecological performance. The nascent industry began to flourish in the first decade of the 21st century as the opportunity to make a fair return on investment through providing restoration results at scale became established. Those needing permits increasingly recognized that purchasing credits from a third party was far more efficient than trying to develop a project themselves and manage it over the long term. At the same time, individuals and companies with an interest in restoration had a clear incentive to develop quality projects.

Following two decades of learning from mitigation project implementation, this specialized industry came together to petition Congress and the key federal agencies to codify industry practices and standards in the 2008 Compensatory Mitigation Rule (the "Rule"). The Rule clarified performance standards, financial assurances, the geographical area within which credits could be sold, and many other technical issues. All mitigation credits are approved and released by the Corps under standards articulated in the Rule. As shown in Figs. 14.1 and 14.2, the decade that followed passage of the Rule has seen exponential growth in the development of mitigation banks and corresponding credits across the country. To track credits and compliance with the Rule, the Corps established an online registry of credits known as "RIBITS," which contains credit ledgers and detailed information about individual mitigation bank performance requirements.

The Rule requires that mitigation plans contain twelve specific elements in order to allow certification of credits that can be sold for Clean Water Act compliance.

These are:

1. Objectives.
2. Site selection.

3. Site protection.
4. Baseline information.
5. Determination of credits.
6. Mitigation work plan.
7. Maintenance plan.
8. Performance standards.

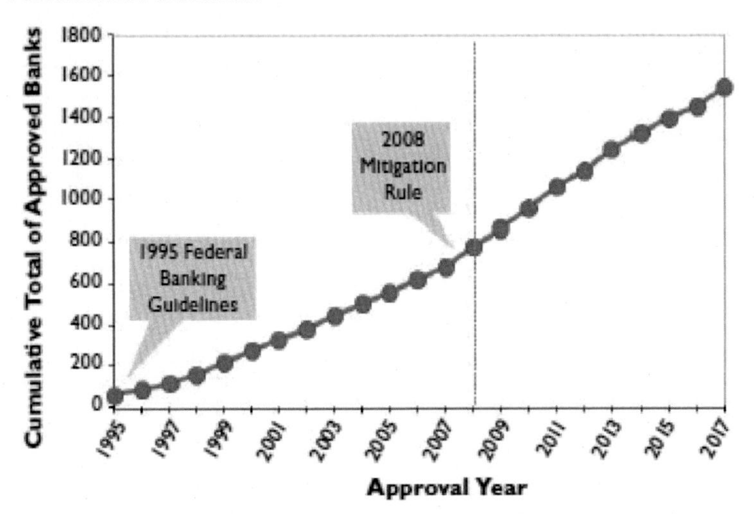

Figure 14.1 Cumulative total of all mitigation banks with Section 404 credits approved 1995–2017. *(Source: Hough and Harrington, 2019).*

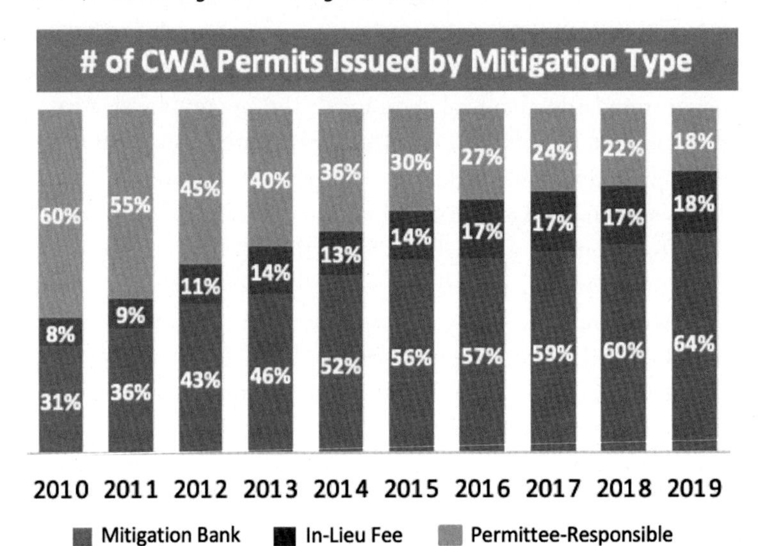

Figure 14.2 Number of CWA permits issued by mitigation type. *(Source: Hough and Harrington, 2019).*

9. Monitoring requirements.
10. Long-term management plan.
11. Adaptive management plan.
12. Financial assurances.

During the 50 years since passage of the CWA, an ecological restoration industry has matured to provide professional and focused development of wetland and stream projects. The combination of the Rule and the presence of mitigation banks and specialized restoration firms making a return from wetland restoration attracts increasingly large investment. Today, private investment finances a significant portion of the restoration done across the United States, from mountain streams, where former coal miners in Appalachia are employed, to the restoration of coastal wetlands in Louisiana, creating jobs for former oil rig workers.

As the industry developed capacity to undertake restoration projects at scale, it began to develop projects to address environmental issues beyond the direct impacts from development covered by the "no net loss" (Section 404) provisions of the CWA. Increasingly, water quality and endangered species-related needs are being addressed through the private investment that follows the same fundamental principles that govern the mitigation banking program. Projects range from floodplain restoration to dam removal to coastal resiliency, and this capacity offers the possibility of large-scale public/private projects to address climate adaptation needs in the future.

While mitigation banking provides an example of one of the most successful environmental markets to date, it is not without challenges worth noting. First, the question of which waters are protected by Federal law and thus subject to CWA permit requirements has been a complex and confusing issue. Over the history of the CWA, the definition of "waters of the United States" has contracted and expanded following Supreme Court decisions and multiple rule-makings under opposing political Administrations. This history reflects the tension between private property rights and the need for economic development on the one hand, and the public need for environmental protection on the other.

Second, inconsistent implementation of certain aspects of the Rule has frustrated investment and implementation of mitigation projects in some areas of the country. There are 38 Districts in the Corps' Regulatory Program, each of which maintains a significant level of discretion about how they implement the Rule. In practice, these Districts use differing standards for land protection, financial assurances, credit approvals, and other requirements. While some discretion to address regional ecological differences is certainly appropriate, these widely varying interpretations of administrative and procedural matters cause unpredictable timeframes and requirements for restoration project approval.

14.3 What is an ecological credit?

A credit that functions as a unit of measure for "success" is a fundamental element of mitigation banking. Credits define ecological success criteria along with legal and financial assurances to provide durability over time and clear target for restoration outcomes. This simple notion enables private investment in a wide variety of public goods that can be defined by ecological success criteria.

As a form of offset, an ecological "credit" allows regulators to measure a restoration project's ecological results in a transparent and accountable manner. This approach ensures that the offset has a nexus with the environmental harm requiring compensation.

The fundamental elements of verifiable credits for restoration are:

Science: which informs the location of restoration, design, and of course measurement of ecological uplift - often through publicly developed functional assessment methodologies.

Additionality: distinguishable from preservation, the restoration must be additional to baseline conditions to achieve "no net loss" or a "net gain" of the resource.

Durability: achieved both through permanent site protection, such as a conservation easement, and an endowment to fund long-term maintenance.

Timing: work is done *before* credits are released: the credits represent permanent conservation and restoration work that is in the ground and performing (i.e., delivering ecological services), not a "plan" to deliver those services in the future.

Functional assessment methodologies score restoration projects on fundamental measures of ecological success like hydrology and the establishment of vegetative communities that only thrive in hydric (wetland) soils. Additionality is an essential requirement; mitigation banking is not about protecting pristine places but restoring ecological function to areas that have been cut off, drained, or otherwise damaged.

Durability is provided both through permanent land protection under a conservation easement, which is a modification of the legal title to the land protecting against future development, and also through a complex but well-established system of endowment and long-term stewardship. While not as visible as the construction that restores wetland function of stream channel stability, this aspect of banking is essential because it provides accountability and funding to make sure a restored site *stays* in good condition.

All ecological restoration projects used for compensatory mitigation must meet rigorous standards through a review process by the Corps, EPA, and State environmental agency representatives. Requirements are codified in a Mitigation Banking Instrument, which is a publicly available document detailing the location, objectives, standards, and requirements that must be met in order to earn credits. Approved Mitigation Banking Instruments are available on the RIBITS website.

Once the restoration is complete and credits have been released from a mitigation bank, the purchase price is established between the buyer and the seller with no government involvement. There is also no compulsion to buy a credit. An entity, for example, a state transportation agency building a new highway bypass near a stream can either do its own mitigation project ("permittee responsible") or purchase credits from a mitigation bank that has earned credits (Gaudioso et al., 2014; Environmental Law Institute & Stetson University College of Law, 2019).[2]

This is an essential benefit of the policy structure that enables mitigation banking: the risk of being able to sell credits and make a return on investment is entirely on the private party doing the restoration, and not on the public. Environmental and performance standards are set by government, and businesses must meet those standards as efficiently as possible. Investment in permanent land protection, ecological success criteria, and long-term financial assurance are required to allow infrastructure, housing, mining, and other developers to buy credits that meet compliance needs. As this investment has expanded, mitigation banks are a steadily growing percentage of the permits required to address aquatic resource impacts under the CWA (Fig. 14.2).

Mitigation bank credits offer both legal liability benefits and time savings to customers as well. Purchase of a credit from an already approved restoration project eliminates the performance risk of a "de novo" project, and

[2] It is also important to note that another option for obtaining mitigation credits, known as an In-Lieu Fee (ILF) program also exists. ILF programs are typically administered by nonprofit organizations or government agencies. For more information on this option, see the following resources: https://www.eli.org/sites/default/files/docs/article_kihslinger.pdf; https://www.eli.org/research-report/lieu-fee-mitigation-review-program-instruments-and-implementation-across-country.

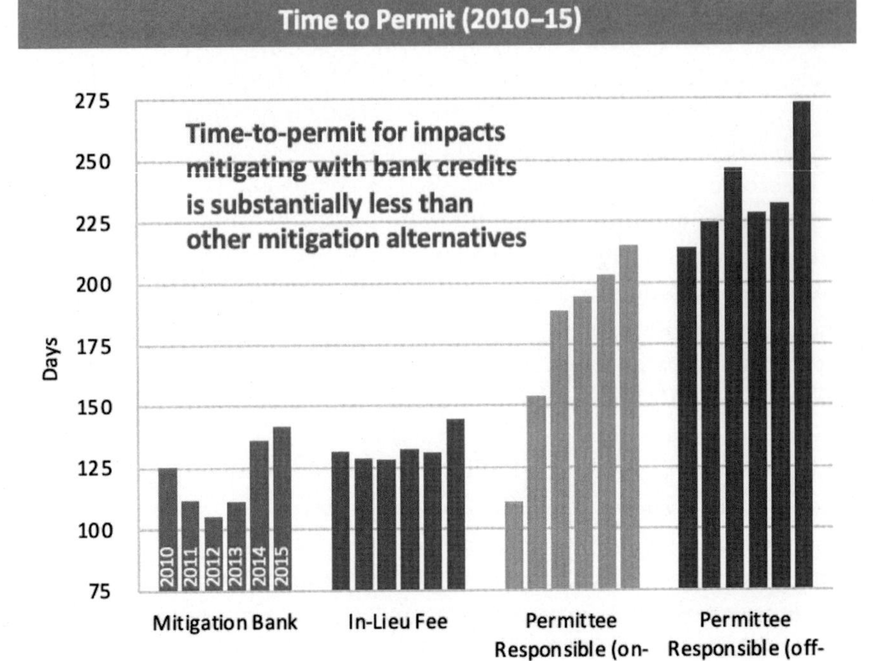

Figure 14.3 Time required to issue a Permit 2010–2015 (After Institute for Water Resources, 2015). (Disclaimer: The appearance of US Department of Defense (DoD) visual information does not imply or constitute DoD endorsement).

the associated legal liability. Purchasing credits from completed restoration projects also means that Clean Water Act compliance can be obtained in roughly half the time it would take compared to obtaining a permit by doing a restoration project from the start (Fig. 14.3).

There are many nuances and complexities to the mitigation bank instruments that codify requirements for individual mitigation banks. However, one worth noting in particular is the notion of a service area. Each mitigation bank has a geographic region defined within the watershed that determines where credits may be sold. This service area is authorized by the Corps and agreed upon by the project sponsor. For example, a highway impacting a stream must be offset by a credit representing restoration of a similar stream type in the same watershed as the impacted stream. If such a credit is not available in the immediate watershed, then the Corps may make an exception and authorize the use of credits from an adjacent watershed but this can require the purchase of credits at a higher ratio, that is, 1 mile of stream impacted is offset by 3 miles of stream restoration.

14.4 How credits incentivize investment: alignment of interests

Because environmental goals like clean water and biodiversity are the very definition of public goods, goals, and criteria for success must be set through a public process and overseen by government. But the work of putting projects into action involves complex real estate transactions, very technical design and permitting processes, and construction management. These skills can be drawn from private firms that can innovate quickly and access investment capital. This is important because in many parts of the world the scope and scale of private capital resources far exceed the capacity of governments to invest in ecological restoration.[3]

First, as in any part of the economy, specialization allows focus. Companies that offer ecological credits from restoration projects *must* focus on meeting government standards in order to earn the product they can sell. While the activities that comprise the daily work of these firms are common enough—real estate control, design and permitting, and construction— the application of these skills to ecological restoration is quite specialized. The expertise required includes detailed knowledge of engineering, hydrology, ecology, regulatory requirements, unique types of construction in sensitive habitats, and ability to address the local concerns and priorities that provide the context for each project. In addition to all this, one must of course be organized to manage and run a sustainable business.

The second element of restoration industry success is scale. Once a firm is focused on delivering restoration results in a specific watershed, the fixed costs for design, permitting, and mobilizing equipment for construction can be quite large. Larger projects can spread these costs over more which of course drives down the cost per credit. In practice, this results in larger restoration projects that have more significant ecological benefits. Scale, after all, is a fundamental principle of landscape ecology, and larger restoration projects have relevance to wildlife migration corridors, water supply and water quality across entire watersheds, and issues like flood storage capacity. As island biogeography shows, larger areas of contiguous natural function are more resilient and diverse than smaller ones (Quammen, 1948).

[3] Consider some 2020 figures for comparison: U.S. EPA's Budget was about USD 9 billion and TNC's Assets were USD 7.8 billion. Versus, total Private Equity Funding at USD 203 billion and U.S. Retirement Accounts at USD 31.9 trillion. Sources: https://www.epa.gov/planandbudget/budget; https://www.nature.org/content/dam/tnc/nature/en/documents/TNC-Financial-Statements-FY20.pdf; https://www.institutionalinvestor.com/article/b1q214krq0zdkn/Private-Equity-in-2020-Not-as-Bad-as-You-Thought; https://www.napa-net.org/news-info/daily-news/retirement-assets-bounced-back-2nd-quarter.

Finally, access to capital provides the resources necessary to control land in locations with the right characteristics for restoration, and to deliver the completed restoration results required to earn saleable credits. This upfront access to capital is essential to enabling ecological restoration firms to invest in sited, performing restoration projects on the ground in advance of a permittee's impacts. As the sector has matured, the range and size of investors have increased; the industry now has nearly 100 ecological restoration companies active in the national trade association called the Ecological Restoration Business Association.[4]

Today, billions of dollars in private investment have become available to address environmental challenges in response to the development of clearly defined metrics that are consistently applied to evaluate the loss or restoration of a regulated environmental resource (Quinson, 2021).

14.5 Case studies

As of this writing, the members of the Ecological Restoration Business Association together employ several thousands of workers in a rapidly growing industry. The case studies that follow provide an overview of selected projects completed or being implemented by members.

14.5.1 Ecosystem Investment Partners

Ecosystem Investment Partners (EIP) is a firm founded in 2007 that invests in mitigation banks and other ecological restoration projects around the United States. Its 2012 Fund, at USD 181 million, was the first private equity fund created for the sole purpose of providing a market rate return to institutional investors through ecological restoration, and the firm now has USD 885 million of assets under management. Since its founding, EIP has restored—and earned credits from restoring—177 miles of degraded stream channel and over 44,000 acres of degraded wetlands. Several specific projects are summarized below.

14.5.2 Mitigation bank

In rural Kentucky, EIP has used investment capital to purchase multiple properties, and earn "stream credits" by restoring natural meandering patterns to artificially straightened channels and restoring stability to the stream beds.

[4] One of the authors of this chapter is the Executive Director.

These mitigation banks address active erosion and degraded riparian buffers that severely impacted the water quality and habitat value of these streams. The work done by EIP makes use of modern techniques for assessment and design that analyze in great detail the natural pattern, profile, and dimension of streams that function in highly altered terrain. EIP is working to reverse impacts including sedimentation, physical alterations of streams that occurred over many years, and nutrient loading, algal blooms, and various pollutants that affect water quality.

The work here results in riffle-pool sequences, riparian corridor establishment, and created wetlands and vernal pools which together reduce downstream sediment loading, improve water quality, and increased habitat for local native flora and fauna.

14.5.3 Performance-based contracting

In the San Francisco Bay Delta, EIP invested in 3400 acres of land in a location suitable for tidal wetland restoration in a project called the "Lookout Slough Tidal Habitat Restoration and Flood Improvement Project." The project is restoring approximately 3000 acres of tidal wetland, creating habitat that is beneficial to native fish and wildlife. Lookout Slough is adjacent to additional tidal habitat, and the work completed here will create a contiguous tidal wetland restoration complex spanning 16,000 acres in the Cache Slough region. Once completed, the Lookout Slough project will be Delta's largest single tidal habitat restoration project to date. In addition to the restoration of important tidal wetland habitat, the project will also meet objectives of the Central Valley Flood Protection Plan to reduce flood risk.

In order to address regional concerns about loss of productive agricultural operation, EIP worked with the existing cattle ranch operation to build facilities that allow them to relocate to an adjacent area. Project design and operation will also require extensive efforts to minimize any impacts on listed species that were present on the site before restoration began.

The site of the project had been completely cut off from tidal influence by levees in the 1920s and 1930s; the work EIP engages in includes construction of over 20 miles of channels to allow movement of water across the site, and three miles of new setback levee that will provide 100-year flood protection with additional height for climate change and sea level rise resiliency. Breaching and degrading the existing levees will restore historical tidal influence on the site, providing food web, and other benefits to Delta smelt and increasing seasonal floodplain rearing habitat for salmonids. This will allow the project to earn credits that represent the habitat needs of the

Delta smelt, a species under threat of extinction and listed under the federal Endangered Species Act.

In a process analogous to the development of a mitigation bank, EIP is obtaining approvals from multiple resource agencies to earn credits, but the credits are committed to the state Department of Water Resources under a fixed-price contract rather than sold on the open market. This kind of public/private partnership, sometimes described as Performance-Based Contracting is distinguished from conventional public sector contracting because results must be achieved through investment before payment is made, and the cost to the state of California is determined by the number of credits actually delivered—rather than paying for the time and effort of the contractor.

14.5.4 Resource Environmental Solutions

Resource Environmental Solutions (RES) also began to specialize in water resources project delivery in 2007 and is now the largest full-scale operational ecological restoration company in the US Originally a regional-based, family firm focused on land-based holdings in Louisiana, early company leadership identified the potential of the ecological credit and regulatory certainty of the Rule to build value in land investments. Considering the highly specialized nature of ecological restoration, the company saw an opportunity to maximize efficiencies and expertise by organizing under an operating model with full vertical integration of services and project delivery. To achieve this operational status, RES underwent multiple strategic transactions to acquire the unique construction and knowledge infrastructure for full-delivery ecological restoration from wetland to species and water quality projects. This focused, highly skilled model attracted substantial private equity investment and in 2016 the company entered a majority stake investment partnership with KKR Green Solutions. As of this writing, RES employs over 900 employees across the country and has annual revenue in excess of USD 300 million for sales of ecological services and credits.

This operating model approach allows for the execution of large-scale restoration projects in support of the development of critical infrastructure for local communities. For example, the North Texas Municipal Water District constructed a surface water reservoir in Fannin County, Texas, to supply a growing regional population; the first reservoir in Texas in approximately 30 years. Environmental impacts from the resulting Bois d'Arc Lake are being mitigated by restoring, enhancing, and preserving approximately 15,000 acres of wetland and upland habitats and 69 miles of ephemeral, intermittent, and perennial streams within the same watershed. Restoration and perpetual protection of these habitats will provide significant benefit to local wildlife populations and migratory birds as well as decrease erosion, sedimentation, and nutrient loads in a significant portion of the Bois d'Arc Creek watershed that drains into the Red River between Texas and Oklahoma.

This project is facilitated through a full-service provider contract that designated one entity with responsibility for design, construction, monitoring, and maintenance of the entire mitigation project. The contract is based on a pay-for-performance model, with milestones tied to each habitat's functional success as dictated by the Corps' issued CWA Section 404 permit and approved mitigation plan. The performance obligations extend for 20 years after construction, anticipated until 2042.

14.5.5 Westervelt

Many companies in the ecological restoration sector started in other land management and resource extraction industries, which already required a certain knowledge of land stewardship and environmental permitting. These companies are diversifying to expand their land investments and services to include ecological restoration. One of the leading examples is Westervelt Ecological Services (WES), which grew out of the Westervelt Company, a privately held forest resource company with roots in the Southeast dating back to the late 1800s. With its history and experience of land stewardship of their 500,000 acres, The Westervelt Company sought additional opportunities to grow their land portfolio. In 2006, shortly before passage of the Rule, WES was established to deliver wetland mitigation and species conservation projects. Their specialized staff comprises highly skilled ecologists, engineers, senior design planners, and landscape architects to manage and act as land steward for over 30,000 acres.

WES works with public agencies and private developers to advance regional conservation and mitigation projects throughout the country. Since contracting with the California High-Speed Rail Authority (CHSRA) in 2016, WES has become the leading expert in developing mitigation solutions in the San Joaquin Valley, fulfilling over 4500 acres of required mitigation on thirteen separate properties in under five years. Through these projects, WES has developed efficient internal methods for communication, document review, and coordination with State and Federal resource agency staff. Their work has expedited permit approvals for their clients, bringing major infrastructure and development projects to shovel-ready with minimal regulatory delay.

14.5.6 Looking forward

Recent years have seen a trend of other major landholding companies also entering the restoration industry from sectors including timber and real estate, at least in part because of the consistency and transparency of the high standards for mitigation under the Rule. As the industry becomes more sophisticated and mature, a trend is observed of consolidation with many smaller bank companies winding down and selling their bank assets, some developed pre-2008 Rule, to larger companies with the capacity and operational skills to responsibly manage the banks for the long term.

While the unique suite of skills cultivated by each of these companies initially grew out of the CWA Section 404 wetland mitigation market, the sector's skills and capital are poised to invest and deliver for other public environmental outcomes as soon as the policy signals are present. Despite the challenges inherent in mirroring CWA restoration incentives in other silos of environmental law, the industry continues to diversify; some companies are now specializing in a variety of projects that provide benefits to water quality, flood and coastal resilience, and natural resource damages restoration. This diversification is reflected in the decision of the industry's leading national trade association to change its name from National Mitigation Banking Association to Ecological Restoration Business Association to more accurately reflect the reach of environmental markets.

14.6 Lessons learned

What are the principles and lessons that come from the US mitigation banking experience, and to what extent can it be used to garner investment in restoration elsewhere in the world?

While the Clean Water Act regulatory framework and specific mechanisms for land control and property rights are unique to the US, the elements of mitigation banking that enable investment certainly have the potential to be applied broadly. One essential element is the basic principle of a credit that is defined by land protection, science-based success criteria, and a financial mechanism for long-term monitoring and maintenance. A second is a notion that private investment in restoration must provide results *before* sales can occur and a profit obtained.

Once credits are earned, however, they do not need to satisfy a legal offset or "compensatory mitigation" compliance obligation to be useful. Government agencies or even international organizations could purchase credits representing restoration success as an alternative to investing in, and taking the risk of, new restoration projects. Private investment can be used for a range of environmental outcomes as long as there is a clear definition of the desired outcome and a contractual mechanism to purchase the credits that represent that outcome.

In most cases, it is helpful for there to be an *explicit goal* for a defined environmental resource, such as the CWA's "no net loss" of aquatic resources. "No net loss" is defined at the watershed scale but exists as a national goal. Water quality goals, while set within the context of national policy, are also typically defined at the level of individual water bodies. For example, the "impaired waters" list maintained by EPA under the CWA 303(d), or goals set for a specific environmental outcome, such as restoration of a tributary to reduce nitrogen loads by a target amount. Goals can be forward-looking, such as the "No-Net-Loss" of wetlands that generally operates with the date of the law as the baseline. Alternatively, goals can account for prior environmental harm, such as those arising from chemical or oil spills. In these cases, the national policy is set by the Oil Pollution Act or Comprehensive Environmental Response, Compensation, and Liability Act and the specific required outcomes are set by councils that include representatives of the various responsible public resource agencies.

In any event, it is necessary to have *a mechanism to fund the restoration work*. There has to be a contractual means to procure ecological credits in order to provide incentive for investment. In the case of CWA "no net loss" or other offsetting programs, the permittee or polluter pays for mitigation credits to offset their environmental impacts, which are measured as "debits" of the regulated resource. This internalizes the cost of impacts that had previously been environmental externalities. As described above in the case studies, government can also be a customer for completed restoration rather than taking on the responsibility for managing projects and delivering results. In this approach, government can purchase credits via contracts awarded through customary competitive procurement processes.

Lastly, and critically, investment requires a *metric of ecological success*. The metric used reflects scientific understanding of desired physical, biological and chemical outcomes, and of course will vary across resources and programs. Once chosen, however, metrics applied in a predictable, consistent manner for a given resource type will best incentive investment. This means that the implementing entity of the government must hold individual restoration projects to equivalent standards and insist on quantifiable results.

Once these three factors are in place, then incentives are aligned in such a manner that verifiable ecological results produce return on investment from successful projects. A structure exists within which "the more good is done, the more money is made." Each verifiable unit of ecological benefit has a value that is set by supply and demand, and results must be delivered before payment is received. This represents an alignment of interests that has the potential to attract significant private capital resources.

Given the increasing pressures of adaptation to climate change, and absolute need for biodiverse natural systems to function in order to meet human needs, there is a worldwide urgency to move from planning to implementation. Plans and goals are helpful of course, but it's time for action. A critical tool for action at scale is public-private partnership wherein government defines ecological goals and success criteria, and the private sector is rewarded *only if* it meets them.

References

BenDor, T., Lester, T.W., Livengood, A., Davis, A., Yonavjak, L., 2015. Estimating the size and impact of the ecological restoration economy. PLoS ONE 10 (6), e0128339. https://doi.org/10.1371/journal.pone.0128339.

Environmental Law Institute, and Stetson University College of Law 2019. "In-lieu fee mitigation: review of program instruments and implementation across the country." https://www.eli.org/research-report/lieu-fee-mitigation-review-program-instruments-and-implementation-across-country.

Gaudioso, L., Kihslinger, R., Woolsey, P., 2014. Establishing In-Lieu fee mitigation programs: identifying opportunities and overcoming challenges. National Wetlands Newsletter 36 (4). https://www.eli.org/sites/default/files/docs/article_kihslinger.pdf.

Hough, P., and Harrington, R. "Ten years of the compensatory mitigation rule: reflections on progress and opportunities" 49 ELR 10018, 10023 January 2019.

Institute for Water Resources (IWR), The Corps, The Mitigation Rule Retrospective: a review of the 2008 regulations governing compensatory mitigation for losses of aquatic resources (2015) (2015-R-03), https://www.epa.gov/sites/production/files/2015-11/documents/mitrule_report_october_2015.pdf.

Quammen, D. 1948. "The song of the dodo: island biogeography in an age of extinctions." New York, NY: Scribner, 1996.

Quinson, T. "The boom in ESG shows no signs of slowing." Bloomberg Green. 2021; https://www.bloomberg.com/news/articles/2021-02-10/the-490-billion-boom-in-esg-shows-no-signs-of-slowing-green-insight.

CHAPTER FIFTEEN

Tapping local capital markets for water and sanitation: the case of the Kenya Pooled Water Fund

Joris van Oppenraaij[a], Roy Torkelson[a], Dick van Ginhoven[a], Maarten Blokland[a], Ng'ang'a Mbage[b], Jean Pierre Sweerts[a] and Eddy Njoroge[b]
[a]Water Finance Facility, Kenya
[b]Kenya Pooled Water Fund

15.1 Introduction

Achieving UN Sustainable Development Goal (SDG) 6—Ensure access to water and sanitation for all—requires major investments in water, sanitation and health (WASH). Traditionally, public sector budget allocations, complemented by Official Development Assistance (ODA) flows, have been the predominant funding source in developing countries. However, there are major shortfalls: according to the UN-Water/WHO, over half of countries surveyed stated that water tariffs are at a level that allows the recovery of only 80% of operating costs (O&M), to say nothing about capital costs ((GLAAS), 2019). In Kenya, sector funding is estimated to be less than 40% of the sector requirements, and the funding gap is approximately USD 10.6 billion.

There are three ultimate sources of funding for WASH capital investments, operation, and maintenance: the "3Ts" are *T*axes in national and local budgets paid by individuals and businesses; *T*ransfers or subsidies/ grants paid by foreign national and/or local governments and/or development partners; and *T*ariffs or user fees paid by households, businesses, and governments. Transfers are generally declining, particularly in countries with Lower Middle Income Country (LMIC) status or higher, requiring a shift in emphasis to tariffs and taxes, which must be adequate not only to finance operations and maintenance but also to mobilize investment, thereby establishing creditworthiness and unlocking commercial financing. Where that is not feasible, commercial financing is possible only in combination with transfers.

This chapter focuses on how developing financing alternatives involving local capital markets to mobilize private-sector, domestic capital investment

Financing Investment in Water Security: Recent Developments and Perspectives.
DOI: https://doi.org/10.1016/B978-0-12-822847-0.00011-9

can help to remedy the financing gap for water and sanitation infrastructure. It also discusses how to identify countries, water providers, and projects that can benefit from this approach and presents the lessons learned from experiences to date in one country—Kenya. The Water Facility Fund (WFF) model proposed for Kenya will be presented here. The strength of that model for developing a local capital market financing mechanism to mobilize investments in pooled water and sanitation loans is based on establishing longer tenor debt (matching the financial obligation to the useful economic life of the assets), accessing financing in local currency (so that there is no foreign exchange risk) and aggregating (or pooling) several loans in larger bond transactions (to a level that makes them interesting for institutional investors). However, tariffs (or taxes, depending on local statutory and regulatory frameworks) have to be well regulated and set at socially and politically acceptable levels in order to cover operations and maintenance as well as the costs of capital investment. At the same time, they need be indexed annually to keep up with inflation.

This chapter explains the background (Section 15.2), the Kenya Pooled Water Fund (KPWF) structure (section 15.3), the challenges met with regard to the establishment of a long-tenor debt financing facility in Kenya (Section 15.4), the KPWF business approach as applied in Kenya (Section 15.5), KPWF's comparative benefits (Section 15.6), and the further development of local capital market financing (Section 15.7).

15.2 Background

On April 20, 2015, the Minister of Foreign Trade and Development Cooperation of the Kingdom of the Netherlands held a High Level Breakfast Event and a Bilateral Meeting at the Permanent Mission to the United Nations in New York to focus on innovative financing strategies in the area of water, sanitation and hygiene in the context of the Post-2015 SDG Agenda and the upcoming Financing for Development Conference in July 2015, in Addis Ababa. The event was significant since the Government of the Netherlands (GoN) had decided to step up and support the development of new approaches to sustainable financing in order to implement the policy objectives of SDG 6. This policy decision, as anchored in the Water, Sanitation and Hygiene Strategy (2016–30) from the Dutch Foreign Ministry's Department of International Cooperation (DGIS, 2016), was inspired by history and the role assigned by the GoN to the Netherlands Water Authorities Bank (NWB) to rebuild the water infrastructure of the

BOX 15.1 The Nederlandse Waterschapsbank

The **Nederlandse Waterschapsbank** (NWB) is a privately incorporated bank that was founded in 1954 to help public entities achieve their policy goals by providing them with loans. Its shares are held by the publicly incorporated water authorities (81%) and national and provincial government authorities (19%). It is a very lean and efficient organisation with a triple A rating. It has successfully handled the banking and private sector lending activities for all water authorities in the Netherlands using pooled financing and revenue bond credit techniques. Over time, NWB has begun to finance the wider Dutch public sector as well.

Netherlands after the Second World War and the devastating floods of 1953 (Box 15.1).

By setting up the WFF, the GoN aimed to collaborate with interested partner governments on the establishment of local-currency-denominated "Water Financing Facilities" in partner countries over a period of four to five years. The thinking was that these facilities would be tailored to each country's legal framework and structured to offer "pooled" investment opportunities to private capital market investors, providing small- and medium-sized local water utilities with access to capital market financing and lowering annual borrowing costs by providing longer loan tenors. This initiative also acknowledged that not every country was in a position to benefit from this approach and that the financing work needed to be accompanied by intensive technical assistance to ensure that the local utilities, which would benefit from these financing facilities, could in fact achieve and maintain creditworthiness. WFF prepared a methodology for the selection of WFF target countries and incorporated it in a policy process to guide the identification of new WFF target countries.

The selection criteria applied include:

- Is the country on the OECD DAC list?[1]
- Is there a sufficient need for water and sanitation investments in the country of interest to justify an intervention by WFF, possibly through the establishment of a national WFF? Does the country have at least 5 million people deprived of access to sustainable, safe drinking water and/or a sanitation service?[2]

[1] The Organization of Economic Cooperation and Development's (OECD) Development Assistance Committee (DAC) "maintains" the definition of development aid funds (Official Development Assistance—ODA) and the list of countries that receive ODA.

[2] This will be judged quantitatively on the basis of UNICEF and World Health Organization's (WHO) Joint Monitoring Protocol (JMP), which tracks progress towards the achievement of Sustainable

- Is there an active debt capital market? Local capital market status and forecast.
- Is there interest and/or any activity from development partners in the water and sanitation sector of the country?
- Description of the country context leading to an overall indicative country risk outline, including current political situation and trends.

WFF produced a target list of 22 countries and one region for serious consideration and one country in addition to Kenya (Indonesia) was selected for a more thorough appraisal. However, it was decided not to proceed outside of Kenya but rather to focus all the attention on KPWF to complete pooled bond financing in Kenya.

To achieve the GoN objective and decide which GoN partner country to focus on first, staff of the Netherlands Water Partnership (NWP) and a team of experts that eventually established the WFF, first reviewed the findings of a 2012 Mission on Private Financing in the Kenyan Water Sector conducted jointly by the World Bank (WB) and the United Nations Secretary General's Advisory Board on Water & Sanitation (UNSGAB). The "Aide Memoire" from that mission described the demand for, and obstacles to, private financing in the water sector in Kenya at that time and identified technical support that the donor organizations could provide to address those constraints (see Box 15.2).

The concept of blending and using a pooled bond financing facility in a local capital market, with that facility aggregating several borrowers' water and sanitation infrastructure needs in a single bond transaction, was introduced in the Aide Memoire. This was also discussed with the relevant senior Government of Kenya (GoK) officials from the National Treasury and the Ministry of Water and Irrigation[3] as a future option for Kenya dependent on the continuing maturation of the bankability of the sector. This possibility was deemed a logical next step and an alternative to other commercial financing approaches that were being piloted at the time, such as the Output-Based Aid programs of the World Bank and SIDA, which catalyzed access to commercial bank loans using the shadow credit ranking

Development Goal 6. The calculation is as follows: the JMP percentage of urban population with no access to a safely managed service multiplied by the urban population. If the resulting figure exceeds 5 million, the criterion is deemed to have been met. This threshold was selected because it corresponds to an investment need of EUR 250 million (based on EUR 45 per capita and 10% of investments not leading to a direct increase in access), which would typically correspond to 5–10 bond issuances.

[3] Now the Ministry of Water, Sanitation and Irrigation (MoWS&I).

BOX 15.2 Recommendations from the 2012 Aide Memoire of World Bank and UNSGAB

- An affirmative statement of support from the Ministry of Water and Irrigation, supported by the Ministry of Finance, that private financing is an important way to augment existing funding for the water sector.
- Policy guidance on private financing in water development to articulate the GoK's commitment to support the water sector. This includes clear information on the legal and regulatory requirements and guidelines on how to address them.
- Regular communication with the general public and the business community on the progress made by the water sector.
- Fast-tracking of the pilot transaction being undertaken by IFC and the Water and Sanitation Program of the World Bank (WSP) to take Water and Sanitation Companies (WASCOs) to the financial market as a way of learning how best to structure long-term private-sector financing instruments.
- The establishment of a national finance entity with the legal authority to pool various WASCO credits as a way to enable access to long-term, low-cost finance from the local capital market.
- Economic-like incentives to creditworthy WASCOs such as longer licences and service provision agreements, automatic tariff adjustment for inflation, and tariff approvals prior to finalising financial agreements.
- Finalisation of the 20-year National Water Master Plan, including cost estimates, and steps to ensure that the plan provides a basis for future financial planning and the identification of commercially viable projects.
- A regular process for National Strategic Financial Planning where capital investment plans are matched by external and internal funding sources for the next five years.
- Encouragement of efficiencies in the operations of WASCOs to facilitate commercial lending.

work of the Water Services Regulatory Board (WASREB, 2019) established by the Kenyan Ministry of Water, Sanitation and Irrigation (MoWS&I) under the 2016 Water Act.

The NWP/WFF team followed up their desk review of the Aide Memoire with a mission led by the Directorate-General for International Co-operation (DGIS) of the Ministry of Foreign Affairs of the Netherlands to Kenya in the second half of 2015 to conduct an on-the-ground assessment. Their findings included:

1. Enabling Environment
    - The institutional design of the Kenyan water and sanitation sector was among the most advanced in Sub-Saharan Africa since the enactment of the 2002 Water Act that had reorganized the sector and, among other things, created independently administered water companies with ringfenced tariffs and an independent regulator (WASREB) that approves the water tariffs.
    - The "Vision 2030" document prepared by the Kenyan Government in consultation with a wide array of local stakeholders, promises that "improved water and sanitation are available and accessible to all" by 2030 in line with SDG 6 (KenyaVision2030, 2008 (annual update 2015)).
    - The commitment from the national budget of approximately USD 420 million annually, or 3.1% of central government spending for the water sector, 40% of which is deemed to come from foreign assistance. Since this USD 420 million, according to Vision 2030, would be able to deliver only 60% of the USD 700 million required annually, the GoK adopted a policy of attracting private-sector financing to bridge the financing gap.
    - A very active, forward, and independent regulator (WASREB), which regulates tariff adjustments (on a full cost recovery principle to include operation, maintenance, and capital expenditure costs) and, on an annual basis, monitors the progress of the sector on the basis of national policy objectives and the operational performance of the Water and Sanitation Services Providers (WSPs), and assigns about half of the 88 water companies a ranking in their Creditworthiness Index Report.
2. Legal and Regulatory
    - Alignment under a proposed new Water Act (introduced and subsequently enacted in 2016) of the water sector's legal and institutional framework, with the recently approved 2010 Constitution (GoK, 2016). Kenya's 2010 Constitution made access to potable water and basic sanitation a human right and assigned (devolved) the responsibility for water supply and sanitation service provision to 47 county governments. Under the new Water Act, the national government is tasked with the management and protection of water resources and the county governments are tasked with the provision of water and sanitation services and the implementation of the national

government policies on natural resources, including soil and water conservation.

- Transfer, under the new Water Act, of every water and sanitation company's (Water Services Providers or WSPs) 100% shareholding from the national government to the county, requiring relicensing by WASREB in order to provide water and sanitation services.
- Reauthorization, under the new Water Act, of the power of the WSPs to ringfence tariff-based revenues from county funds and allow those revenues to pay for WSP operations and to borrow for their infrastructure needs. In fact, a very small number of the better managed WSPs had successfully taken on commercial bank loans under the Water Act of 2002 to finance their infrastructure, backstopped with a guarantee provided to one of the commercial banks by the United States Agency for International Development (USAID) or an OBA grant component provided by development partners. The payment status of all those loans is currently good.
- Authorization under the new Water Act of a change of the name of the Water Services Trust Fund, which comes under the MoWS&I, to the Water Sector Trust Fund (WSTF) and the empowerment of that fund to provide conditional and unconditional grants to counties in addition to the Equalization Fund[4] and to assist in financing the development and management of water services in marginalized areas or any area which is considered by their Board of Trustees to be underserved.

3. Capital Markets
    - A relatively well-developed capital market and a strong Capital Markets Authority (CMA), established in 1989 by an Act of Parliament (Cap 485A, Laws of Kenya) as the regulatory body for the capital markets.
    - A large number of pension funds and institutional investors interested in diversifying their portfolios with investments in debt instruments other than those issued by the National Treasury. These could include water and sanitation investments if properly structured with credit enhancements and CMA approval.

[4] Please find background information about the Equalization Fund on this website: https://www.klrc.go.ke/index.php/constitution-of-kenya/147-chapter-twelve-public-finance/part-1-principles-and-framework-of-public-finance/373-204-equalisation-fund.

- Verbal support from all governmental and non-governmental stakeholders indicating that the timing for developing an innovative program for access to private-sector financing using a non-governmental facility was favorable.

These findings helped the team determine that significant groundwork had already been done by multiple development partners—World Bank, KfW, GIZ, the USAID and others—by providing technical assistance to the Water Services Regulatory Board (WASREB) and the Water Service Trust Fund (WSTF) in order for these agencies to promote the development of the capacity and creditworthiness of the WSPs so that they could access commercial financing. These development partners had some success in getting some WSPs to take on commercial bank loans but with a number of conditions and limitations. These included short tenors, variable rates, the parking, and control of system revenues at the lending bank, one-time credit support in the form of a partial guarantee to one bank from a donor, and a grant funding program to pay up to 60% of a loan from a commercial bank if the WSP's construction project met donor grant conditions. Given this focus on commercial bank financing, the GoN decided to support a more traditional bond financing program that would allow multiple creditworthy WSPs to access longer-term, fixed-rate financing using the local capital market for their capital investments. Stakeholders were in agreement that the timing might be right.

On the basis of these positive findings, the Embassy of the Kingdom of the Netherlands (EKN) boldly decided, with agreement from the DGIS, to allow the Kenya Innovative Finance Facility for Water program (KIFFWA) to seed fund the development of multiple strategies to access private financing for the water sector in Kenya, and so assist the "aid to trade" transition of the Kenya–Dutch partnership. KIFFWA's earliest initiative was to fund the establishment of a local capital market facility for water financing known as the KPWF. The team that was responsible for the initial feasibility work to get to this point then formally established itself in 2017 as the WFF, a subsidiary of Cardano Development Foundation (see Box 15.3). The overall set-up cost was estimated at approximately EUR 3 million.

The WFF team began its work in Kenya during 2016 by:

1. Preparing a comprehensive Business Plan for KPWF.
2. Establishing a private, non-profit financing entity under the Kenyan legal framework. The founding "members" of the company—limited by guarantee—are two Dutch non-profit entities, namely NWP and Cardano Development.

BOX 15.3 Cardano Development

Cardano Development is an incubator and fund manager that initiates and develops innovative financial risk management solutions, for application in complex environments in developing countries. Cardano Development facilitates inter-company learning with startups such as WFF and more mature companies like The Currency Exchange Fund, GuarantCo, and Frontclear.

3. Developing a "request for funding" application by WFF to DGIS for the work in Kenya with the intention of raising more funds from other donors and scaling up the model in other countries.
4. Initiating a project development process that encouraged water companies to express interest in obtaining local currency financing for their capital investments because donor-based financing was dwindling at that time and commercial bank funding costs exceeded that of the capital market, resulting in exposure to interest rate risks (variable rates) and costing more on an annual basis due to short tenor loan terms.
5. Identifying Kenyan candidates for the Board of Directors of the newly established Kenyan management company as well as identifying staff to take charge of the activities in Kenya with professional support from WFF.

KPWF was operationalized in 2017–18 by appointing a Kenyan Board of Directors and recruiting a local CEO and CFO/COO and transferring responsibility for all contracts with local service providers (such as legal firms, tax advisors, placing agent, financial advisor, and project development consultants) that are needed to arrange the capital market financing for the water sector and prepare the investment projects. With WFF's continuous support, KPWF has developed into a solid company with strong corporate governance, local management, and professional staff. WFF continuously provides a full range of technical assistance to KPWF management as they work to achieve their goal of issuing a first bond against a pool of creditworthy water companies that are spread across several counties in Kenya.

WFF and KPWF, with leadership from the Ambassador of the Kingdom of the Netherlands to Kenya, were able to establish a GoK Steering Committee in 2016 to support the KPWF initiative. At that time, the membership included the Cabinet Secretaries of the National Treasury and the MoWS&I, the Netherlands Ambassador, and the Chairman of KPWF. This high-level committee was established as a visible demonstration of the GoK's support for the KPWF. This group met several times and decided to add the Council of Governors to the Committee once the new Water Act of 2016 was passed.

During the early stages of the development of KPWF, the Steering Committee was able to obtain parliamentary approval for a supplemental appropriation (in the 2016–17 budget) to the MoWS&I through the WSTF of USD 2.5 million out of a planned USD 5 million in order to support the development of private-sector financing for water and sanitation. At that time, the thinking of EKN and KPWF was that the funding could best be used to capitalize a debt service reserve account (DSRA) for the first bond transaction in order to subsidize the WSP borrowers in the first bond transaction because reserve funds are normally capitalized by borrowers to credit-enhance their own revenue bond transactions. In 2019, the Kenyan government decided it could not legally invest these funds, which were housed with a private bond trust (KPWF), to credit-support a pooled bond transaction benefiting WSP borrowers. Without this GoK budgetary support for the KPWF private financing initiative, alternative ways were needed to show investors that the Kenyan government's support for private capital market financing was real.

In the latter part of 2019, The Steering Committee, therefore, agreed to formalize the bilateral cooperation of GoK and the GoN to develop a private financing initiative on the basis of a Memorandum of Understanding, which was later elaborated into a Framework Arrangement for signing by the Cabinet Secretary of the National Treasury and the Netherlands Ambassador to Kenya. In addition, the Treasury wanted an Implementation Agreement to be signed by the MoWS&I, the WSTF, and KPWF to demonstrate that the implementing entities for this initiative were cooperating. As of mid-2021, these agreements have not yet been signed. They are still the subject of negotiations and official signing will require clearance from the Attorney General. This explicit government support is needed because institutional investors are typically risk-averse and not used to investing in the water and sanitation sector. The reason for the delay in the signing of the Framework Arrangement and Implementing Agreement is the transformational nature of the KPWF initiative. Large-scale financing of sector investments from the capital market could fundamentally reorganize funding flows (off and on budget). Careful consideration by those responsible is therefore needed.

15.3 KPWF structure

On the basis of the findings from the original feasibility study in late 2015 and the work of various donors to activate commercial bank financing in Kenya, the Dutch government decided to commit itself to work on unlocking the local capital market as a way to establish more sustainable, long-tenor revenue-based debt financing of the water and sanitation sector in Kenya. To that end, WFF established a private Kenyan management company—the KPWF, a nonprofit company limited by guarantee—to act as the financial arranger for a pooled bond financing structure (which has been used internationally for over 80 years in international debt capital markets) that would enable small- and medium-sized creditworthy water and sanitation companies to access the Kenyan debt capital market. The decision to set up a private company was based on feedback from stakeholders in the investment community that indicated its strong support, trust, and comfort to invest in projects prepared by a private, professional financial intermediary rather than a government institution.

One of the first steps taken by WFF and KPWF was to determine precisely whether KPWF could directly access the debt capital market with a pooled water and sanitation bond. KPWF retained a leading Kenyan law firm to help it establish KPWF as a financial intermediary. That firm advised KPWF that it could not be the issuer of a publicly issued pooled bond transaction because it had not been in for existence long enough (the minimum requirement is 5 years) and did not have a positive balance sheet, even though it intended to issue debt on a nonrecourse basis to itself[5]. KPWF was also advised that the only way to access the market was for KPWF to create an special purpose vehicle (SPV) Trust as a conduit issuer and develop the pooled bond financing structure in a way that would comply with the CMA's guidelines for using an Asset Backed Securitisation (ABS) in order to achieve KPWF's goal of financing the water sector with a public bond offering.

KPWF, therefore, overhauled the components of the internationally recognized Bond Bank model for pooled financing and established an asset-backed structure that would use an SPV Trust, instead of KPWF, as the bond issuer. This means that KPWF can create a Trust as a SPV that will act as the

[5] A nonrecourse debt (loan) does not allow the lender to pursue anything other than the collateral pledged for repayment of the debt obligation. For example, if a borrower defaults on a non-recourse home loan, the bank can only foreclose on the home.

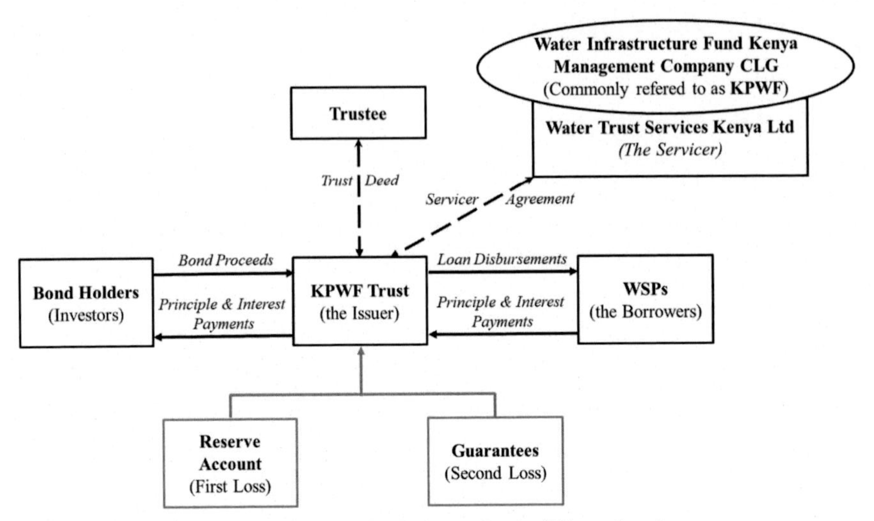

Figure 15.1 KPWF trust structure. KPWF, Kenya Pooled Water Fund.

actual issuer of the pooled bonds. The assets needed for the repayment for the bonds will be the various loan agreements signed by the Trust and the WSP borrowers that are deemed creditworthy by KPWF (see Fig. 15.1).

Following discussions with the law firm, a Kenyan placement agent and a financial advisory firm were retained by KPWF to assist in the structuring, marketing, and sale of the bonds. On the basis of international experience with brand new bond issuances, the financing structure of at least the first bond transaction needed to include a specific de-risking instrument to provide bond investors with comfort about repayment. Moreover, the loan obligations of the WSPs will be corporate debt obligations for each company on their balance sheets, not obligations for the county or the national GoK. The proposed credit enhancements have proven valuable in other emerging markets to help stimulate the appetite of domestic investors by mitigating their concerns about the bankability of various subnational infrastructure endeavors.

Some of the techniques KPWF will use involve:

1. The establishment of a DSRA. The DSRA will be an account of the Trust to provide emergency liquidity to the trustee if there is a shortfall in interest payments or principal repayments from the borrower(s). In this scenario, these reserve funds will be used to pay bond holders on time and in full in line with the bond amortization schedule. This facility provides bond investors with first-loss mitigation. It is cash funded and consists of a percentage of the par amount of the bonds issued, the maximum

annual debt service on the bonds or the average annual debt service on the bonds to achieve an investment-grade credit rating. The account is held in escrow and managed by the trustee on behalf of the bond holders. Normally, revenue bond issuers capitalize this reserve requirement by bond financing the required amount and amortizing its cost over the life of their loan. In Kenya, a decision was made to subsidize this requirement with funding from the GoK, but that subsequently proved unfeasible. The Dutch government then decided to provide the funds needed to capitalize the reserve for the first bond transaction.

2. The purchase of partial guarantees (one or more) from external private or donor entities. These serve as the second loss-risk mitigation if the reserve fund is fully utilized and/or if a borrower in the pooled program defaults in full. The costs for obtaining guarantees will be passed on to the individual borrowers on a pro-rata basis.
3. There is a legal requirement for each WSP borrower to establish a secured escrow account for repaying their debt obligations. The monitoring of payments into the escrow account allows for the early identification of problems affecting the borrowers. KPWF will therefore require WSPs to set up a special bank account to collect and make monthly repayments to the SPV Trust funded from the revenues received for water services. Moreover, the amounts that have to be deposited in that account will be specifically identified in each WSP loan agreement.
4. A requirement to make full annual payments to the trustee in nine equal monthly installments. This credit technique allows the Trust to identify repayment problems early before a debt service payment to bondholders is due and gives the trustee and servicer time to cure the problem and/or arrange access to the loss mitigation instruments mentioned above. This technique is a credit enhancement and gives investors more comfort because it over collateralizes the semi-annual interest payment obligation of the borrower.
5. A check by the trustee on the disbursement of bond proceeds. On the basis of investor concerns in Kenya that bond proceeds might be misused by WSPs during the construction period, WFF and KPWF decided to use a project financing technique to provide investors with comfort on this issue. In addition to KPWF's requirement that an independent resident engineer must be contracted by the WSP to supervise and verify the construction of the assets, KPWF will therefore retain an independent lender's technical advisor to work on-site to review payment invoices from the WSPs, to pay their contractors, to determine that the

constructed works have been completed in accordance with the approved designs and to approve each invoice for payment by the trustee.

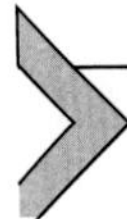

15.4 Challenges to establishing a revenue-based long-tenor debt financing facility in Kenya

Despite the positive findings described here that led to the decision to establish a private financing facility in Kenya for creditworthy WSPs, many challenges have since surfaced that need to be overcome before capital-market-based investment financing can be a realistic and sustainable financing option.

These challenges include:

1. A lack of expertise in capital improvement and financial planning at the water companies, which is an area that the World Bank identified as a critical component for technical assistance.
2. The limited enforcement by the GoK of the requirement to implement or adjust findings in WSP audit reports.
3. The creditworthiness of only a relatively small number of WSPs (less than 20 of 86 licensed water companies), primarily due to a low level of operational efficiency and inadequacies in the governance of the water companies.
4. Water tariffs are decided by WASREB after stakeholder consultations. The tariffs are set to cover the full cost of operation, maintenance, and capital investments. In some cases, the implementation of the revised tariffs by the WSPs has been opposed by county government authorities, resulting in a political risk.
5. Limited financial support from national government for the development of a financially viable water sector; lack of a national water financing strategy and a costed SDG6 investment plan; failure to commit national funds to co-finance portions of investment projects as a grant/subsidy or as loans with a low (subsidized) interest rate and blend them with a commercial loan at a market rate; and/or an unwillingness to credit-enhance proposed WSP's debt obligations.
6. The WSP practice of allocating budget to plan, design, obtain permits and licences, and execute capital investment projects, only after the funding of the investments has been secured.
7. The complete shutdown of any decision-making throughout the government for almost one year before, during, and after national elections because of the requirement that candidates for office must resign (that is

to say, vacate) their positions well before the new elections, and because of the time required after the elections for the newly elected officials and their appointees to be in position and assume their responsibilities.

8. Another government decision-making installment occurred as a result of the COVID-19 pandemic.

9. A debt capital market that has not been fully developed, as evidenced by bond yields based upon the price of Treasuries with no interest rate differentiation for debt issues based upon their ratings, structure or tenor.

10. A perception by local investors that the water and sanitation sector is a risky investment because it is perceived to be poorly managed, and subject to undue interference by political interests. The only other sector that was successful in mobilizing investment finance from the capital markets was the power sector. It did so through KenGen, Kenya's electric utility before 2016.

11. No investor confidence that any publicly owned national government entity is in existence or could be organized to develop and manage a financing facility in which confidence is possible and on which it is possible to depend as an issuer of debt.

12. Lack of a coherent financing policy for accessing private financing and using blending techniques to lower the high-interest costs for Kenyan commercial and capital market loans.

13. The absence of an institutional champion in Kenya for the private financing of investments in the water sector.

14. Legal and regulatory growing pains in the water and sanitation sector as a result of the devolution of responsibility for water-sector investments (which used to be made through the eight regional Water Boards) and the ownership of the WSPs by the 47 county governments.

15. Legal and regulatory constraints on the creation of a financial structure that meets the capital market requirements.

16. Real and perceived competition from DFI concessional loans and/or donor grants (many water companies and their county government officials still hope to be financed by these historical means, despite the reality that, being a LMIC now, Kenya is no longer eligible for many of these sources of international financing).

17. The absence of any requests from WSPs for the approval of annual tariff adjustments (using indexation) to maintain the coverage requirement for long-tenor private financing from the regulator (a matter which the Kenyan regulator can approve). Coverage means that a WSP must raise, or commit to raise, revenue over and above the revenue needed to meet all of its obligations (opex and capex) in order to obtain and maintain an investment-grade credit rating.

18. The long period required for moving projects, and especially the associated financing, through the mandatory approval processes from the WSP, local government, and the national government.

19. Frequent changes in the leadership of national ministries that oversee and review funding matters, and, to a lesser extent, in water company managerial staff and Boards of Directors.

15.5 WFF/KPWF business approach to successfully accessing private-sector financing

As mentioned above, there are many obstacles that need to be overcome in order to mobilize revenue-based, long-tenor debt in Kenya, and to build a bridge between the water and sanitation sector and the capital market. The KPWF business model, therefore, includes a rigorous and disciplined process for delivering technical assistance to the WSPs in the areas of project planning, engineering, and financing, and to the capital market parties with the aim of demystifying and de-risking the proposed investment in the sector.

It is important to note that the KPWF financing model focuses on analyzing a water company's overall financial condition and creditworthiness as a basis for determining the ability of a WSP to service its prospective debt obligation. This credit assessment is not based solely on the expected revenues and/or savings associated with a specific investment project, as is common in project financing. The loans that KPWF will structure for the capital market investment will commit a WSP to repay its capital market debt from its system revenues rather than specific project revenues, meaning that the investment project is not necessarily the source of debt repayment. Capital market fixed-rate investments result in borrowers incurring interest payment obligations on the whole amount of the loan, starting immediately after financial closure, whether or not the project is constructed on a timely basis. So, the WSPs need to get the project constructed as soon as possible after financing to avoid paying interest on idle funds. This can only be done if the project development process has been finalized before financing occurs. This was new to WSP managers because the practice in Kenya is that the detailed design phase for an investment project commences only after a WSP has obtained a commitment to finance the project from a financier, often a bilateral donor or IFI.

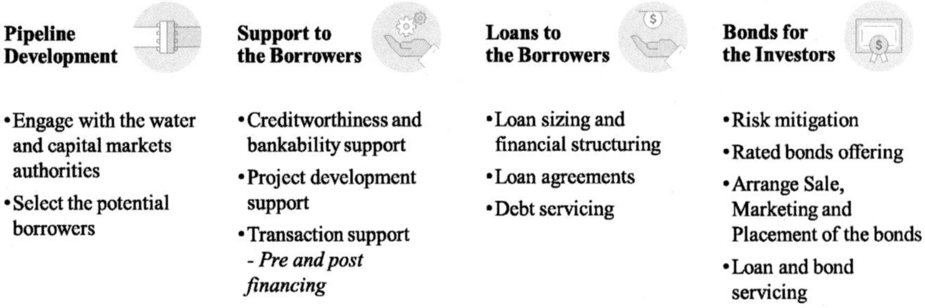

Pipeline Development

- Engage with the water and capital markets authorities
- Select the potential borrowers

Support to the Borrowers

- Creditworthiness and bankability support
- Project development support
- Transaction support
  - *Pre and post financing*

Loans to the Borrowers

- Loan sizing and financial structuring
- Loan agreements
- Debt servicing

Bonds for the Investors

- Risk mitigation
- Rated bonds offering
- Arrange Sale, Marketing and Placement of the bonds
- Loan and bond servicing

Figure 15.2 Summarized Overview of KPWF's Services. KPWF, Kenya Pooled Water Fund.

By combining financial structuring skills and project development skills with support from donors, development finance institutions, and guarantors, KPWF's focus has been on developing loans to creditworthy water and sanitation service providers, organizing an approved capital market financing structure for issuing debt, obtaining a single "A" national credit rating for the bond structure and de-risking the bond issuance with credit enhancements to a level where domestic institutional investors become confident to invest in this sector. Fig. 15.2 provides an overview of the KPWF services model.

15.5.1 Project and loan development support activities

KPWF engages with the water utilities in an advisory role throughout the project development period, in other words from project conceptualization to the completion of the tender documents for the capital investment projects to be funded by the loan. Project development was financially supported by development partners, allowing the WSPs to contract third-party consultancy firms to help them prepare detailed engineering designs and tender documents for the proposed investment projects. Financial and legal transactional support is regularly provided by KPWF staff to ensure the WSPs have developed sound capital investment and financial plans so that they can commit to capital market loan agreement terms and pay back their loans.

The project and loan development phases are monitored by a KPWF Investment Committee (IC). The IC consists of four experts who assess the WSP proposals and advise the KPWF Board about the proposed loans. The IC holds intermediate progress meetings as necessary to expedite the process. The outcomes of the IC meetings are presented to the KPWF Board of Directors for their information and approval.

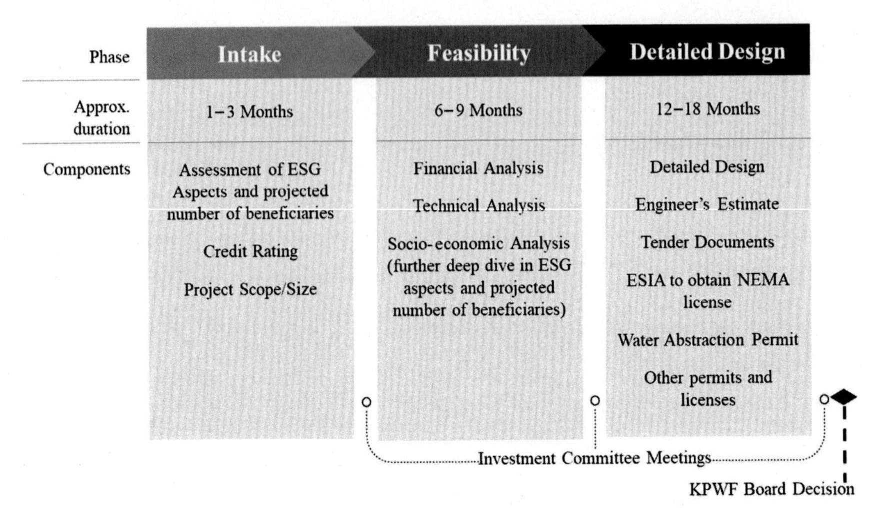

Figure 15.3 KPWF's project development process. KPWF, Kenya Pooled Water Fund.

KPWF follows a robust and disciplined project development process that includes the following elements (summarized in Fig. 15.3).

15.5.1.1 Preproject development phase

This phase consists of the pre-selection of a group of well-performing and creditworthy or near-creditworthy Kenyan WSPs (assessed using the Key Performance Indicators and Credit Worthiness Index published annually by the Water Services Regulatory Board) and an informative workshop for these WSPs and other stakeholders. WSPs interested in pursuing project and loan development with KPWF are invited at this time to submit an Expression of Interest to KPWF. These Expressions of Interest are assessed based on Environmental, Social and Governance (ESG) information provided by the WSP and on the initial description of the project to which the possible corporate loan to the WSP will be tied. It should be noted that KPWF does not make an assessment about which project a WSP will finance. Those decisions remain the purview of the Board of the WSP, its shareholder, the county, and its capital improvement plan. In order to finance any proposed project, a WSP must incorporate it in the overall WSP's tariff. WASREB regulations require proposed tariff adjustments to be subject to local stakeholder consultations, for example, to determine whether they are necessary.

15.5.1.2 Intake phase: preparation of the intake document

This phase takes between one and three months and comprises an initial analysis of ESG aspects, creditworthiness, and consultations with the WSP to conceptualize and describe the proposed project in terms of its tentative scope, size, and benefits. This phase results in the preparation by KPWF of an intake document that is submitted to the IC for review and comment.

The IC determines which WSPs/projects will move to the feasibility phase and typically provides conditional approval for some of the WSPs pending a written response from KPWF/WFF Project and Loan Development Team to specific IC questions or concerns. The written response to the IC has to be submitted within one month to avoid delays.

15.5.1.3 Feasibility phase: preparation of the project proposal

The feasibility phase takes between six and nine months and comprises a highly detailed analysis of the WSP by KPWF, including due diligence for the WSP's legal, commercial, financial, and ESG status, socioeconomic and technical surveys, the scope and size of the proposed project, and financial projections and analysis, among other things.

In terms of loan development, the following steps are taken during this phase:

1. The establishment of further awareness about the requirements for commercial lending and KPWF's product. During a workshop, the WSP Management Team is informed about the details of KPWF's product and the steps to be taken before financial close. Most WSPs have very little experience with commercial lending and so this workshop and the hand-holding that follows from it is key to successful loan development. The workshops are tailor-made for each WSP.
2. A letter to the WSP about the prospective KPWF loan outlining the expected loan terms, which will be discussed in person with WSP management as needed.
3. The introduction of KPWF's due diligence questionnaire and the loan conditions precedent (CP) to the WSP requesting that all financial, commercial, operational (including but not limited to ESG operational procedures) and legal information be readily available so that KPWF can check whether a loan can be provided. After the questionnaire is returned to KPWF, an iterative process of reviewing questions and answers, and the sharing of additional documentation, takes place as needed.
4. KPWF retains an external international accounting firm to conduct a financial due diligence review of several years of audited and unaudited financial statements for each WSP, obtaining clarification of findings, as necessary.
5. KPWF uses its own externally and independently validated credit assessment model to calculate the WSP's overall financial viability. It is

a financing and project evaluation model capable of generating profit and loss, cash flow and balance sheet projections, and key financial ratios (including, *inter alia*, debt service cover ratios, operating cost cover ratios, and internal rate of return calculations) for the WSP on the basis of a set of operational, financial and socio-economic assumptions and information provided by the WSP. On the basis of this analysis, a WSP may be found not to be creditworthy or be given advice about the steps necessary to achieve creditworthiness. Ultimately, using the output from this model, a projected loan repayment schedule can be forecasted for each WSP using the prevailing market-pricing assumptions.

6. After the completion of the feedback sessions described above, KPWF formally sends a preliminary, draft, loan term sheet to the WSP on the basis of which the WSP is expected to respond with a letter of intent to proceed to financing.
7. The IC determines which WSPs/projects will move to the detailed design phase and typically provides conditional approval for some of the WSPs pending a written response from the KPWF/WFF Project and Loan Development Team to specific IC questions or concerns. In several cases, the IC has advised staff to suspend work with a WSP pending the resolution of particular issues that would otherwise prevent the WSP from becoming a credible borrower.

15.5.1.4 Detailed design: preparation of the complete project documents

This phase involves the preparation of detailed engineering reports and designs, the engineer's estimate, the tender documents, the Environmental and Social Impact Assessment that is required to obtain a licence from the National Environment Management Authority, any hydrological study required to obtain a water abstraction permit from the Water Resources Authority and any other work to obtain other permits and licences. This phase takes between twelve and eighteen months and, once completed, satisfies all project development requirements by KPWF and signals the start of the borrowing process to get to financial close.

After the completion of the detailed design and cost estimate, the management of the WSP submits the proposed project, and its financing with a loan from KPWF, for approval to the WSP Board of Directors. The KPWF team shares a draft board resolution to borrow with the WSP that is drafted by the KPWF external legal advisor on the basis of the outcome of its legal and external accounting firm's financial due diligence. The legal counsel of the WSP will be instructed to prepare a legal opinion on the board resolution in advance of the board meeting in question. Pursuant to, or simultaneously with, the approval of the board resolution, a letter of no objection to the proposed project is also expected to be signed by the county government authority in question to mitigate the political risk of interference in, or the retraction of, the loan commitments made by the WSP. In any case, both the board resolution to borrow by the WSP and the letter of no objection from the county government are needed for KPWF to initiate the documentation phase leading to financial close.

Subsequent to the execution of the board resolution to borrow and the county letter of no objection, the KPWF team will share drafts of the loan agreement, the proceeds account agreement and the loan performance monitoring agreement with the WSP. As far as is possible, these drafts will include any additional requirements set by the IC and/or any requirements the KPWF team deems necessary as a result of the due diligence outcome. It should be pointed out for the sake of clarity that any such requirements can be brought up by third-party stakeholders, including but not limited to the trustee, guarantors or other providers of a different type of assistance to the trust. Moreover, it is expected that any such third-party stakeholders will also conduct additional due diligence through visits, access to virtual "rooms" with all information required for due diligence or otherwise. The WSP is expected to cooperate with these requirements. KPWF also expects each WSP to obtain their own legal advice from an external legal counsel to ensure that the WSP is prepared to provide the legal representations and warrants for the KPWF loan and that the WSP management has the authority to execute the loan agreement as well as any ancillary loan-related documents needed for the financing.

All project development phases have been built into KPWF's service model to help develop the capabilities of the WSPs that are seeking private sector financing for the first time. The oversight and continuous involvement of KPWF staff with the management of each WSP is critical to changing the perceptions of investors in the local capital market about the water supply and sanitation sector. The detailed attention to financial and technical matters by each WSP will build confidence in the management of the sector and help KPWF to build a sustainable financing facility for Kenya's water sector. However, KPWF is not only providing expertise and paying close attention to the front end of the financial preparation process, it also has detailed operational guidelines for its staff to provide ongoing monitoring and surveillance of the borrowers after financing has been arranged.

15.5.1.5 Final approval for a WSP to take out a KPWF loan

As a WSP shareholder, a county government authority will be required to confirm that it has been informed about the loan commitments to be undertaken by the WSP. KPWF will provide the WSP with a draft letter of support that the county can use for this purpose. The letter of support will be needed not only to provide comfort for investors that the county, as a shareholder, acknowledges and understands the terms of the loan: it will also include a statement from the county that it will not interfere with the WSP's compliance with the loan agreement during the lifetime of the loan. This is of particular importance for the implementation of future tariff adjustments required by borrower WSPs to safeguard their debt service capacity. When the county government has approved the letter of support, it will then be required to obtain approval from the MoWS&I and the National Treasury as required under the State Corporation Act.

15.5.2 Lessons learned about project development

Several important lessons have been learned from the project development process in Kenya since 2016 and these factors are continuing to cause delays in getting to the first financing. They include:

1. Many WSPs, with the possible exception of the very large ones, have difficulty freeing up capacity for matters other than regular operations and the related administrative activities. They lack the experience and professional capacity relating to long-term financial planning and to project development and financing because, prior to the 2016 Water Act and, to a lesser extent, even at present, infrastructure investment projects have been prepared and implemented by Kenya's eight regional Water Services Development Agencies, and then handed over for operational execution to the WSPs. Rather than being active borrowers, WSPs were merely the recipients of new infrastructure and the accompanying debt service payment obligations that had been arranged by others.

   The recommendation here is to provide the WSPs with technical assistance targeting knowledge and skills development in the areas of financial planning, and project development, and financing.

2. Another problem is the slow pace of decision-making in the WSP's Board of Directors and the county government. This can also, at least in part, be attributed to the changes brought about by the 2016 Water Act in which the counties were given responsibility for water and sanitation in their jurisdiction. Counties are now the owners of the water supply

companies and they appoint several board members. This all played out in the 2017 general elections that resulted in the new county governors and other elected officials, which in turn led to changes in the appointed board members and senior staff of the water companies. Moreover, the new government in some counties, in line with their mandate, sought to restructure the water companies in their jurisdiction. The initiatives for these mergers severely delayed project and loan development, if only because the merger process was often improperly executed (in other words, it failed to comply with WASREB regulations) and, moreover, generated uncertainty at KPWF and among prospective lenders about the legal identity of the borrower who would be the counterparty to a loan agreement.

The recommendation is that major sector reform of the kind implemented in Kenya in 2016 should be accompanied by intensive consultations, training, and other forms of technical assistance to ensure that the implications of the reform are clearly understood by the stakeholders concerned and also that these stakeholders have adequate capacity and resources to assume their new roles and responsibilities.

3. The uncertainties described here relating to leadership and decision-making delayed the start of the project preparation work by the water companies' professional staff. In response, KPWF was able to source donor financing in 2017 for several of the KPWF pipeline candidates. The donor funds covered the cost for the preparation of the detailed designs and tender documents as the WSPs were not accustomed to include these in their own budgets. Currently, in late 2020, five of the six potential WSP borrowers for the first bond transaction have finalized their project scope, cost, detailed designs, and tender documents. The sixth WSP wants a major revision of the project after the WSP Board decided to re-prioritize its investment policies. In the other cases, the cost estimates need to be updated given that the CPI has averaged about 5% annually since the projects were first costed in 2018.

   The recommendation is to promote the understanding and ownership of the commercial financing concept by the local and national stakeholders concerned, to monitor progress made with the project and the loan development process, as well as with the approval processes, and to undertake remedial action where there are misunderstandings and where progress is slow.

4. The WSPs have to provide the funds needed to make the final adjustments to the investment projects, whether in respect of the revised

cost, an additional study, or a revised design. This will most likely cause new delays as the water companies are suffering revenue setbacks due to the COVID-19 pandemic and because some of them do not have a specific budgetary allocation to take the project preparation process to its conclusion.

The recommendation here is for WSPs to engage in investment planning and to include foreseeable project and loan development costs in their annual budgets and also to include these costs in their application for tariff adjustment.

5. Each WSP was advised to register the disbursement schedule of the loan for their proposed project with WASREB for it to become part of their application for the regular tariff adjustment or be incorporated in their tariff that had already been approved. Only one WSP has followed through on this, even though the rest have agreed to do so. This example serves to demonstrate why, over the life of the loan term, the financing entity will need to monitor the compliance of WSP management to ensure they fulfill their loan obligations. This includes, among other things, the drafting of full cost-recovery budgets and financial statements, the timely payment of the debt, and applications for future tariff adjustments. These tariff adjustments are regulated and therefore should not encounter politically motivated reluctance to implementation at the county level. The challenge at the political level is to explain to the users and general public that a sustainable water and sanitation delivery model for all Kenyans (Water Vision 2030) depends on a full cost-recovery model based on regulated cross-subsidized tariffs that take into account affordability and consumer willingness to pay.

   The recommendation here is to promote understanding among key stakeholders that a full cost-recovery tariff is necessary to sustain water and sanitation services. This requires communications targeting politicians, national and local governments, WSP boards, and the general public.

6. Over the past few years, KPWF has worked in good faith while being assured that each of the six WSPs were committed to arranging their financing through the KPWF program. However, some of the WSPs' senior professionals and some of their board members, as well as county officials, have consistently highlighted a concern about the high-interest cost of capital market financing as a barrier to making a commitment. Searching for a least-cost financing option is a natural and responsible thing for utility managers to do. Moreover, this is not unexpected

behavior since these water companies have generally had access to free grant money, soft loans from donors and the possibility of lower interest rates on loans from the commercial banks in Kenya for a number of years (the latter was occasioned by the Banking (Amendment) Act of 2016, which regulated interest rates in Kenya at a statutory cap of 4% above the central bank rate). This law was rescinded in late 2019. Nevertheless, there was no commercial bank lending to the sector or other sector when it was in force.

The recommendation here is to raise awareness among the members of the WSP boards that the investment funding landscape for the water sector is changing, that the opportunities to obtain grants and soft loans are decreasing, and that commercial financing is the only way forward, with limited prospects for the blending of commercial financing with grants and soft loans.

7. Cheaper options are virtually unavailable in Kenya at this time. WFF and KPWF have approached donors and others for funds with the aim of establishing a blended-rate program to cushion the high interest costs of the market. Unfortunately, no cofinancier has stepped up. Blending with government subsidies through WSTF has also proved impossible. There is, however, some expectation that a small World Bank water financing program may require blending with commercial financing. If that happens and it can be done efficiently, financing will move ahead. Without waiting for that to happen, KPWF staff is continuously educating WSPs about the nuts and bolts of capital market bond pricing, in other words how the initial interest rate for a bond is tied to the price of treasury bonds issued by the national government, over which KPWF has no control. It has also been modeling and showing WSP management that longer-tenor, fixed-rate loans from the Kenyan capital market, even with high-interest rates, are more affordable on an annual basis than shorter-term loans with highly variable rates from commercial banks. While it is unclear what commercial banks would charge for water company infrastructure loans now that the loan cap rate is no longer statutorily mandated, it can be assumed that the rate will, all in all, be as high or higher than rates on the capital markets and very short-term.

The recommendation here is that the senior management of the WSPs, members of WSP boards and elected officials and staff in local government should be trained to understand commercial financing, to assess the pros and cons of the various commercial financing options, and have the capacity to make informed choices about commercial options for financing investments.

8. Corporate management and operational efficiency improvement programs, like reducing levels of nonrevenue water and implementing energy efficiency programs, may be instrumental not only in improving service delivery but also in boosting revenue and therefore the debt-service cover ratio of the WSPs.

The recommendation here is that WSP senior management and WSP board members should focus more sharply on cost control and revenue enhancement, positioning the WSPs better for commercial investment financing. In addition to investing in the expansion of systems to enhance service coverage, WSPs should roll out high-yield efficiency improvement projects, in particular in the areas of non-revenue water and energy management.

15.5.3 Transaction-related activities for issuing the first pooled bond

Once KPWF receives each WSP board resolution and the county letter of no objection, it will begin the extensive preparations for issuing a pooled bond transaction. There will be a number of steps that need to be taken to start the CMA approval process. They will involve the production of final or close-to-final transaction-related documents. These types of activities and documentation requirements are not unique to Kenya but are fairly standard for capital market financing in any country.

First and foremost, KPWF's board will, on the basis of the IC recommendations, move to formally approve the creditworthy WSP borrowers that will be included in the first bond transaction. Staff will follow up this approval by listing and obtaining all the information and requirements that the WSPs need to deliver or adhere to prior to financial closing, otherwise known as CP. This list may differ slightly depending on the WSP. In any event, the list will contain requirements such as providing all appropriate licences and permits, and the signed board resolution to borrow, as well as the county letter of support, the approval of MoWS&I under the State Corporation Act and the gazetted notice of the latest tariff adjustment from WASREB. Although these deliverables are conditional upon the finalization of the loan agreement, it is important for all preparations to be made to ensure that these conditions are met by each WSP in good time. Key tasks for each step of the process are detailed in the Annex.

15.5.4 Lessons learned about transactional services

15.5.4.1 Responsiveness of WSPs

It is important to understand that the water sector is going through a multifaceted transition that involves both water sector reform and Kenya's newly acquired status as an LMIC. The 2016 Water Act followed up on the 2010 Constitution of Kenya that devolved power and responsibilities from the national government to the 47 elected county governments. The Water Act devolved responsibility for water and sanitation services to the counties and also made them the owners of the WSPs operating in the county. Since its enactment, project development and financing have been the responsibility of the WSPs and their boards, with the county governments playing an oversight role as the owners of the WSPs. This is in sharp contrast with the situation before 2016, when the development and financing of investment projects for the WSPs was firmly in the hands of the eight regional water services boards, in concert with the national government and its agencies. Not surprisingly, one challenge is that the capacity of the local actors to handle their new tasks has yet to be fully developed, with hesitation and delays as a result.

The other major change is the transition of Kenya to an LMIC in 2014, which is slowly but surely leading to a change from an almost unique reliance on conventional grants and soft-loan programs to a situation in which there is much more need for the commercial financing of investments.

A primary lesson learned to date in Kenya is that the responsiveness of local utilities to the demands of the project preparation phase should be assessed before engaging in costly financial and transaction preparations. If utilities are responding in a timely fashion, the finance facility can consider starting on the early preparation of the financial documents needed to access private financing. If they are not, the early preparation of the legal, compliance, and capital market documents will, at least in part, need to be repeated, driving up the related costs as the utilities' governance, financials, operational performance, and project information change over time.

On several occasions, KPWF, working in concert with WSP management, has set deadlines that were not met. This may be attributed to the WSP but it must be seen, in part, as a result of delays in obtaining support from county governments, and the absence of encouragement, and long delays on the part of the national government and its agencies. The consequence was that KPWF engaged in significant capital outlays to prepare the myriad of transactional legal and financial documents with the aim of expediting readiness to access capital market financing, including steps such as establishing shadow ratings and the independent assessment of WSP financials by a Tier 1 audit/professional services firm, only to be disappointed by the failure of the WSPs and their county governments to comply with time schedules.

The consequence is that shadow ratings and audit due diligence had to be repeated. Moreover, the severe delays affecting project approval at the WSP board, and at local and national government, levels meant that initial engineer's estimates had to be reviewed and updated to account for inflationary adjustments. Engaging legal, regulatory, guarantee and other market experts at the outset can engender deal fatigue and a heavy cost run-rate without actually generating capital market financing.

15.5.4.2 Testing investment appetite

Furthermore, testing the investment appetite of capital market investors also produced challenges regularly as market pricing evolved in response to movements on the supply and demand sides. It is therefore apparent that the pipeline preparation phases should be concluded well before any meaningful transactional development expenses are incurred by a financing facility. On a positive note in the case of Kenya, however, the numerous legal documents, the credit assessment model, and the bond structuring model which were developed early on are all excellent and valuable templates that can still be used, not only in Kenya but also with some modifications in other emerging market countries that might be looking for support in their efforts to access capital market financing.

15.5.4.3 Phasing of the contracting of external service providers

Another lesson learned includes limiting, or keeping to a minimum, early discussions about transaction preparations with external parties, especially if the potential borrowers do not stick to time schedules and delay decisions. Talking about a concept that takes a long time to mature leads to scepticism and makes stakeholders less keen to be involved or provide support.

15.5.4.4 The essence of political support

Political support at the national and local levels is critical to a successful outcome in developing countries. That is particularly true of support from the Ministry of Finance (National Treasury) and water-related ministries as well as local government authorities. This is a specific issue where WFF/KPWF could have done more in the beginning. Despite catalyzing a high-level steering committee, its establishment was not codified in any formal documentation which would have solidified GoK's formal and irreversible commitment to the KPWF initiative. In the early stages of the project, KPWF/WFF had too little political acumen, lean staffing levels, and a huge workload but it was unable to get a strong, committed central government on board convincingly. Unfortunately this takes much longer than anticipated and forms a real risk to the initiative.

Political support at the county level for the unhesitating implementation of the regulated and agreed tariff adjustments, as well as the annual indexation, is mandatory for the development of sustainable, revenue-based, long-tenor debt financing for the water and sanitation sector with access to safely managed water supplies and sanitation for all Kenyans by 2030 (Water Vision 2030).

15.5.4.5 The essence of support from development partners such as the multilateral development banks

Support from the leading international finance institutions such as the World Bank and African Development Bank is also essential because of their expertise, political, and financial clout, and strong physical presence in emerging countries. UNSGAB wanted more commitment from IFIs to helping developing countries to develop an active local capital market with the capacity to finance the broad infrastructure needs of those countries, particularly in the fields of water and sanitation, once countries are left to their own devices. This type of technical assistance is still critically needed at the national and local government levels because the requirements of private sector financing and particularly capital market financing are difficult to digest for those lacking experience with the associated requirements. Unfortunately, there would seem to be no incentives for IFIs to provide this type of technical assistance to the extent that a gap is clearly visible in developing countries where the ready availability of conventional donor finance has not cultivated an environment that encourages expertise about how to leverage private financing for infrastructure investment in local currency.

Finally, it is also clear that a government champion for the development of sustainable financing is important, as evidenced in emerging countries that have successfully implemented innovative private financing initiatives in the past. Unfortunately, a champion of this kind, other than the regulator, has not yet come to the fore in the public sector in Kenya. WSTF (public), working together with KPWF (private), could possibly step in to fill this void.

15.6 KPWF's comparative advantage

By comparison with other financial initiatives, KPWF's product is appealing for the water and sanitation sector because of:

1. The support for project development and the fact that domestic private sector investment is available for all water and sanitation sector infrastructure needs. The average cost of project development including KPWF's assistance is approximately 5% of the related investment cost.
2. The project development effort in the context of private financing and long-term debt-servicing obligations requires an unusually thorough analysis of the costs and benefits of the proposed investments, resulting in projects that are affordable, sustainable, technically sound, and well balanced in terms of their scope, size, and benefits for WSPs and user communities.
3. The long tenor, which stretches the repayment obligation and more closely matches the useful life of the assets (20–30 years for civil works, and 7–8 years for mechanical and electrical works), lowers the annual debt service, even if interest rates are high.
4. The discipline required to access private financing leads to performance improvement: readily available funding from capital markets for borrowers with good governance, uninterrupted loan repayment history, and bankable credentials (credit rating and debt service coverage ratio of at least 1.3) and a transparent publication of annual audited accounts.
5. The fact that local currency borrowing eliminate the need for a GoK guarantee or the purchasing of a forex hedge to mitigate the currency risk that donors demand as a result of their creditor status.
6. Legally ring-fenced revenues generated by the borrower to leverage private investment do not add to the national debt because the WSP loans are corporate debt obligations that will be paid from system revenue without recourse to the national or county governments.

Despite the numerous benefits of local currency investment, the cost of borrowing from local commercial financing sources cannot compete with the subsidized interest rates of concessional loans, and so water utility managers are hesitant to take on commercial loans at higher interest rates. This deleterious crowding-out effect is being corrected by donors who are now trying to find ways to blend their concessional financing facilities with commercial financing facilities to reduce the overall cost of capital for borrowers in the sector. The reshaping of these financing options will bring about the symbiotic operability between concessional and commercial finance that is needed to accelerate the development of water and sanitation infrastructure in Kenya toward achieving Vision 2030 and SDG 6.

To achieve this, private nonprofit financial intermediaries could be made eligible to receive those concessional funds and direct them on a low-interest-rate basis toward blending for qualifying public utility investment projects. The long-term benefit of blending is that it weans water utilities away from free subsidies, crowds in repayable financing, and moves water utilities more toward assuming 100% responsibility for financing their operations and for infrastructure financing from their own revenue.

By comparison with other investments, the KPWF product appeals to institutional investors because:

1. The pooled water and sanitation loans are a new creditworthy debt product with SDG/ sustainability relevance for diversifying lenders' investment debt portfolios.
2. The bonds have been deemed eligible for "green" bond status, which could help to attract a larger buying universe, and particularly impact investors.
3. The debt is structured in a way that is acceptable for purchase by long-term investors, such as pension funds and insurance companies
4. The local capital market and Kenya's water regulator, international development agencies, and development partners are interested in enhancing the mix of available financing options for the sector to achieve the country's Vision 2030.

Donor agencies are interesting in partnering the WFF mission in Kenya because of KPWF:

1. Contributes to the achievement of SDG 6 and other SDG goals.
2. Leverages donor financial support with private sector investments, making the donor financial support more catalytical and "sector transformational."
3. Improves the management and creditworthiness of water & sanitation utilities.
4. Develops local capital markets to give creditworthy water & sanitation utilities "sustainable" access to finance as long as they take steps to raise the revenue needed to support local capital market debt financing, eliminating the currency risk.

15.7 Further development of local capital market financing

The development of domestic capital markets enhances the real potential for access to local currency financing for infrastructure investments, not only for the water and sanitation sector but also for many other government sectors (roads, health care, housing, airports, etc.). Commercial borrowing opportunities are basically limited only by the creditworthiness of the sectoral enterprises and their ability to service the debt. To achieve the required level of creditworthiness, the leadership, and the managerial and operational capacity of, in this case, the water companies, a clear regulatory framework is needed that is backed by a strong water regulator and supplemented by ongoing capacity development (with external funding and assistance). All this is necessary to establish a well-functioning water and sanitation sector that can finance its operations and investments from the tariffs received for their services. Access to capital markets, therefore, gives the creditworthy water companies the opportunity to provide more people with access to better and safer water and sanitation services. The rigor required to access commercial borrowing also encourages the water companies to become better operators and more creditworthy.

Those water utilities that cannot become totally creditworthy in order to access commercial financing for their infrastructure development need the national government to provide them with direct financial assistance that helps them mature and develop basic capabilities for sound financial management. To achieve this, the government needs to develop alternative financing programs that gradually weans the non-creditworthy WSPs off exclusive dependence on free funding to a situation where grants and commercial funding go together. These programs should always include a component targeting the development of a loan with minimal interest rates to teach small and intermediate, noncreditworthy borrowers what they need to do in order to access the commercial financing sector for some of their needs.

There are no appropriation risks for, or partial financing required from, national or local government budgets associated with borrowing from domestic capital markets unless the government decides to become involved in this way. Encouraging local currency borrowing brings some budget relief for government authorities because no government guarantees are needed and there is, therefore, no contingent liability to cover foreign-exchange risk. The establishment of local capital markets generates enormous benefits for governments who are attempting to finance development internally. It also goes hand in hand with the transition from a low-income to a middle-income country. This is a development, which all IFIs and donors should help develop, as the Dutch are doing in the WFF/KPWF initiative.

The transition for governments and for water companies toward private financing using their domestic capital markets for their water and sanitation infrastructure needs has been seen in many countries. Obviously, this requires the presence of a domestic capital market that operates adequately and the investors have to be actively approached and encouraged to be interested in investment in the water and sanitation sector.

Active leadership and support from the Ministry of Finance are required for this new financing route for social infrastructure in general, building on the successful experiences in many countries which have made this transition a success for water and sanitation infrastructure as well as for other social infrastructure, including energy, housing, and transport. Acceptance followed by leadership is also required from the Ministry in charge of water and sanitation.

The regulator must have unequivocal support to maintain the full-cost recovery principle in setting the water tariffs. In Kenya, WASREB has the responsibility under the Water Act 2016 to evaluate and recommend regular and automatic tariff adjustments to the water service provider and approve the imposition of such tariffs in line with consumer protection standards. A tariff adjustment application should be initiated by the WSP, six months prior to the expiry of the prevailing tariff. Citizens and county governments are key stakeholders and the WSPs tariff adjustment proposals must be shared with and agreed upon by these parties before submission to WASREB.

The county government shall establish water services providers and in doing so, shall comply with the standards of commercial viability set out by WASREB. The county government needs to ensure that the water service provider fully implements the tariffs gazetted by the Regulator WASREB. These tariffs shall be sufficient to cover the reasonable full life cycle cost of providing the services, maintaining the facilities, and to meet any other costs such as providing new facilities, debt servicing on new loans related to the improving operations, and extending coverage, asset renewal, and development.

In summary, collective and strong support is needed by the Ministry of Finance (National Treasury in Kenya), the Ministry in charge of water and sanitation (Ministry of Water, Sanitation, and Irrigation in Kenya), the water regulator (WASREB in Kenya), and the water companies and the county governments, in order to achieve domestic capital market financing for water and sanitation infrastructure. When mechanisms that allow for such an additional financing stream are in place it will be possible to implement the Kenya Water Vision 2030 and to meet the SDG 6 targets of universal coverage by 2030 for all Kenyans.

Annex Detailed steps in transaction-related activities to support issuing the first pool bond

As discussed in Section 15.5.3, there are a number of key steps required to begin the extensive preparations for issuing a pooled bond transaction. First and foremost, KPWF's board will, on the basis of the Investment Committee recommendations, move to formally approve the creditworthy WSP borrowers that will be included in the first bond transaction.

Other staff activities will focus on:

1. Initiating a procurement process to select a Tier 1 Kenyan bank to hold the Trust's account.
2. Finalizing agreements with the trustee and the servicer.
3. Creating a variety of accounts at the Trust bank consistent with the trust deed, identifying their purpose and establishing controls on access to those accounts.
4. Commencing discussions with the relevant banks or banks that hold each WSP's general water revenue accounts to set up an escrow repayment account and help them to activate a cash-sweep mechanism for debt service repayment.
5. Engaging with any third-party lenders to the WSP to establish the Trust's debt obligation as the senior obligation and, if that is not possible, establishing an agreement (intercreditor) to make the Trust obligation *pari passu* with those other debt obligations. This may be the most difficult challenge to resolve but only one potential borrower's commercial bank is expected to be a problem.
6. Continuously reviewing, with WSP management and their local counsel, the form of reports and monitoring obligations to ensure that the WSP has a good understanding of the covenants it enters into and confirming that it is able to comply with the related reporting requirements.

Together with the need for frequent interactions with the WSPs about the CP process, KPWF staff will also need to complete numerous other tasks expeditiously to get CMA's approval completed in time to issue the ABS-listed securities in the bond market. These tasks include:

1. Finalization of the particulars of the financing transaction—amortization schedule, flow of funds, capitalized interest, if any, identification of the cost of issuance, etc.
2. Negotiation and finalization of one or more guarantee agreements (USAID+ SIDA + GuarantCo[6]). In the past, these three guarantors have worked together and shared risk in bond traxsnsactions.
3. Creation of a data room for, among other things, all WSP CPs & signed agreements and the financing structure.
4. Receipt of the third-party international auditors' financial accounting review reports on 3 years of audited financials, and up to 6 months of unaudited financials for each borrower prior to financing.
5. Initiation and final receipt of a tax opinion.
6. Incorporation of the Kenya Pooled Water Trust 2019 as an SPV issuer.
7. Obtaining an updated domestic shadow rating from the Global Credit Rating Company's (GCR) Public Finance Group for each of the proposed borrowers (not published) and a published domestic rating from GCR's Structure Finance Group on the ABS pooled bond structure.
8. Confirmation of the availability of funds to capitalize the DSRA.
9. Final review of all transaction & legal documentation by the SPV Trustee: MTC (Information Memorandum (IM)); bond trust deed (which sets out the flow of funds and legal recourse mechanisms in the case of late payments and/or default); agency agreement, placing agreement. servicing agreement; account bank agreement; loan agreement(s); fee agreements with the trustee, financial advisory firm, securitization manager, registrar and accountants. These documents have been shared and discussed with WSP management and their outside legal counsels.
10. Bond counsel final review and legal opinion.

When all the preparations above have been completed, the placing agent can submit KPWF's structure to the CMA for their approval to issue pooled "green" bonds on the Kenya capital market using an SPV Trust. On the basis of numerous meetings with CMA officials over the past few years and the opinion of the placing agent and bond counsel, the expectation is that the CMA review and approval process should take approximately 6 weeks.

[6] This involves providing partial financial guarantees to support projects and companies in order to raise debt financing for the development of infrastructure; see Guarantco's website: https://guarantco.com/who-we-are/.

As the CMA process proceeds, other preparations for the next phase to get to financing will be accelerated but only once it is clear that CMA appears likely to approve the Trust issuance. These steps include:

1. Marketing, education of investors, and building of the book of interested investors.
2. Initiation of the development of servicer reconciliation & monitoring processes (8 weeks are required for development, testing, and implementation).
3. WSP tariff Implementation.
4. CMA approval received & printing of pricing supplement.
5. Bond offer opens.
6. Green Bond certification received.
7. Servicer application finalized.
8. Offer closes.
9. Listing approval from NSE.
10. Tax exemption received on bond structure.
11. Closing of the bond issuance.
12. Funding of WSPs loans from bond proceeds & activation of loan proceeds account.

References

(GLAAS), U.-W.g., 2019. National systems to support drinking-water, sanitation and hygiene: global status report 2019. Geneva: World Health Organisation.

Directoraat Generaal voor Internationale Samenwerking (DGIS, the international cooperation arm of the Netherlands Ministry of Foreign Affairs), 2016. WASH Strategy 2016–2030. The Hague: DGIS.

Government of the Republic of Kenya (GoK), 2016. Water Act. Nairobi: GoK.

KenyaVision2030, V.D., 2018. Kenya Vision 2030 Popular version and sessional paper nr 10. Nairobi: Kenya Vision 2030.

WASREB, 2019. Impact - a performance report of Kenya's Water Services Sector - 2018/19. Nairobi: WASREB.

CHAPTER SIXTEEN

Investing in catchment protection: The Water Fund model

Bert De Bièvre[a] and Lorena Coronel[b]
[a]Fondo para la Protección del Agua (FONAG), Quito, Ecuador
[b]AquaNature, Quito, Ecuador

16.1 Introduction

Water Funds (WF) are to date one of the most widely replicated water conservation strategies, with 41 operating worldwide and more than 35 in development (The Nature Conservancy, 2020a). This ability to replicate is due to the fact that they have managed to adapt to local realities, and due to the evolution of the concept itself.

The initial concept of a WF was a stable, transparent, and long-term financial mechanism, with a special focus on obtaining contributions of the main users and stakeholders in a specific basin who derived benefits from improved water management. Some of them, like FONAG, were linked to strategies for the financial sustainability of protected areas (Jepson et al., 2018). Later on, WFs adapted and, while maintaining some focus on financing, they included governance and management as decision-making elements for their design and implementation. Despite the similarities, an evolution and adaptation of the original concept have been seen, moving from a financing approach to an approach of creating organizations that promote water security. Bremer et al. (2016) point out that WFs share three primary organizational components: funding, governance, and management. Each individual fund has a different approach to these components.

This chapter analyzes the case of FONAG as a demonstration of the importance of financial robustness in order to achieve an impact on the governance, catchment protection, and achievement of its main objective: improving water security.

16.2 Principles and rationale

At the very origin of WFs is the need to establish a relationship between water sources and their users. This relationship is consolidated

Financing Investment in Water Security: Recent Developments and Perspectives.
DOI: https://doi.org/10.1016/B978-0-12-822847-0.00008-9

through the creation of a vehicle, in some cases a trust, that allows efforts to be united to solve a common problem around water. The *co-responsibility* around water makes users contribute to the preservation, restoration, and maintenance of their source water catchments, under the assumption that this will maintain the water availability in the long term.

The problems around water are diverse, as well as the possible solutions. The protection and restoration of hydrological services requires an investment in the *nature-based solutions (NbS)*[1] that promote not only strict conservation but countless interventions in catchments that maintain or increase the capacity of ecosystems. WFs are based on the use of NbS as an integral part of water management and became the mechanism to channel the co-responsibility of the users to connect with nature that provides the ecosystem services.

The original *financial model* of a WF was based on a trust, with financial capital that would be added to, by a recurring, sustainable mechanism. The trust would be public or private, depending on the current legal framework. However, in many cases, legal complications were encountered to add public funding to a private trust, and more recent WFs in Ecuador, for example, are now managed by a public fiduciary[2]. A trust is a legal act, in which one or more constituents (water users) irrevocably transfer assets and contributes to an autonomous endowment. A third party, the fiduciary or trust company, will act as its legal representative and will fulfill the specific purposes established in the constitution contract of the trust, either in favor of the constituent itself or of a third party referred to as a beneficiary. The main characteristics of a trust are: fulfilling a specific purpose, having an autonomous endowment (long term, protected from seizure and irrevocable) and allowing the leverage of other resources. This financial vehicle has shown to be appropriate because it enables several water users with a common mission, source water protection, to join efforts. In the cases where a considerable capital has accumulated in the trust, it has been clearly shown that the trust provides a solid basis, resilient against economic and political shocks, and in general good stability and sustainability.

[1] Nature-based Solutions (NbS) are defined by IUCN as "actions to protect, sustainably manage, and restore natural or modified ecosystems, that address societal challenges effectively and adaptively, simultaneously providing human well-being and biodiversity benefits."

[2] Whether public and private funds can be combined into the same trust depends on the specific legal framework. For example, in Ecuador, a public fiduciary can manage private and public funds, but public funds must be managed by a public fiduciary.

On the other hand, there are several examples of WFs that actually do not have a trust. In places where the trusts are not a legal possibility, there are still financial contributions of several parties, but the creation of the WF can be established as a corporation, a civil association, among others. The possibilities and limits arise from specific legal framework in place. Examples include corporations like Vivo Cuenca in Colombia; civil society associations like Agua Capital in Mexico City, a foundation like FUNCAGUA in Guatemala. They operate with the funding they are able to obtain and channel this directly into different projects. In small funds, the administrative cost of the fiduciary limits the possibility to establish a trust, some use bank accounts.

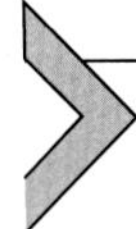

16.3 Application to catchment protection

16.3.1 The case of Fondo para la Protección del Agua (Quito's WF—FONAG)

After the creation of several protected areas in Ecuador in the 1990s, such as Antisana Ecological Reserve, Cayambe Coca National Park, and Cotopaxi National Park, it was estimated that 80% of the water supply for the city of Quito had its origin in these protected areas or their buffer zones. The importance of sound management and the financial challenges of doing so were the starting point of establishment of the first WF. In 1997, the Antisana Foundation and The Nature Conservancy (TNC) proposed Quito's public water utility to create a joint fund that would include voluntary financial participation of water users. Key elements like political support, technical leadership, and a high-level champion (the mayor at the time) were aligned. A major challenge to overcome was the consolidation of funding by public institutions that by law could not invest in private financial mechanisms. This obstacle was eliminated as of 1999, when a stock market reform allowed public funds to be contributed to private Trusts and therefore the Public Water Utility and Public Electric company were allowed to allocate resources to this autonomous financial mechanism.

Quito's WF, FONAG, was established in 2000, by the Public Water Utility of the city of Quito, and TNC, as a mercantile trust with a duration of 80 years. Soon, other public and private constituents joined: Quito Electric Company (a public hydropower and electricity distribution company), Cervecería Nacional (a beer brewer), and Tesalia Springs Co (a mineral water and other beverages bottler), as well as a national NGO. By 2007, there were six constituents, each of them sitting on the Board of the Trust, the highest decision-making body. The legal representative is a private fiduciary company, and under Ecuadorian law, FONAG reports to a "Superintendency of Companies," as do the majority of private companies in the country.

Contributions of the constituents to the trust are those they committed to upon foundation of or adherence to the fund. The initial capital endowment was USD 20,000 from the water utility and USD 1 000 from TNC. The power company committed to and contributes USD 45 000 annually, while Cervecería Nacional and the Tesalia Springs Co. USD 6000 and USD 7000 annually, respectively. In addition, the water utility committed to an annual contribution derived from a share of its income from water and sewage fees. This was initially set at 1%. As a Public Municipal Company, this contribution was formalized in 2007 in a Municipal Ordinance, which also established the role of FONAG as responsible for source water protection for the Metropolitan District of Quito. The ordinance also established a gradual increase of the percentage of the utility's income contributed, from 1% eventually reaching 2% in 2011. It has remained at this level since then. All of these contributions strictly capitalized the endowment until 2011. The funding for implementation of projects was derived only from financial revenues on the still-limited endowment, and leveraged resources. In 2011, a reform to the trust contract established that up to 30% of new contributions could go to the direct implementation of projects, while at least 70% should continue to capitalize on the endowment.

In addition, other contributions increased the scope of assets held by the trust. TNC contributed a 900 ha private reserve it owned to the trust in 2016, making FONAG a landowner. Further, the Electrical Company made special contributions totaling almost a million dollars, conditional on the purchase of another 3000 ha property relevant for the protection of the catchment of highest strategic importance for the company.

Over the years, FONAG has leveraged additional financial resources from a variety of sources, notably development partners. This included a multiyear cooperation agreement with USAID allowing field action while local contributions were still exclusively dedicated to the endowment. Also, there were projects financed by the French Institute for Overseas Development Research IRD, the German cooperation GIZ, the Inter-American Development Bank IDB, among others. In more recent years, FONAG has managed other cooperating entities' contracts, and became an implementing partner of a major REDD+ project managed by the United Nations Development Programme (UNDP) for the Ecuadorian government.

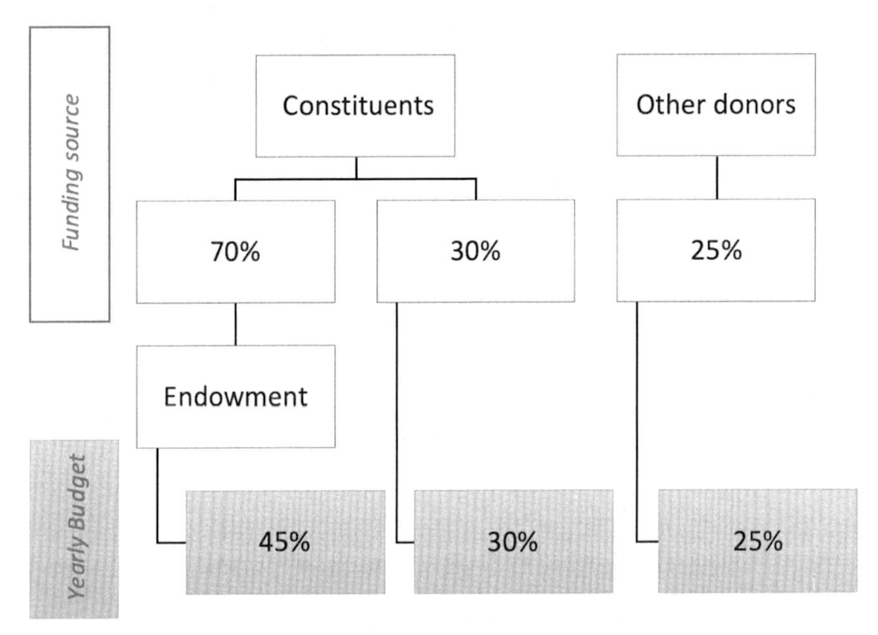

Figure 16.1 FONAG's yearly implementation budget (all actions in catchments, administration, salaries). *(Source: authors).*

However, the main financial motor of FONAG is without doubt the 2% of water fees contributed by the water utility, which is responsible for the 87% share of the utility in the endowment of about USD 21 million as of 2020. As the power company is also a public entity, the share of the public constituents in the endowment is 96%.

A yearly implementation budget of FONAG is currently around USD 2.5 million, with income from financial revenues of the endowment accounting for approximately 45%, with the 30%–70% rule of new contributions accounting for 30%, and leveraged income from other donors for another 25%. The flows of funding and how they provide a current annual budget are given in Fig. 16.1.

Financial revenues from the endowment have benefited from the first 20 years of Ecuador being a "dollarized" country. Without an own currency, the endowment has not suffered from any devaluation, while benefiting from relatively high-interest rates in the country. As for financial regulations for trusts with a majority of public contributions, most of the endowment is invested in Ecuadorian internal debt, while a smaller part is invested in shares of the major national companies. The average performance of the investments in the last few years has been over 8%.

In terms of the focus of activities, in the early years, there was a strong emphasis on biodiversity conservation in combination with source water protection. Actually, in the startup phase, the "Condor Bioreserve" concept was used, which consisted of efforts to consolidate a large protected area by managing Antisana and Cayambe Coca National Parks and their buffer areas in an integrated way. However, the trust was created with the clear mission of "Conserving and recovering the water sources for the Metropolitan District of Quito." After 20 years, FONAG now covers the whole of source water areas, most of which are situated in the *páramo* ecosystem, the high altitude tropical grass- and shrublands. FONAG applies a variety of strategies, detailed in the following paragraphs.

FONAG manages "own" protected areas, i.e., areas owned by the water utility since 2011, or by FONAG itself since 2016. They now total 20 000 ha with 15 *páramo* rangers in charge of surveillance and monitoring.

On private and community-owned land, FONAG pursues the negotiation of conservation agreements. Community *páramo* rangers (currently 8) are on FONAG's payrole. Additionally, it is able to offer, in a flexible way, compensation in the context of such conservation agreements. Examples include the conversion of productive activities that put a burden on water conservation toward less environmentally impactful activities. For example, this may be conversion from livestock farming to ecotourism. More recently, this may also include improving water security for these communities or owners themselves. This part of the portfolio of interventions could be considered a Payment for Ecosystem Services (PES) scheme, since it connects the provision of a service, i.e., protection of sensitive source areas, with a compensation that can take different forms. However, FONAG never provides cash payments in this context.

FONAG restores ecosystems, mainly *páramo*s, and promotes *active* restoration in heavily degraded areas, and drained wetlands, reforestation in Andean forest, but also a lot of *passive* restoration, i.e., effective elimination of pressure factors and threats, such as reducing cattle and feral dogs, enforcement of the ban on motorized sports in the *páramo*, and most importantly fire prevention and *páramo* fire combat with its own rangers who are equipped and trained as wildfire fire brigades. Although natural fires in páramo do occur, burning is mostly caused by hunters, tourists, and communities looking for expansion of grazing land, increasing the burning frequency beyond ecosystem recovery capacity. Passive restoration, seen as natural regeneration when these pressures are reduced, is by far recognized as the most cost-effective measure. FONAG normally does not intervene in hydraulic infrastructure, but in the case of extreme degradation, physical erosion control works can be included in the intervention package, and rural sanitation can be funded where its absence jeopardizes the quality of water sources downstream.

FONAG has run a large environmental education program since its very beginning, with rural schools in source water areas, targeting children, but nowadays also their teachers, i.e., the educational system itself, in an extensive awareness-raising effort. It also coordinates with other entities on campaigns for the responsible use of water.

Through the scientific station "Water and *paramo*," and extensive hydrometeorological monitoring, FONAG actively promotes, partially funds, and orients relevant research that feeds into its own decision making, e.g. optimization of conservation and restoration projects, as well as into decision making of other stakeholders in the catchments, through making data and information products easily available in information systems.

Finally, FONAG devotes a lot of effort to monitoring the impact of the interventions. Impact monitoring focuses on the quantification of the benefits that contribute to the mission, notably in terms of water quality and quantity. The impact monitoring provides feedback on the design of the portfolio and as an input for return-on-investment studies.

An overview of relevant indicators of FONAG as of 2020 is given in Table 16.1. Over its 20-year history, FONAG has invested in capacity strengthening using leveraged funding sources as well as its own resources. Continuous education is provided to its technical staff, the *páramo* rangers and a very close relationship is maintained with researchers in different fields that are important for the effectiveness of the portfolio of interventions in the catchments. The relevant stakeholders express an overall very good perception of the performance of FONAG investments, which contributes to the trust of current and new contributors. This trust is further reinforced and consolidated by the monitoring of impact of interventions, which allows for the quantification of benefits, and informs studies on the return on investment. This trust is probably the overall basis of financial sustainability when all is said and done.

Furthermore, over time, FONAG has adapted to different socio-cultural and political conditions, including those that oppose the commodification of natural resources (Kaufmann, 2014). Remarkably, Ecuador passed new legislation in the period 2007–2014, with notable importance. This included a new constitution that recognizes the rights of nature, a Water Resources Law that establishes catchment councils, and other environmental legislation that prohibits commercialization of ecosystem services. By sticking

Table 16.1 FONAG: key statistics documenting achievements as of 2020.
Financial information

Original constituents of trust	2
Constituents adhered to trust	4
Value in endowment 2020	USD 21.4 M
Annual budget 2020	USD 2.7 M
Average annual financial return on endowment	8 %
Implementation	
Own conservation areas	19 870 ha
Area under formal conservation agreements	16 734 ha
Restored area	21 068 ha
Impact Monitoring sites	3
Signed conservation agreements	20
Persons that have participated in education and awareness raising processes	49,058
Information platforms	2
Ongoing projects	25
Number of meteorological stations operating	33
Number of hydrological stations operating	36
Staff	
Technical staff	21
Páramo rangers	20
Educators	7
Communicators	2
Researchers/impact monitoring	7
Logistics	3
Administrative	6

Source: authors.

consequently to its mission, FONAG kept strengthening its role within the changing context. According to Joslin and Jepson (2018), the WF somehow bypasses the government administrative system, but the state itself may use the arrangement as a platform to exert power within a territory.

16.3.2 Water Funds in Latin America: state of play

Currently, there are 30 WFs in Latin America (Annex 16.1) and more than 20 under development globally[3]. According to the methodology of the Latin American WFs Partnership, FONAG is the only one to have reached

[3] The Latinamerican experience has been disseminated globally. There are 13 water funds in North America, there are specific examples such as Nairobi and Cape Town Water Funds in Africa. There are also developments in Asia, such as the Longwu and Qiandao Lake & Xin'an River Basin Water Funds, but these are currently in feasibility and negotiations phases.

Annex 16.1 Operating Water Funds in Latin America.

Country	Water Fund	Location
Mexico	Fondo de Agua Metropolitano de Monterrey (FAMM) a	Monterrey
	Agua Capital	Mexico
	Fondo Semilla de Agua b	Chiapas
	Fondo de Agua de Guanajuato: Cauce Bajío.	Guanajuato
Guatemala	Fondo de Agua de la ciudad de Guatemala (FUNCAGUA)[a]	Guatemala
Costa Rica	Agua Tica[a]	San José
República Dominicana	Fondo de Agua Yaque del Norte[a]	Yaque del Norte
	Fondo de Agua Santo Domingo[a]	Santo Domingo
Colombia	Fondo de Agua de Cartagena[a]	Cartagena
	Fondo de Agua de Santa Marta y Ciénaga[a]	Santa Marta
	Alianza Bio Cuenca[a]	Cucutá
	Cuenca Verde[a]	Medellín
	Fondo de Agua Bogotá Región[a]	Bogotá
	Fundación Fondo Agua por la Vida y la Sostenibilidad[a]	Valle del Cauca
	Madre Agua[a]	Calí
	Vivo Cuenca[c]	Chinchiná
Ecuador	Fondo para la protección del Agua (FONAG)[a]	Quito
	Fondo del Agua para la conservación del río Paute (FONAPA)[a]	Paute
	Fondo de Agua de Guayaquil para la Conservación de la Cuenca del Río Daule (FONDAGUA)[a]	Daule
	Fondo de Páramos Tungurahua y Lucha Contra la Pobreza[a]	Tungurahua
	Fondo Regional del Agua (FORAGUA)[d]	Región sur del Ecuador
Perú	Fondo Regional del Agua (FORASAN)[a]	Cuenca Chira Piura
	Fondo de Agua para Lima y Callao (Aquafondo)[a]	Lima
	Fondo del Agua del Río Quiroz Chira[e]	Quiroz Chira
Brasil	Programa Reflorestar[a]	Espiritu Santo
	Produtores de Água e Floresta Guandu-RJ[a]	Río de Janeiro
	Fondo de Agua de Sao Paulo[a]	Sao Paulo
	Produtor de Água do Rio Camboriú[a]	Camboriú

(continued on next page)

Annex 16.1 Operating Water Funds in Latin America—cont'd

Country	Water Fund	Location
	Projeto Produtor de Água no Pipiripau-DF[b]	Formosa/GO
Chile	Fondo de Agua Santiago-Maipo[a]	Santiago
Argentina	Fondo de Agua de Mendoza[a]	Mendoza

Source: Latin American Water Funds Partnership (2020). Note: This Water Funds have the support of the Latin American Water Funds Partnership (LAWFP).Source: Bremer et al. (2016).Source: Corpocaldas (2019).Source: Foragua (2020).Source: Contreras (2017).
[a] Source: Latin American Water Funds Partnership (2020). Note: This Water Funds have the support of the Latin American Water Funds Partnership (LAWFP).
[b] Source: Bremer et al. (2016).
[c] Source: Corpocaldas (2019).
[d] Source: Foragua (2020).
[e] Source: Contreras (2017).

maturity phase[4] and this largely responds to its financial robustness that has allowed it to become a technical entity with a wide spectrum of solutions to catchment protection. WFs will continue to be replicated. As Kauffman (2014) points out, they were easily adapted to different sociocultural and political conditions and provided an innovative model for providing sustainable financing for watershed conservation. However, it is necessary to go beyond their own establishment to achieve stable financial mechanisms that allow them to reach their objectives and contribute significantly to the governance and management of water.

16.4 Economic benefits and return on investment

WFs have started operating in an environment where they were considered generally as part of social and environmental responsibility management. In some cases, including FONAG, studies on willingness to pay of final users contributed to the design of the funding scheme, but their influence on final agreements was minor, since agreements were with large institutional users and not with the final users in the general public. At the early stage, anything resembling results-based payments was not an option. However, recently, a few studies became available that estimate the return on the investment of contributors (Kroeger et al., 2019; Ochoa-Tocachi et al., 2018). These studies have basically two components: the quantification

[4] The Maturity Phase is a determination that assures the long-term viability of the water fund to create significant and lasting impact that positively contributes to water security(TNC 2020).

of hydrological benefits of the fund's action, and the economic analysis of benefits and costs for those who contribute.

The quantification of hydrological benefits of nature-based solutions remains a huge challenge. While many WFs devote significant efforts to the task (Bremer et al., 2016), few have been able to describe the impact of individual interventions or a combination of them, on water quality and quantity. Many hydrological modeling studies are performed in the stages of feasibility studies and prioritization of interventions and their geographic locations, but models in general clearly fall short when it comes to quantifying impact on specific relevant parameters. Therefore, FONAG and others set up impact monitoring sites, within a strategy to identify impacts, as much as possible with a "Before–After" combined with "Control–Intervention" approach (Bremer et al., 2016). "Control–Intervention" has been promoted by a group of partners under the Hydrological Monitoring of Andean Ecosystems Initiative (iMHEA by its acronym in Spanish) (Ochoa-Tocachi et al., 2018), as the most rapid way to generate information on how different interventions improve the hydrological regime, but the combination with monitoring "Before–After" interventions makes this information more robust. FONAG participates in this kind of regional Andean efforts to quantify impacts of recurring interventions. Apart from providing feedback on the optimal design of interventions, the data provided by this type of monitoring feeds into a new generation of hydrological modeling, calibrated by the monitored ground-truth, and subsequently into return-on-investment studies. Overcoming the bottleneck of the sound description of hydrological benefits is key to the feasibility of return-on-investment studies. Cobenefits such as biodiversity conservation and carbon benefits could be added to this type of analysis, although they require different strategies than the primary water benefits, and considering only those would in any case be a conservative approach on benefits.

Furthermore, a detailed analysis of what is considered a benefit by the contributor, and how these generate financial return, is the other pillar of such studies. Benefits can be of diverse nature (e.g., sediment load reduction, overall flow increase, dry season flow increase, improvement in specific chemical water quality parameters, to name a few), and not all of them necessarily reduce costs or need for grey infrastructure investments. In its first comprehensive study of the financial return on investment the water utility makes in FONAG, covering the complete set of contributing catchments to water intakes for Quito, the return on Investment was estimated at USD 1.7 for every dollar invested (Ochoa-Tocachi et al., 2018).

The technical feasibility of these kinds of studies is currently causing a shift from a "social and environmental responsibility" motivation of contributors, toward a more economically-motivated one, although it has to be said that this trend has not been observed within the private partners of the fund. As WFs strengthen their technical capacity and can demonstrate effective interventions, we foresee their sustainability will be increasingly based on hydrological performance in the future. Therefore, it cannot be stressed enough that the basis of sustainable finance for WFs is technical soundness, that builds trust among the contributors.

16.5 Trends: cobenefits, water funds beyond source water protection?

In almost every case, apart from the direct hydrological benefits for constituents, interventions of WFs generate other cobenefits. Biodiversity is probably the most significant one, and also the most visible one in the startup or early years of many initiatives, given that many initiatives receive start-up support from environmental organizations such as TNC. Many funds have been able to attract funding from sources with an interest in supporting biodiversity. More recently, carbon and climate finance has been gaining more attention. Wherever funds conserve and restore ecosystems, carbon is captured, and in many cases they are able to demonstrate benefits in terms of avoided emissions. Therefore a group of Ecuadorian WFs is already an implementer of a major Green Climate Fund project run by UNDP and the Ministry of Environment. A major study by the Earth Innovation Institute (Earth Innovation Institute, 2019) of the potential of obtaining avoided emissions by the array of implementers of this project, ranked the WFs among the most effective ones. In addition to this contribution to climate change mitigation, WFs also clearly contribute to climate change adaptation, for example when restoring hydrological regulation in wetlands, while this same regulation is being lost with glacier melt in the same catchment.

WFs in general have not yet been able to leverage significant funding from markets for carbon credits. In the specific case of funds like FONAG that are active in the high Andean non-forest ecosystems, knowledge gaps on the carbon dynamics (mostly related to soils in these ecosystems) and the overwhelming emphasis of international carbon markets on forest-related carbon, has been a major obstacle. However, FONAG is now preparing itself for an implementing role of a national carbon footprint compensation mechanism, recently established by the Ecuadorian government. We consider that this mechanism is today the one with the greatest likelihood of attracting significant private funding to WFs. Private partners could obtain carbon neutrality certification, while at the same time contributing to water security and biodiversity conservation, subjects that are highly appreciated in public opinion.

In relation to the Sustainable Development Goals (SDGs), the contribution of WFs to SDG 6 "Clean Water and Sanitation" through source water protection is clear. Moreover, the cobenefits generated contribute to a range of other SDGs such as healthy ecosystems (SDG 15) and climate action (SDG 13). Their contributions to goals on reducing inequality (SDG 10) and sustainable cities and communities (SDG 11) are also recognized (see for example UNEP, 2016), and they are, by definition, an example of Partnerships (SDG 17).

TNC has recently begun promoting a role for WFs in water security in general, including filling the huge gap in access to water supply, sanitation, and hygiene services. This would require major changes in the concept, and in the case of FONAG, a fundamental change in its mission, inscribed in the trust contract.

16.6 Conclusions

Funding is a corner stone for sustainable WFs. A solid funding scheme is built on the commitment and trust of the constituents of the fund, with complementary fundraising strategies, such as with international development cooperation or climate financing. As for the constituents, in an ideal situation, both public and private funding contributes. Based on experience, public funding has proven to be recurrent and long-term. We consider of utmost importance that the WF be clear and sincere on its main mission objective. Cobenefits can join along the way with great potential of increasing financial strength, but should be streamlined in accordance with the main mission.

There are different possible arrangements to create a WF and this adaptability has allowed for the replication of WFs to different legal settings in different countries. There is no unique recipe due to distinct legal contexts, as well as biophysical priorities. The success of each depends, in part, on its institutional architecture and suitability for the context. In the case of FONAG, the solid financial base has contributed to the positive impact in water management and governance.

Monitoring and information generation allow for adequate decision making and effective implementation as well as adaptive management and institutional learning. One very important aspect is impact monitoring to demonstrate the long-term impact and financial returns. This is only possible with rigorous data collection in collaboration with constituents and scientific partners.

The conservation and preservation of natural ecosystems to protect water is so important that it has become "obvious." Many concepts have evolved, like ecosystem services (Millenium ecosystem assessment- 2000), Nature-based Solutions (2006), green infrastructure, ecosystem-based adaptation, and others. But at the end of the road, what prevails is effective interventions in the field, in the catchments.

WFs are powerful mechanisms that contribute significantly to addressing water-related problems, but they cannot solve them all. An important part of the design of a WF is to identify the specific role in facing the challenges of water security.

References

Bremer, L., Auerbach, D., Goldstein, J., Vogl, A., Shemie, D., Kroeger, T., Nelson, J., Benítez, S., Calvache, A., Guimaraes, J., Herron, C., Higgins, J., Klemz, C., León, J., Lozano, J., Moreno, P., Nuñez, F., Veiga, F., Tiepolo, G., 2016. One size does not fit all: natural infrastructure investments within the Latin American Water Fund Partnership. Ecosyst. Services 17, 217–236.

Bremer, L., Vogl, A., Petry, P., De Bievre, B. (Eds.), 2016. Bridging theory and practice in hydrological monitoring in Water Funds, Latinamerican Water Funds Partnership, https://naturalcapitalproject.stanford.edu/publications/bridging-theory-and-practice-monitoring-water-funds, Accessed February 2022.

Contreras, L., 2017. El Fondo Quiroz- Chira. Un mecanismo de gestión para los ecosistemas de Piura, Perú. Programa Bosques Andinos de la Agencia Suiza para el Desarrollo y la Cooperación (COSUDE), Naturaleza y Cultura Internacional. Perú.

Corpocaldas, 2019. Fondos de Agua: estrategias para la seguridad hídrica. http://www.corpocaldas.gov.co/publicaciones/1599/2019-03-27/14-Fondos%20de%20Agua-AndresFelipeBetancourt-VivoCuenca.pdf, Accessed May 2021.

Earth Innovation Institute, 2019, Evaluación del Impacto de políticas públicas destinadas a reducir la deforestación y degradación y acciones destinadas a la gestión sostenible de los bosques en Ecuador, Technical Report, http://proamazonia.org/wp-content/uploads/2020/01/EIIProductoDos-Evaluacio%CC%81n-de-Impacto-Ecuador_P2-min.pdf, Accessed 28 Feb 2022.

FORAGUA, 2020. Fondo Regional del Agua y Fondo Ambiental, http://www.foragua.org, Accessed February 2022.

Joslin, A., Jepson, W., 2018. Territory and authority of water fund payments for ecosystem services in Ecuador's. Andes. Geoforum 91, 10–20.

Kauffman, C., 2014. Financing watershed conservation: Lessons from Ecuador's evolving water trust funds. Agricultural Water Manag. 145, 39–49.

Kroeger, T., Klemz, C., Boucher, T., Fisher, J., Acosta, E., Targa Cavassani, A., Dennedy-Frank, P.J., Garbossa, L., Blainski, E., Comparim Santos, R., Giberti, S., Petry, P., Shemie, D.,

Dacol, K., 2019. Returns on investment in watershed conservation: Application of a best practices analytical framework to the Rio Camboriú Water Producer program, Santa Catarina, Brazil. Sci. Total Environ. 657, 1368–1381. ISSN 0048-9697 https://doi.org/10.1016/j.scitotenv.2018.12.116 .

Latin American Water Funds Partnership, 2020. Water Fund Maps. https://www.fondosdeagua.org/en/the-water-funds/water-fund-maps/, Accessed February 2022.

Ochoa-Tocachi, B., Buytaert, W., Antiporta, J., Acosta, L., Bardales, J., Célleri, R., Crespo, P., Fuentes, P., Gil-Ríos, J., Guallpa, M., Llerena, C., Olaya, D., Pardo, P., Rojas, G., Villacís, M., Villazón, M., Viñas, P., De Bièvre, B., 2018. High-resolution hydrometeorological data from a network of headwater catchments in the tropical Andes.

Ochoa-Tocachi, B., Buytaert, W., Ochoa-Tocachi, E., Torres, S., Vera, A., Osorio, R., De Bièvre, B., 2018. Calculating the return on investment of nature-based solutions for water security. American Geophysical Union Fall Meeting 2018, 12–16 December 2018.

The Nature Conservancy, 2020a. Investing in nature to improve water security. Fact Sheet. https://waterfundstoolbox.org/library.

The Nature Conservancy, 2020b. Toolbox: project cycle overview. https://waterfundstoolbox.org/project-cycle.

UNEP, 2016. Sustainable Development in Practice, United Nations Environment Programme, Panamá.

CHAPTER SEVENTEEN

Leveraging private finance for landscape-level impact: the growing role for bankable nature solutions

Dean Muruven Nalandhren
Global Policy Lead Freshwater, WWF, Switzerland

17.1 Introduction

Water is the world's most precious resource, but it is invariably undervalued relative to its wide range of uses and benefits. Globally, water resources are not managed in a way that reflects their full values—and this pattern of neglect has serious consequences. The stakes are high. Historically, poor management of water has contributed to the decline of civilizations and continues to threaten the vitality and viability of communities, cities, and countries today (Barbier, 2019). Our freshwater ecosystems and the species that inhabit them are under increasing threat. Approximately 35% of the world's wetlands were lost between 1970 and 2015 and the rate of loss has been accelerating since 2000 (Gardner and Finlayson, 2018). WWF's 2020 Living Planet Report noted that the populations of various freshwater species monitored throughout the world have declined by 84% on average since 1970 (Almond et al., 2020). Almost one in three freshwater species are threatened with extinction.

Decades of mismanagement, underinvestment, and weak policies—often due to a low appreciation of the diverse values of healthy river systems or the proper management of groundwater—have placed water resources under increasing strain. As a result, it is estimated that by 2025, 60% of the world's population will live in regions potentially experiencing moderate to extreme water resource-based vulnerability (WRV) (Kulshreshtha, 1998).

With climate change and demographic pressures set to worsen, water crises will jeopardize food production, put business value at risk from water shortages and flooding, pollution, and degraded natural ecosystems. It is time for novel solutions.

Financing Investment in Water Security: Recent Developments and Perspectives.
DOI: https://doi.org/10.1016/B978-0-12-822847-0.00018-1

Every crisis presents opportunities. The world's worsening water crisis provides opportunities for the private sector (which includes corporates, private and institutional financiers, impact investors, and others) to create financial value, while also supporting more sustainable management of water. A review of recent estimates of the costs of achieving SDG 6 concludes that the global costs exceeds USD 1 trillion annually, while only a fraction of this amount is currently being spent to close the service gap (Hutton, this volume). Indeed, private sector involvement is essential in order to finance this gap.

A range of policies, institutional mechanisms, and approaches are available to make private sector participation financially attractive, such as blended finance, along with opportunities to reduce costs and risks, and increase revenues. It is unlikely that the gap in water-related investments can be filled without the involvement of return-generating projects and companies. This chapter will refer to these investments in, and loans to, projects and companies as "bankable projects." For the purposes of this chapter, bankable projects are projects which have the ability to create positive environmental returns that lead to improved biodiversity and climate mitigation and/or adaptation, while also being attractive for financial institutions to invest in (WWF, 2020).

A wide variety of projects exist that offer financial returns and positively impact freshwater ecosystems. While projects need to be realized individually, it is important they happen as part of a broader development effort, including strengthened policies and governance. This chapter argues that there are benefits in assessing and planning investments at landscape level. It illustrates the requirements to combine project-level analysis with plans and financing strategies at the landscape level of aggregation. This will significantly enhance their overall impact on freshwater ecosystems.

17.1.1 The opportunity for bankable water solutions

For companies and private sector investors, who have become increasingly aware of the scale of water risk in recent years as alarm bells have been repeatedly rung by a range of organizations (World Bank, 2016; World Economic Forum, 2019; OECD, 2019; WWF, 2018), this dire scenario presents an opportunity to improve sustainable water resource management while also generating solid financial returns. There are three main factors behind this:

- Governments, businesses, and individuals are increasingly aware of the value of water, which is driving more stringent regulation and potentially greater willingness to pay for water services.

- As competition for water resources intensifies and supplies become less secure, users are more receptive to projects that can reduce the quantity of water they consume and improve the quality of the wastewater they discharge. These projects at the same time boost the resilience of their business case and ensure the companies can retain their licenses to operate, especially in water-stressed areas.
- Companies are starting to appreciate the benefits of water-related risk mitigation. In a world where floods and droughts are becoming more common and severe, this will include initiatives that ensure sufficient water to operate facilities or meet citizens' needs. But mitigation will also limit reputational risks (e.g., a public backlash from polluting or overusing freshwater sources) and regulatory risks (e.g., fines for noncompliance as regulations tighten).

To illustrate the scale of the opportunity, a 2020 report from UNCTAD estimated that an additional investment of USD 260 billion a year is needed for wastewater treatment, water supply plants, and supply networks alone. Some of these projects and other essential investments are not bankable propositions and will depend on the backing of the public sector, in particular in emerging and developing economies where affordability issues are more pronounced and where initiatives are designed to meet basic human needs.

However, a significant proportion of the USD 260+ billion investment need is made up of projects with the potential for strong business models and revenue profiles that are attractive to the private sector; for example, industrial wastewater treatment plants or clean technology investments in water-reliant businesses.

17.2 Support for bankable water solutions

While some water projects have clear and strong business cases, others can be more challenging due to inadequate governance, poor pricing, or weak creditworthiness—often in developing countries that lack good water-related infrastructure. Fortunately, the United Nations Sustainable Development Goals (SDGs) —ambitious objectives to end poverty, foster prosperity, and protect the planet by 2030—are creating a momentum for change. Given the size of investment required to realize the SDGs, the private sector is expected set to play a pivotal role. In the water sector, philanthropic donations and grants make up only a fraction of the level of investment needed. To gain further insight into the magnitude of the financing needs, most countries in Asia will need to allocate between 1% and 2% of GDP on water supply and sanitation infrastructure over the period 2015–2030 (Leckie et al., 2021) (see Fernandez, Cardascia and Leflaive, this volume).

Clean, fresh, and accessible water is key to almost all SDGs that companies, governments, and development agencies have committed to tackling within the UN's time frame. However, water resources are limited so we have to manage the land and water resources we have sustainably and for multiple stakeholders and benefits. Consequently, these different stakeholders are ready to provide a range of support for players participating in projects that have a net positive impact on water resource management.

For example, leading governments are providing financial incentives, streamlining investment approval procedures, and developing more effective regulations and penalties that encourage investment in water projects. An interesting example is the case of the Senegal River Basin Development Organization (OMVS), which showed how a specific legal framework can reduce some of the existing risks and pave the way to finding water funding solutions in the transboundary Senegal River Basin (Padt and Sanchez, 2013). In the development community, a variety of organizations—from Development Finance Institutions to foundations and impact investors—offer financial support and technical expertise, including helping businesses to develop project ideas, create strong business cases, and structure financing deals. Examples are the Dutch Fund for Climate and Development (DFCD).

One of the key trends enabling private sector participation is the growth in blended finance as noted in Gietema (this volume). A notable example of the successful use of blended finance is the Partnership for Cleaner Textiles in Bangladesh—PaCT (Huq and Karim, 2017). PaCT is a cooperation between the International Finance Corporation (IFC, part of the World Bank Group), apparel brands, textile manufacturers, international and local NGOs, governments, financiers, and many other stakeholders. Together, they drive investment in cleaner production methods in the local textile industry. In this partnership, the government of Bangladesh provides subsidies to the textile industry, minimizes internal disputes among the exporters, and has withdrawn the withholding and sales taxes (Rahman et al., 2018). These actions allow manufacturers to attract commercial finance in order to invest in their production facilities. This leads to a substantial improvement in the investment's profitability, as well as to a very positive impact on the environment.

17.3 Bankable projects case studies

Bankable water projects come in many shapes and sizes, but they share a common characteristic: they combine environmental sustainability with financial sustainability. Firstly, to be sustainable, bankable freshwater projects must have a positive effect on the environment: some improve the management of water resources or mitigate negative impacts, while others prevent the fragmentation and degradation of watersheds, rivers, and wetlands. Secondly, the design of bankable water projects should consider the whole basin as well as trade-offs and/or synergies with other projects in the basin. Thirdly, projects should generate a sufficiently large positive cash flow, through increased revenues or cost savings, and/or mitigating risks. For (external) funders, such as clients, borrowers, corporates, etc., the bankable project needs to be sufficiently creditworthy or at least have guarantees or backing from a creditworthy partner (e.g., a large buyer or government); the project (possibly an aggregation of many smaller projects) needs to be sufficiently large to merit the required due diligence and resource investment, and the political and regulatory environment should be relatively stable.

So, while bankable projects come with different levels of risks, risk appetites vary depending on the type of financial institution. Working with the right partners can help bring different types of investors and other funders together, which could help offset or address those risks through on-the-ground expertise, guarantees, aid, and more in-depth due diligence (e.g., financial modeling of the whole project lifecycle and a stress test of this model, and the project's legal contract, licenses and permissions, and more) (Hampl et al., 2011).

We have highlighted four different types of water projects below that have the potential to be successful, bankable investments: (1) agricultural and industrial water usage projects; (2) water supply networks and leakage prevention to curb nonrevenue water projects; (3) wastewater treatment plants, and water and resource reuse technologies projects; and (4) restoration of freshwater ecosystems projects. For each of these four types of water projects, we have added a specific case study highlighting the benefits derived from investments and how that can be monetized to translate into financial returns, who reaps those benefits, and how that shapes incentives to act, and the timeframe for return on investments. Although this is not intended to be an exhaustive list, each project type has a proven and positive impact on the environment, a clear revenue model, and can generate robust cash flows.

17.3.1 Agricultural and industrial water usage

The agricultural and industrial sectors are large users and polluters of water, as many of the processes are both extremely water-intensive and require substantial amounts of chemicals. Consequently, these sectors depend heavily on the availability of water of a sufficiently high quality and quantity. For the industrial sector, high water quantities are needed for cooling purposes for instance. A restricted availability of water is, therefore, a cause of concern. For the agricultural sector, both water quantity and quality are essential. The absence of adequate water quantity and quality can directly cause agricultural land to be abandoned and industrial assets to be shut down. Because of these risks, and the related costs of dealing with used water (e.g., discharge of polluted water), many agricultural and industrial facilities can benefit financially from investing in better water management (e.g., improvements in factory-internal management of water flows and engaging in the management of the water resource outside of factories at the basin level).

One example is the use of sustainable practices on farms close to Australia's Great Barrier Reef, which were introduced to curb river pollutants that are significantly damaging the reef (Thorburn et al., 2013). These projects have significant financial and environmental benefits. Indeed, the Great Barrier Reef credit scheme has already successfully attracted global banks such as HSBC, which are paying farmers to improve their practices. These reef credits are sold by farmers or project developers to organizations and companies looking to offset their environmental footprints. Those sales help fund improved land management practices. But instead of removing or avoiding carbon in the atmosphere, reef credits go toward helping improve water quality in this very specific area to protect the reef. As with any offsetting the usual caveats related to monitoring apply. However, a prerequisite is that water allocation needs to be properly managed—otherwise it does not lead to a reduction in water usage. Another example is how textile manufacturers in Turkey are reducing their costs by implementing cleaner production methods (see Büyük Menderes case study later on).

17.3.2 Water supply networks and leakage prevention to curb nonrevenue water

Investing in technology to improve timely asset management, including identifying leaks from water pipelines or building water supply facilities that provide drinkable water for residential use can yield significant benefits and returns, especially where robust water pricing is in place (with benefits also in terms of water and energy savings). In Lisbon, Portugal's capital, EPAL, the city's water company, is using a leakage-monitoring computer program to improve the supply of water to residents (Climate-ADAPT, 2016). It identifies potential leaks by comparing information about water usage in real-time with anticipated usage levels based on average household consumption. From 2005 to 2015, nonrevenue water was cut from 25% to below 15% resulting in accumulated savings of EUR 68 million. To achieve this outcome, the city invested an initial EUR 2 million followed by EUR 500,000 per year over the same period, implying a heathy financial return. Over these 10 years the annual rate of return was a stunning 946%! This kind of leakage control programs is well mainstreamed also across many developing countries but there is still room to replicate and scale.

17.3.3 Restoration of freshwater ecosystems

Improving the condition of rivers, wetlands, and other freshwater ecosystems enhances their natural capacity to mitigate the impact of floods, droughts, and storm surges on communities and cities. They also create value for the wider community, see de Bievre and Coronel (this volume). For instance, since healthier ecosystems are better at filtering pollution, they generate more plentiful and higher-quality water for residential and commercial end-users. Protecting these ecosystems also provide numerous other benefits thanks to the various ecosystem services that they provide. Upkeep costs for the public authorities that manage them and for taxpayers are also reduced.

In 2017, UK utility Anglian Water raised GBP 250 million in a green bond arranged by ING (Anglian Water Services Limited, 2018). The bond will mature in August 2025 with a return to investors of 1.625%. At one point the order book peaked at GBP 800 million with nearly 80 investors participating. To attract investors, all activities financed by the bond are fully transparent, and rigorous M&E and reporting have been put in place. The proceeds will be used to tackle a broad range of ecosystem-related issues, including resilience, drought, and water recycling. For instance, the Chalton Water Recycling Center in Bedfordshire has been transformed into the biggest sand filtration site in Europe. This allows for tertiary treatment of wastewater (removing ammonia) thus ensuring less polluted effluent discharges and safeguarding the nearby Flitwick Moor. The funds raised through this bond are kept separate from the regular budget of Anglian Water to ensure resources are allocated to eligible projects and to enable proper reporting of the impact achieved. It is noteworthy that Eau de Paris considers catchment protection a second core business, in addition to water supply.[1]

In California, privately funded forest restoration programs—the removal of excess vegetation to return the region's forests to a thinner and healthier state—are cutting the risk of wildfires. They are also enabling more water to flow through to reservoirs and farms rather than evaporating from excessive vegetation. By reducing costs, these programs benefit both forest managers and downstream water users. The work is funded by a 'Forest Resilience Bond', which pays out to investors provided projects meet preagreed goals (for other examples related to privately financed large-scale ecological restoration, see David & Johnson, this volume).

Another example comes from the Haringvliet, a now closed-off arm of the Netherlands' river delta, where the local authority has converted an agriculture area into a wetland (Marks et al., 2014). It has reserved a small part for housing and is using the proceeds of property sales to fund ecosystem restoration for the area as a whole. Rising land values have helped make the project bankable. Indeed, there is a huge untapped opportunity to invest in green infrastructure projects that support freshwater ecosystems, including the protection of headwaters upon which industrial and drinking water supply systems depend.

There are many other types of projects—some that are even outside the water sector—that can have an equally positive impact on water resources and freshwater ecosystems. For example, since hydroelectric power plants disrupt the natural flow of rivers, where feasible solar or wind farms could be installed in their place, avoiding degradation of the ecosystem while still providing extra energy and a financial return to private investors. Enhancing climate resilience by protecting and restoring healthy wetlands and floodplains, which mitigate the impacts of extreme floods and storms and so reduce insurance losses, is already often bankable as the world warms and extreme weather events increase. Meanwhile, any investment in water-dependent projects will encourage local and central governments to tighten the regulations that govern river basins, further improving the conditions in those basins.

[1] See http://www.eaudeparis.fr/les-metiers/preserver/ in French.

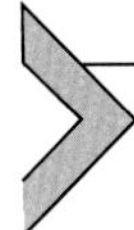

17.4 Case study: Büyük Menderes river basin

17.4.1 Background

The Büyük Menderes river basin in southwestern Turkey is a vital source of water for the region and an area of rich biodiversity. It is home to two globally important wetland protected areas—Lake Bafa and Büyük Menderes Delta National Park. The Büyük Menderes basin is of key importance to the textile industry, accounting for about 60% of Turkey's textile exports. Yet, the textile manufacturing industry near Denizli causes severe water pollution, threatening flora and fauna, and local livelihoods. In addition, downstream industries such as cotton farming and fishing are negatively affected by the water pollution, posing additional economic risks to the area.

17.4.2 Investment context

Turkey's textile industry is booming. In 2017, the textile industry amounted to USD 10.5 billion—roughly 16% of Turkey's total exports—and this number is expected to rise given predicted demand, production, and export. Denizli shows a strong performance as a leading textile manufacturing city in the country. While the Ministry of Environment and Urbanization has shifted its focus toward the textile sector, incentives to implement clean production methods have been limited due to the absence of strong environmental regulations. This led to the implementation of an innovative project to support garment and cotton production factories to adopt better water management (WWF, 2020).

17.4.3 Project description

The project was developed to significantly improve water quality in the Büyük Menderes river basin and ensure a sustainable and clean water supply for businesses, people, and nature. The project does this specifically through supporting a group of small and medium-sized textile (dying) companies to adopt cleaner production processes that use less water, chemicals, and energy, and reduce solid waste and wastewater. A multistakeholder negotiation platform has been created to develop this project at the basin level and thorough work has been conducted to align a wide range of stakeholders with diverse priorities and agendas, resulting in widespread support for the project. This multistakeholder involvement mechanism has supported the aggregation of small interventions (in each SME) into a collective project that delivers environmental benefits at scale. Funding for these interventions mainly comes from purchase guarantees from clothing brands to garment manufacturers to finance more efficient and cleaner production practices. SMEs are also incentivized to participate based on these purchase guarantees, the additional bank loan finance they can access based on these guarantees, increased profitability from improving production practices, and achieving compliance with environmental regulations. The local banking sector is incentivized to participate by the expanded customer based they achieve for their loans products.

These interventions range from small alterations, such as changes in chemicals and improved water management, to large investments in equipment. Investing in "grey infrastructure" helps to minimize the impact of the industry on "green infrastructure" (or nature-based solutions), that is, freshwater resources and the health of the basin.

The long-term goals are to:

- Raise the water quality from a low to a good status, especially in highly polluted locations.
- Create a basin-wide partnership with civil society, the Turkish government, and (international) private sector companies to set agreed conservation targets in key biodiversity areas, reduce resource use in industry production, and establish an effective monitoring system.
- Enforce effective wetland management and restoration in order to protect freshwater habitats and species.

17.4.4 Investment structure

WWF supported textile dyers to attract grants to fund feasibility studies for about 40 processing facilities at a cost of EUR 400,000–800,000. The largest sum was granted by the South Aegean Development Agency. The additional financing requirement for the feasibility studies was met by textile buyers and brands. In addition, WWF created protocols with participating banks to facilitate the process of obtaining loans worth EUR 3.6–8 million (EUR 90–200k per facility) for investment into cleaner production processes. These loan's repayment periods are in the range of 6–48 months. For commercial reasons, exact interest rates could not be disclosed. Seven textile manufacturing companies have already invested EUR 6.5 million in cleaner production methods and 12 other producers are committed to invest an additional EUR 3 million. The total cost of this basin-scale project is estimated to be between EUR 5–12 million and will generate cost savings of between EUR 4–12 million annually.

The project's main financial risk is related to default on repayment by the garment factories. Several instruments have been put in place to mitigate this risk, including offtake guarantees by global brands (their motivation is partially related to protecting their brand and reputation through a cleaner supply chain), first-loss guarantees by development finance institutions, and provision of collateral in loan agreements.

17.4.5 Business model and revenue-generating activities

The business model is built upon cleaner production processes that simultaneously help to lower the cost of production and sustain the business, for example, by improving compliance with environmental legislation and regulations, alignment with the demands of international brands, and increased brand value. Financial returns are specifically generated through:

- Reducing the use of water (the improved 40 production facilities will generate a combined saving of 1.5 million cubic meters of water annually) chemicals and energy—cutting production costs.
- Feasibility studies point out that the interventions resulted in significant savings—EUR 4–10 million/year through an investment of EUR 5–12 million—with payback periods ranging from 6 months to 2 years. There have not yet been follow-up studies on the feasibility studies forecast. Exact financial saving values will be confirmed in the future.

17.4.6 Risks and safeguards

A potential risk is related to labor conditions, as the global trend of cost reductions drives cheap labor and long working hours. To mitigate this, all manufacturers who wish to become part of the project need to provide assurance of fair working conditions. In addition, brands have indicated that they will only buy from suppliers who meet the minimum standard as measured by the Higg Index (WWF, 2019). WWF has moreover developed a safeguards manual to identify and manage social and environmental risks and opportunities. Stakeholder engagement is a key aspect of this manual. To ensure that activities do not harm certain groups, various stakeholders have been engaged throughout the project

17.4.7 Lessons learned

Arguably the first lesson is that in order to protect the wetlands of the Büyük Menderes river basin there must be a recognition that the interventions needed to happen upstream to address the pollution by the textile mills. The model is innovative, since it involves different actors throughout the supply chain and along the river basin, and since it aligns the goal of reducing water pollution in the river basin with the commercial goals of the garment industry. The project is supported by the private sector with seven brands having already invested and an additional 12 brands having committed to invest as of 2019. It is anticipated that the remaining factories will soon follow suit once they see the tangible financial and environmental benefits observed by already participating factories. Key lessons learned include:

- The importance of multistakeholder platform negotiations when developing a project at basin level. WWF Turkey managed to align a wide range of stakeholders with diverse priorities and agendas, resulting in widespread support for the project.
- Challenges regarding finding the required financing. It proved to be difficult to provide rates and loans that are attractive enough for small and medium-sized enterprises. WWF is looking into blended finance mechanisms that can help lower the interest rates. It is thereby important to understand the desired real interest rate borrowers are seeking.

The project was based on lessons from similar projects across the globe. Other projects and activities in the region were reviewed to optimize models and approaches for the specific landscape of the Büyük Menderes basin. Having a good grasp of local and international finance proved to be key in unpacking certain issues and identifying solutions. The core components to create an enabling environment are summarized below:

- Capacity to pool small projects.
- Capacity to value and monetize the benefits (in that case, through a global value chain).
- Water regulation in place (quality standards and allocation regime).
- An intermediary active on the topic (WWF).

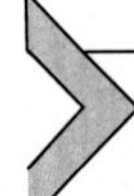

17.5 The roles of the stakeholders in financing partnerships

17.5.1 Corporates

Multinational companies often feel the cost of water misuse in their supply chains. This can be felt in many different ways: financial sanctions from water regulators due to their misuse of water, reputational damages due to media or press releases exposing their water misusages, or even pressure from their clients who increasingly want to buy products whose supply chains are more sustainable (this includes where water is better managed). Corporates would therefore benefit significantly by promoting better use of water resources, by identifying bankable freshwater projects within their own operations and supply chains, and by quantifying potential benefits. To start, this requires a good understanding and monitoring of the key issues and challenges. When a project is identified, they should be actively engaged by providing the financial commitment needed to enable the project to be realized, through seed funding, guaranteeing to meet any initial losses, or off-take agreements. Finally, corporates should advocate for improved water-related regulations and enforcement with local authorities

By proactively supporting investments in bankable projects, they can gain a competitive advantage and benefit from lower costs, more resilient supply chains, and a better reputation with customers and regulators. Companies can also ring-fence water-related investments, making it easier for them to raise funds for specific water projects and improve their sustainability performance so that they can achieve better financing terms. Indeed, financiers particularly like ring-fenced water projects because it ensures that the financing only supports targeted areas. The International Finance Corporation is an example of a financial institution, which is very active in this form of environmental investing (Le Houérou, 2018).

Companies need to set aside budget that helps to de-risk their supply chains and leverage other investments in the basins from which they source. We recommend that each company set so-called leverage targets. Indeed, in discussions about financing for the SDGs, there is a fascination with high leverage ratios and structures where a small amount of development finance can unlock a large amount of commercial finance. As a result, the development of financial structures that provide such high leverage solutions is gaining significant attention (ECDPM, 2018). Developing new water-related infrastructure—such as wastewater treatment plants and nature-based solutions—is another clear potential source of returns for utility and infrastructure companies. By addressing their water-related risks, companies can eventually improve their credit rating and source capital at favorable rates. However, for this to occur credit rating agencies need to integrate water and other environmental parameters into their actual rating methodologies.

17.5.2 Financiers

Financiers will find valuable business opportunities in water-rated investments by collaborating and pooling expertise to develop projects and mobilize capital at scale. Development finance institutions, multilateral, and national development banks, and government aid agencies can work proactively with private funders to combine different types of available funding to ensure financial deals work for all parties. These organizations also have valuable expertise that they can share about which projects are needed and where, and how to meet projects' financial and technical requirements. It is important, however, that development institutions do not 'crowd out' commercial financiers by participating in projects, or parts of projects, that the private sector can fund.

Commercial funders also need to work alongside players in the field to identify bankable freshwater projects. An innovative approach to financing takes into account new financial products, which can be divided into investment, banking, and insurance products. The predominant financial instruments in green finance are debt and equity. However, to meet the growing demand, new financial instruments such as green bonds and carbon market instruments, have been established, along with new financial institutions, such as green banks and green funds. Renewable energy investments, sustainable infrastructure finance, and green bonds are areas of most interest within green financing activities.

Other finance mechanisms include blended finance (as mentioned earlier in this chapter) and cooperation amongst many stakeholders at basin scale. In addition, new financial products, such as landscape bonds and sustainability improvement loans, can support and catalyze investments in bankable water solutions. They will need to be consulted early in a project's life to ensure it is on course to becoming 'investor ready' and attract additional private capital. By working with the companies, they are financing to identify and tackle water-related risks, funders can reduce their own exposure to these risks. Depending on the individual project, they may also have to take a longer-term view of when project returns will be delivered (this is especially the case for impact investment funders and Ultra-high Net Worth family offices, who tend to prefer longer-term investments as opposed to short term high return investments). They also need to adjust how they find comfort in lending.

In short, in order to unlock the full potential of sustainable finance, financiers must look at innovative and new funding mechanisms (in addition to the more traditional ones). In addition to using new funding mechanisms, financiers can also partner with effective NGOs and other stakeholder and experts, who are closely linked to the project, in order to ensure a more successful delivery of the project's environmental and financial potential.

17.5.3 Governments and regulators

Governments have a key role to ensure that private funds go to projects that contribute to wider policy objectives (sustainable growth, water security, etc.). They need to strike a balance between crowding in private finance and a fair allocation of risks and revenues between public and private actors. To do so, their role is to develop robust enabling environments, which combine water regulation (to make water pollution costly and to reflect the opportunity cost of using water) and financial regulation (to channel domestic private finance where it creates value for the community). Indeed, governments in developed and developing countries and their national water regulators play a crucial role in creating the right conditions for effective water management and investment.

But there is more to be done. With supportive policies and regulation, governments can incentivize companies and other project sponsors to develop sustainable freshwater projects. Governments can also provide additional incentives to encourage sustainable investment by means of tax benefits or grants. However, they will need to ensure that the processes that sponsors are required to follow to obtain these benefits are not overly bureaucratic. Governments can also create a positive investment climate by enforcing existing regulations and so create a level playing field for all stakeholders in the river basin. A solid regulatory environment is a good place to start for governments hoping to attract private investments into their basins. Developing-country governments that get the conditions right—by creating a competitive and resilient business environment that is attractive to investors—will benefit from stable project pipelines.

A specific example of this is the strategic investment pathways that is currently in the process of being implemented in the Zambezi Basin (Dominique, 2020). Ensuring that water resources continue to contribute to Zambia's economic, environmental and social development will require significant investment. In this case, scaling up financing for investment will require a pipeline of bankable projects. This pipeline of projects will be situated within a strategic investment pathway, which takes a long-term and basin-wide approach in order to ensure that investments deliver benefits for the environment, the economy, and society over the long term. In this case, an enabling environment for investment will require: (1) Strong legal and policy frameworks and implementation; (2) Improved coordination to improve water management; and (3) Better enforcement of key water regulations (notably related to abstraction). In addition to a strong enabling environment, WWF is working with the Ministry of Water Development, Sanitation and Environmental Protection on developing the landscape finance plan (LFP) for the lower Kafue. The LFP will outline both bankable and nonbankable projects and is being developed with the support of AB InBev and the

Zambian government. To further support the rationale of the LFP, WWF will be piloting a number of bankable projects through the DFCD, which will be integrated into the overall vision for the Kafue. Through the LFP and bankable projects, WWF will create an enabling environment to ensure that the Zambian people and economy are better adapted to the threat of climate change, which ultimately impacts Zambia's water security.

17.5.4 NGOs

To effect change, both in the water sector and other areas, nonprofit organizations need to think outside the box by identifying ways to encourage private-sector investment using innovative approaches. Those organizations that can demonstrate an ability to deploy their philanthropic funding to leverage private capital will improve their relevance for donors. NGOs will have to use their networks and their expertise to identify the freshwater projects needed to restore ecosystems and safeguard freshwater for local communities. A key strength: they have the capability to take a broad, basin-level view of what projects are required and can determine which of these are bankable. They can also use their 'brand' to create awareness about the need for sustainable projects and act as an intermediary to bring together the right people and organizations, from the private and charitable sectors, to make projects happen. This effort goes hand in hand with working with the local authorities to promote adequate regulation and enforcement.

By giving their support to the private sector, NGOs can place a stamp of approval on projects, helping to promote those with social and environmental benefits. In addition, when individual private sector-backed water projects succeed, this leads to the development of other projects and ultimately creates pressure on regulators and governments to improve regulation and enforcement. It also creates a momentum for funding to support broader conservation goals.

17.6 Conclusions

The only way to achieve the required level of investment in the water sector is by significantly leveraging the private sector. The following are key considerations:

- A supportive enabling environment needs to be put in place before the finance needed to meet water security and/or environmental goals will flow.

- Focus on income-generating activities. To effectively leverage the private sector, engagement needs to focus on more than channeling private funding to a project since this does not automatically make the project bankable. Therefore, income-generating activities need to be implemented as these enable the project to provide financial return for investors.
- The landscape approach enables progress between different interests and drivers. It has been promoted and used mainly within the agriculture and forestry sector, and sustainable water management has so far has not been a key focus. Integrating water management in the landscape approach discourse will make it even more effective. Adopting a landscape approach is key to ensure that investments can trigger integrated positive impact. Some direct venture development investments may appear to be sustainable but can sometimes be questionable when looking at the landscape as a whole. At the same time, projects that involve 'grey' infrastructure, may also result in highly beneficial outcomes for the wider landscape.

Solving the world's water-related challenges will need a proactive approach, innovative structures that enable greater cooperation between different players, and the use of blended finance mechanisms. These steps are essential to make freshwater projects a bankable, or investable, proposition and secure the participation of the private sector. Corporates, financiers, governments, and NGOs need to urgently step up their efforts to jointly drive sustainable and bankable freshwater projects.

References

Almond, R.E.A., Grooten M. and Petersen, T. (Eds). WWF, Gland, Switzerland.

Anglian Water Services Limited, 2018. Green Bond Annual Report 2017–2018. https://www.anglianwater.co.uk/siteassets/household/about-us/pr19-10c-green-bond-annual-report.pdf, accessed 10 December 2020.

Barbier, E., 2019. *The Water Paradox: Overcoming the Global Crisis in Water Management*. New Haven: Yale University Press. https://doi.org/10.12987/9780300240573.

Climate-ADAPT, 2016. Private investment in a leakage monitoring program to cope with water scarcity. In Lisbon. https://climate-adapt.eea.europa.eu/metadata/case-studies/private-investment-in-a-leakage-monitoring-program-to-cope-with-water-scarcity-in-lisbon, accessed 10 December 2020.

Dominique, K., 2020. Strategic investment pathways: the Zambezi Basin case study OECD 2020. https://www.oecd.org/water/OECD-(2020)-Strategic-investment-pathways-Zambezi-case-study.pdf, accessed 10 December 2020.

ECDPM, 2018. Leveraging private investment for sustainable development: https://ecdpm.org/wp-content/uploads/Great_Insights_vol7_issue2_Leveraging_Private_Investment.pdf, accessed 10 December 2020.

Gardner, R.C., Finlayson, M., 2018. Global wetland outlook: state of the world's wetlands and their services to people 2018. Secretariat of the Ramsar Convention, Switzerland.

Hampl, N., Lüdeke-Freund, F., Flink, C., Olbert, S., Ade, V., 2011. The myth of bankability-definition and management in the context of photovoltaic project financing in Germany. Goetzpartners & COLEXON (eds.) (2011).

Le Houérou, P., 2018. Opinion: a new IFC vision for greening banks in emerging markets: https://www.devex.com/news/opinion-a-new-ifc-vision-for-greening-banks-in-emerging-markets-93599, accessed 10 December 2020.

Huq, S.N., Karim, I.U., 2017. Greening of textile industries through public and private governance: an explorative research on the garment industry in Bangladesh. Wagenigen University.

Kulshreshtha, S.N., 1998. A global outlook for water resources to the year 2025. Water Resour. Manage. 12 (3), 167–184.

Leckie, H., Smythe, H., Leflaive, X., 2021. Financing water security for sustainable growth in Asia and the Pacific. In: OECD Environment Working Papers, No. 171, OECD Publishing. Paris.

Marks, P.K., Gerrits, L.M., Bakker, S., Tromp, E., 2014. Explaining inertia in restoring estuarine dynamics in the Haringvliet (The Netherlands). Water Policy *16* (5), 880–896.

OECD, 2019. Making blended finance work for water and sanitation: unlocking commercial finance for SDG 6. OECD Studies on Water. OECD Publishing, Paris https://doi.org/10.1787/5efc8950-en.

Padt, F.J., Sanchez, J.C., 2013. Creating new spaces for sustainable water management in the Senegal river basin. Natural Res. J. 53, 265–284.

Rahman, M.H., Muzib, S., Chaity, R.A., 2018. Ready-made garments of Bangladesh: an overview. Barishal University Journal Part 1, 5(1&2): 59–122.

Thorburn, P.J., Wilkinson, S.N., Silburn, D.M., 2013. Water quality in agricultural lands draining to the Great Barrier Reef: a review of causes, management and priorities. Agriculture, Ecosyst. Environ. *180*, 4–20.

World Bank, 2016. *High and dry: Climate change, water, and the economy*. The World Bank. License: Creative Commons Attribution CC BY 3.0 IGO.

World Economic Forum, 2019. *The global risks report 2019*. Geneva, Switzerland.

WWF, 2018. *Banking on financial solutions to save our basins*. https://d2ouvy59p0dg6k.cloudfront.net/downloads/banking_of_financial_solutions_to_save_our_basins_1_1.pdf, accessed 10 December 2020.

WWF, 2019. Conservation Investment Blueprint: Cleaner Production in the Textile Sector-Büyük Mendered. http://cpicfinance.com/wp-content/uploads/2020/12/Cleaner-Textile-Production-Blueprint_WWF.pdf, accessed 10 December 2020.

WWF, 2020. Bankable nature solutions: blueprints for bankable nature solutions from across the globe to adapt to and mitigate climate change and to help our living planet thrive. https://wwflac.awsassets.panda.org/downloads/bankable_nature_solutions_report.pdf, accessed 10 December 2020.

Index

Page numbers followed by "*f*" and "*t*" indicate, figures and tables respectively.

Printed in the United States
by Baker & Taylor Publisher Services